Global Warming and the Climate Crisis

Bruce E. Johansen

Global Warming and the Climate Crisis

Science, Spirit, and Solutions

 Springer

Bruce E. Johansen
University of Nebraska at Omaha
Omaha, NE, USA

ISBN 978-3-031-12356-6   ISBN 978-3-031-12354-2   (eBook)
https://doi.org/10.1007/978-3-031-12354-2

This Springer imprint is published by the registered company Springer Nature Switzerland AG
The registered company address is: Gewerbestrasse 11, 6330 Cham, Switzerland

To Pat Keiffer, wife, keeper of the hearth, sage editor

Preface

Watching the Skies

Since the age of 6 or so, I have been an avid weather watcher, fascinated by clouds and storms, rain and snow, how weather systems come and go. Moving with my siblings and parents halfway around the world (Dad was a senior officer in the U.S. Coast Guard; one might be surprised just how many places that the Coast Guard has bases). The onset of the monsoon in the Philippines with its dark, fat-bellied clouds, was a sight to behold, and not unlike a Puerto Rican tropical storm. The suburbs of Washington, D.C. sometimes received heavy snow and ice storms. I also love shoveling snow and chipping ice. I relished watching the first cirrus tendrils of a snowstorm move in from the southwest. I had a special hill with a wonderful view of the incipient inclemency.

Dad negotiated with the bureaucracy and got me on the subscription list for the U.S. Weather Bureau's daily weather map, which folded out to about 3 ft by 2 ft, the tool that the real weather forecasters used. I wanted to be a weatherman until I discovered that most of their time was spent indoors, in offices, with clacking tele-type machines. Watching the sky from suburban hills had gone out of fashion among the professionals.

My family continued to move. The advent of frequent rains during in the Pacific Northwest got a bit boring during my 17 years there although we did have an occasional heavy, sloppy snowstorm, providing some entertainment. Watching the snow line slide down Olympic mountainsides from Seattle also gave me hours of pleasure.

Following Ph.D. work at the University of Washington, the time came to pick my professorial working location. A major qualification for adult life in Omaha, as I saw it, was *real* weather: thunderstorms, the occasional tornado (as long as it wasn't *too* close). As I write this, memory of an F-4 tornado, with winds of 170 miles an hour, 100 miles east of Omaha, is less than a week old.

Nature has kept its promise to me over the last 40 years. Deluges, droughts, snow and ice, bitter cold snaps, and stifling heat waves all have captured my sense of wonder and danger. An F-1 tornado rolled over a house that our family owns a few miles east of Omaha as my stepson Shannon was folding laundry. He watched a load, previously folded, rise and wildly dance, so it seemed, out a broken window.

All of this Midwestern mayhem comes to us between a few really nice, blue-skied late-spring and early fall days when we could watch spring foliage burst out or leaves turn brilliant color.

I have learned to look for clues in the sky, and not always where one might expect. The Sandhill Cranes, for example, pass through Nebraska each late winter on their way north, along with dark geese, trumpeter swans, and bald eagles. Years ago, the migration northward peaked in mid-March. By 2022, their stop in central Nebraska occurred in the middle of February, responding to temperatures and food availability (Ducey, 2022, B-1). Our climate world is changing around us—for human and the Sandhill Crane. The clues are before our eyes. As a teaching tool, this book should help explain these changes, and what they portend.

Welcome to the Anthropocene

By the last half of the twentieth century, something very important was happening to humankind's relationship with the Planet Earth: the weight of human beings was passing that of every other creature on the planet. At the same time, each of us, on average (an important distinction) was using a larger proportion of the Earth's resources—wood, oil, soil, and much more. We had, by the turn of the twenty-first century, reached what many scientists (and other literate people) were calling the Anthropocene, a new epoch, in which human beings exercise control of Earth's future—no matter whether that influence is malign or benign. As the human race continues to squabble over resources as well as nationalistic differences based in religion and ethnicity that have plagued us since the time when wars were waged with sticks and stones (when humanity as a whole had very little influence on the course of the planet as a whole), we may ask whether our role in the Anthropocene is a blessing or a curse.

The term "Anthropocene" was so new that as of this writing (February, 2022) that the spell-checker on my brand-new Mac cannot find it. We, as human, about eight billion of us, have assumed control of Earth's future at short notice, on borrowed time, whether we are ready or not to maintain a sustainable Earth that meets our energy needs without combusting fossil fuels, because their consumption of them will cause our atmosphere to heat beyond humankind's ability to withstand it. I do not say *"may."* I say *"will,"* unless we banish fossil fuels from our energy equation, and do so very quickly, perhaps in 10–20 years. Not many textbooks start with such an emphatic theme. There exists no way to objectify ourselves out of this. Thus, this is a textbook with *attitude*.

Reference

Ducey, M. (2022, February 20). Sand hill cranes are flowing into central Nebraska. *Omaha World-Herald* B-1, B-2.

Bruce E. Johansen
Omaha, NE, USA

Acknowledgments

Editor Niko Chtouris, of Springer, in Frankfurt. Germany: friend, advisor, and a great editor. I wish I had met you a few decades ago.

Yvonne Schwark-Reiber, for very good advice on a manuscript being born.

Everyone at Springer, an extraordinary group.

Joy Porter of Hull University, UK, for many years of support and advice.

Wife Pat Keiffer, keeper of the hearth for our extended family in Omaha, woman of the world, and sage editor.

Contents

Introduction

Climate Crisis, Science, Spirit,
and Survival

Contents

© The Author(s), under exclusive license to Springer Nature
Switzerland AG 2023
B. E. Johansen, *Global Warming and the Climate Crisis*,
https://doi.org/10.1007/978-3-031-12354-2_1

1

We spoke about the land;
If we lose the land, we lose the culture;
Lose the culture, lose the peace,
lose the peace,
lose the community.
Lose the community, lose our way of life.
Forever.
The Oloiboni Kitok
Kenyan elder, about his world (Owour, 2021, 98).

⊜ Learning Objectives

1. To achieve an overall view of climate change as a subject.
2. Why study of climate is important for the future of humanity and the Earth as a whole?
3. To introduce the subject matter of the book and to introduce some of the terminologies (such as "thermal inertia") that are very important to the subject.
4. Introduce a number of subjects that, taken together, comprise an introduction to facts and points of debate attending climate change.
5. Why does climate change "incubate a future catastrophe?"
6. To present synopses of the book's chapters vis-a-vis each other.
7. To provide a brief introduction of political points of view on the issue.
8. To introduce some of the issue's terminologies and their importance, such a jet stream.
9. To consider broader debates (such as nationalism) that have an important bearing on climate change's usefulness or lack thereof in discourse over the subject at hand.

1.1 Introduction to Climate Change

This book contains work on a number of subjects that, taken together, comprise an introduction to facts and points of debate attending climate change. It has been written for anyone from first-time visitors to this exciting and vital new field of knowledge but also for specialists who will enjoy a conversation with another person with similar interests and concerns about the future of all in our precious habitat. The two photos in ◗ Fig. 1.1, side by side, show the Muir Glacier in Alaska in 1941 and 2004, illustrating dramatic losses about 63 years apart. The two space photos of the Arctic icecap (1984 and 2020) in ◗ Fig. 1.2 illustrate shrinking coverage.

1.1.1 The Eclectic Nature of Climate Science

Climate change (global warming) has been cobbled together from other areas of knowledge. An example is provided by Bob Leverett and Monica Jakuc Leverett,

Fig. 1.1 Muir Glacier in Alaska in 1941 and 2004. Source: NASA; Public Domain

who began by researching new ways to measure the heights of old-growth trees along the Mohawk Trail in Upstate New York and ended up discovering an important new source of knowledge about how older trees remove carbon dioxide (CO_2) from the atmosphere vis-a-vis younger ones. While climate change scientists had long assumed that younger trees removed CO_2 faster, the Leveretts found the opposite among White Pines, which can grow to 170 ft and live as long as 450 years.

The Leveretts then were drawn into the use of the White Pine as a national symbol by the Haudenosaunee (Iroquois) Confederacy for at least 1000 years. The Leveretts were then became attentive to the Jake Swamp White Pine, the tallest tree in New England, named for the late Mohawk leader on the Akwesasne Mohawk (St. Regis) reservation. For many centuries, the Mohawks have been one vital part of a connection that brought Haudenosaunee into contact with Benjamin Franklin and helped to shape the framework of US democracy during the eighteenth century. By that time, the White Pine now named for Jake Swamp probably was about 200 years old. Thus, climate science has been cobbled from several other fields of scholarly endeavor (Diamond, 2022, 21–33).

1

□ **Fig. 1.2**   Arctic icecap (1984 and 2020). Source: NASA; Public Domain

1.1.2 Climate Change and Presidential Politics

Debate about climate disasters has exploded in the past decade, along with accompanying studies, in line with an abundance of evidence that it has become an existential threat to the future of the Earth and its flora and fauna. The issue, barely considered a major problem by much of the general public 10–20 years ago, was a staple of debates between presidential candidates Joe Biden and Donald Trump during the 2020 US election, with the candidates holding starkly different positions. Biden's position stressed reducing fossil fuel emissions with an emphasis on generating wind, solar, and other renewable sources of energy, along with a phase-out of oil, coal, and methane (natural gas), among other nonrenewable sources (□ Fig. 1.3).

Biden agrees with the scientific consensus that climate change is a worsening, destructive force, evidenced by intensifying storms (especially cyclonic storms such as hurricanes and typhoons), drought, fires alternating with deluges, and wildfires (as witnessed in Australia, Brazil, Siberia, the western United States, and other places). These catastrophic events clearly indicate that human-generated climate change is no longer solely a prospective issue meant only for discussion, not action. It is a *very* real, everyday reality worldwide. At the same time, the proportion of carbon dioxide (the major greenhouse gas) in Earth's atmosphere has been increas-

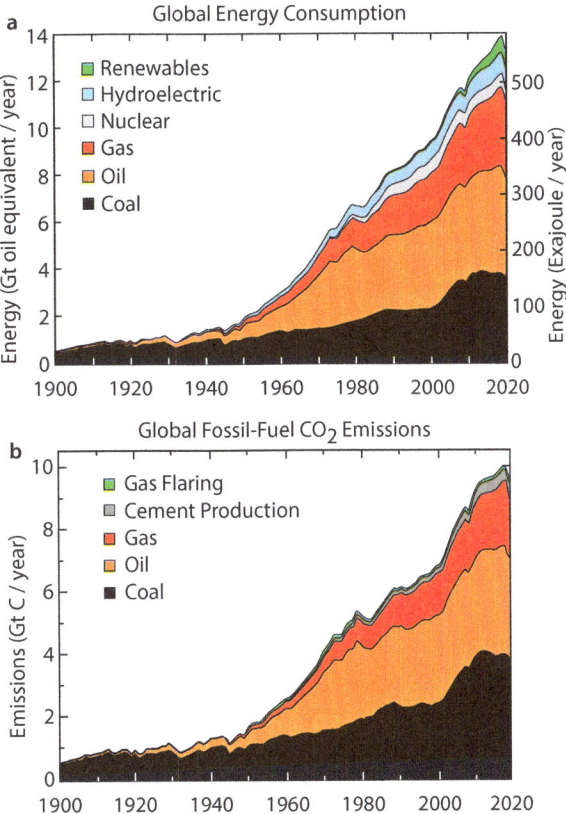

Fig. 1.3 Global energy consumption and fossil fuel emissions (1900–2020). **a** Fossil fuel consumption. **b** Global energy consumption. (Reproduced with permission from Hansen and Sato (2021))

ing rapidly, passing a level of 360 parts per million (which is the highest level considered safe by many scientists) to almost 420 p.p.m., a level that, with further increases, is laying a foundation for world-widen catastrophe (**Fig. 1.4**). As readers will learn in this text, this is a level at which ocean levels may increase by about 25 ft, which will wreak havoc on some of the world's largest coastal urban areas.

Donald Trump's position on this issue was simple. So simple, in fact, that his mental apparatus seemed questionable. He usually referred to climate change as a "hoax," perpetrated by the Chinese, Democrats, and other "radical leftists," to harm industries, such as coal and oil, and to destroy American jobs. In other words, Trump thought that global warming is a politically motivated myth or an outright lie. As president, Trump's standard stump speech included references to his belief that climate change was a Democratic Party conspiracy to obliterate jobs in the coal mining industry, although by that time (2020) wind and solar power were employing four times as many people as coal.

Politics is one profession in which competing candidates can make the most outrageous claims that affect millions of people without a shred of supporting evidence. This mishandling of debate happens so often that many people who

1

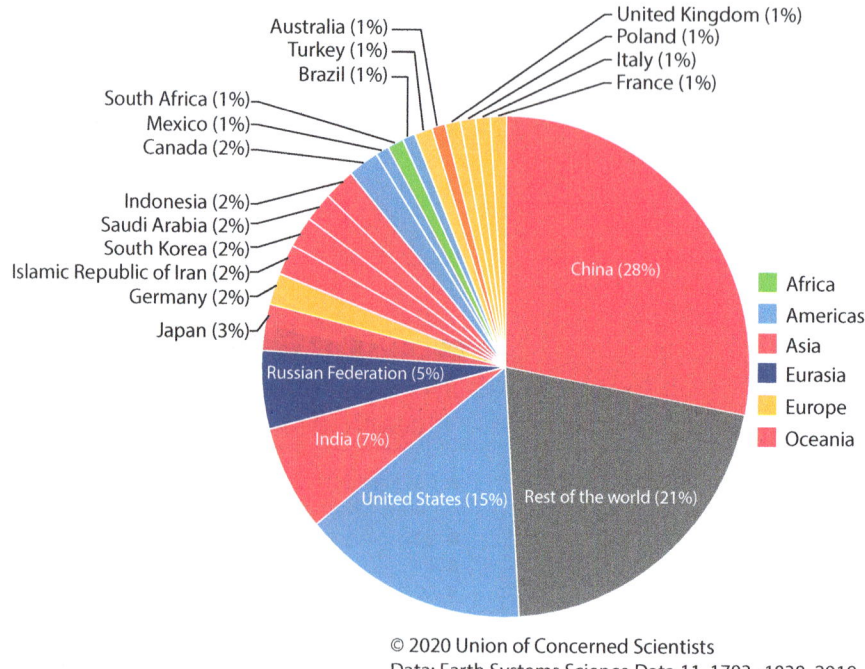

© 2020 Union of Concerned Scientists
Data: Earth Systems Science Data 11, 1783–1838, 2019

☐ **Fig. 1.4** Share of world's carbon dioxide emissions by country and region. (Reproduced with permission from the Union of Concerned Scientists (2021))

patronize politics do so with an engrained prejudice against its factual basis, that is, other than the true believers who demand no standards of proof, who cater to true believers, neglecting basic science. Trump's version of reality is so thickly Orwellian that Big Brother might have trouble recognizing it. He repeats his lies so often and so emphatically that he has perhaps 30% of adult US citizens believing them. Trump's standard riff years after he lost the 2020 election is that Joe Biden hijacked it from him. Hard to believe? Some opinion polls taken during the 1930s had about the same percentage supported Adolf Hitler, for what they thought to be abundantly good reasons.

1.2 Chapter-by-Chapter Descriptions

Big lies non-withstanding, we have reached the point where global warming science has been accepted by most thinking people as a threat involving every living being on our precious Earth. Many people may not understand much of the science behind this belief, but let the daily weather convince them. The year 2021 spilled over with severe weather that eventually becomes climate for an evolving, violent world climate. Humanity has always witnessed natural disasters. So many, over so short a time, provides new perspectives.

▶ Chapter 2 explores the basics of climate change, such things as why and how carbon dioxide, methane, and other "trace gases" play an important role in creating and maintaining a steadily warming lower atmosphere. Witness, please, the role of thermal inertia, which delays the effects of our emissions of carbon dioxide for approximately 50 years on land and 100–150 years in the oceans. The degree of warming and weather instability that we bear today is a result of 1970s' greenhouse gas emissions. The importance of this chapter lies in its ability to explain the science—why and how scientists can interpret the future, within a range of possibilities.

▶ Chapter 3 is a descriptive scorecard of day-to-day meteorological events during the last half of 2021 and early 2022. As weather watchers, we are cautioned by scientists not to mistake a few events (weather) for a much longer trend (climate). This is true, until the evidence is overwhelming, and it continues, with breaks that are relevant in any chaotic natural system. Weather is the story, and climate is the plot. The story in 2021, which had just closed as these words are being written, has been full of weather drama—record-strength hurricanes, record size, heat, and sheer power of wildfires, flooding rains, droughts, and some of the strongest- and longest-lived tornadoes in history (in *December*, no less). Do not forget the Super Bowl on February 14, in Los Angeles, during a heat wave fueled by Santa Ana winds. For a weather watcher, 2021 and early 2022 became quite a feast. So where and when does the story become the plot? Given the fact that heat drives weather severity, and temperatures are rising, betting against wild weather is probably a bad idea, even though all of it will not happen at once.

▶ Chapter 4 develops indigenous (in this case Native American) attitudes toward nature and, by and by, weather and climate. Some observers might be surprised to find Native points of view included in a book such as this, although they kept rough records in their memories and, in the case of peoples who lived in tipis, on the skins of their dwellings. The belief systems of Native peoples are an important counterpoint to dominant systems of belief in today's world. The same goes for ▶ Chap. 5, a present-day description of today's Navajo and Hopi when faced by the worst droughts in their traditional histories. Both have adapted well to drought interrupted by occasional cloudbursts, but it all seems to be a battle that they are losing.

The last chapter, numbered ▶ 6, describes attempted solutions. Facing an existential threat in which thermal inertia guarantees that results follow their initiation by several decades, not to mention that our penchant to argue over weather trends delays real, basic change. Habit is comfortable, as well, and so is comfort, as well as an ability to profit off the status quo. In the face of all of this, change is taking place, with the development of wind and solar power, for example, now outpacing that of oil.

However, many of us have not a clue as to how to solve this problem. Such thinking (and doing) goes beyond solar and wind power, electric cars, and flying jets with used cooking oil; it has to do most basically with how we think and relate to each other as human beings not only on a personal level but also as world citizens. Thus, the first requirement necessary solving this problem is a large-scale acceptance of science-based truth. The road to solutions also is paved with shysters an easy-money hustlers.

1

1.2.1 Cooperation Is Required

To solve the existential problem posed by climate change, we as world citizens must get past individualism as an acceptable, international means for conducting international relations. On a broader level, we must cultivate a spirit of forgiveness between nations and cultures in order to cooperate. To face a basic change in the modus operandi for human relations on this planet in the face of an Earth-wide threat, we must cooperate in ways never heretofore expected or accepted. This is not rabbit food or a hippie frolic. It is essential for survival. We should live for achievements that further a common good for our children and theirs. If not, the entire net of life on Earth will eventually collapse. The world's vast and beautiful carpet of plant and animal life may die. As has been said of Earth-wide nuclear war, the last of the living will envy the dead. Whomever and whatever survives do will do so on a spare and frightening planet.

Today, the human race faces a new kind of adversary; carbon dioxide, methane, and other greenhouse gases do not have a grudge against humans or other living things. They do not care if you drive a Cadillac or ride a bicycle. They do not care whether you ignore its influence in atmospheric chemistry. All that carbon dioxide, methane, etc. do, in this instance, is to hold heat. At the current rate that we are generating it, a rising level of CO_2, methane, etc. will make our planet a very miserable (and, given enough time, uninhabitable) place. One graph and a map illustrate the recent state of greenhouse gas emissions by countries and regions. The other shows greenhouse gas emissions as of 2020 by countries and regions.

Greenhouse gases share some resemblance to the COVID-19 virus, which also harbors no grudges. It does not care what your name or nationality may be. All it seeks is human bodies (often lungs) in which to breed. Masks and vaccines inhibit this spread, which, over time, deprives it of an ability to spread. The virus has no notions of human political freedom or debates over the inhibition of freedom by wearing masks or taking vaccines.

1.2.2 Early Evidence of Extreme Warmth: The "Heat Dome"

Carbon dioxide, methane, etc. are already leaving calling cards across our habitat. The western half of North America, for example, has already been under a "heat dome" which has been a frequent atmospheric pattern for much of the last two decades. The jet stream, a river of air 30,000–40,000 ft above the Earth's surface (at the altitude of jet aircraft), steers weather systems over the Earth's surface. For roughly the last two decades, it often has been swinging northeastward over the northeastern Pacific Ocean to Alaska and then moving south-southeastern to the Gulf of Mexico. After another sharp turn to the northeast. This river of wind often then rolls up the US East Coast toward the Arctic. This pattern can wiggle east-west or north-south, but its basic pattern is relatively fixed, instead of moving east to west in an oscillation, giving everyone a mixture of precipitation, sun, and showers.

1

1.2 · Chapter-by-Chapter Descriptions

During the last week of June 2021, the US Pacific Northwest, where people are accustomed to mild summers, broke all-time highs by substantial margins. In Seattle, for example, the temperature hit 108 °F in an area with very little air-conditioning. In Portland, Oregon, the high on the same day (June 28, 2021) also set an all-time (record) high of 116. On June 28, Lytton, population 300, northeast of Vancouver, hit the heat wave jackpot with a high of 121. That was the highest recorded temperature in the history of weather record-keeping in Canada and not by a small amount. This scalding high was accompanied by a wind-whipped wild-fire that razed about 90% of the town.

A half-million lightning strikes blitzed British Columbia and nearby Alberta. At the US Olympic Games' trials in track and field in Eugene, Oregon, competition was suspended until after the sunset, as the ambient air temperature reached 110, and the air at track level broke 140°. A 110° afternoon temperature inflicted third-degree burns on anyone unwise enough to walk on an asphalt surface barefoot. Wiser people carried umbrellas as shade against the sun, not the area's more familiar gentle rain.

1.2.3 The Power of Denial

Some human beings' sense of denial can be severe enough to be frightening. Climate activist Bill McKibben visited Arizona in the midst of the same heat wave that gave Portland, Oregon, a 116-°F afternoon. With Phoenix carving out highs in the 113–118 range for almost a week, McKibben's amazing sense of climatic irony found a group organizing to pursue its self-supposed right to burn up water on golf courses: "AZCentral reports that some golf-course managers near Phoenix are "▶ pushing back" against a plan that would cut their water use by just 3%. Never mind that reservoirs across the West are ▶ falling to record-low levels (with insufficient snowpack to replenish them, and with that constant evaporation); representatives of the golf industry have formed the Arizona Alliance for Golf, which has met with state officials and launched a Web site urging residents to "speak up for Arizona golf" and "'protect our game'".

I could not help but wonder whether these golfers have air-conditioned carts. Just how hot and dry does it have to get before some people worry about more than their golf scores?

A few days after the heat wave from hell broke, hundreds of people were believed to have died in it, including 20 in Washington; 50 in Portland, Oregon; 70 elsewhere in Oregon, and uncounted hundreds in British Columbia. What is more, salmon (a cold-water fish) were dying en masse, and many beaches in and near Puget Sound were inundated with bacteria, not a pleasant place to spend a sizzling afternoon.

1.2.4 The Importance of Jet Stream Movements

An anchored jet stream pattern (described above) favors warmth or heat, with generally dry weather to its south and cold to cool and stormy to its north. The area in

1

which the jet stream makes its turn (movement southeastward to northeastward) favors climate chaos. Watch the number and intensity of tornadoes.

Thus, the US West also endured drought and heat from California inland roughly to the Missouri River valley. "The Southwest is getting hammered by climate change harder than almost any other part of the country," said Jonathan Overpeck, a climate scientist at the University of Michigan. "And as bad as it might seem today, this is about as good as it's going to get if we don't get global warming under control".

In the US East, this pattern favors storminess—floods during late spring to mid-fall. Otherwise, it favors ice and snow, sometimes in record amounts. Watch the news, and you will see this pattern inflicting nasty weather patterns relentlessly with flooding rains or snow as the West withers in continual heat that is ruining what used to be one of the world's richest agricultural areas.

1.2.5 **Thermal Inertia and Ocean Rises**

Dr. Overpeck might have added that even if our output of greenhouse gases fell to zero instantly (extremely unlikely during today's rapid rise of global human CO_2 emissions), it would take about 50 years over land and 100–150 years over the oceans to restore anything close to climatic stability. The longer we wait, the more entrenched this general pattern becomes.

When the oceans rise a foot and then 2–3 ft, Pacific Islanders' homes will be among the first to be swept into the ocean. As oceans rise as much as 25 ft to catch up with today's greenhouse gases via thermal inertia, large parts of coastal cities such as Miami; Manila; Tokyo; Washington, D.C.; Mumbai; Kolkata; Shanghai; New York City; London; and others will be flooded. Twenty-five feet in less than one human lifetime is rapid in our time and barely a flick of an eyelid by geological standards. This also provides an idea of how quickly Earth's oceans may rise if emergency measures are not taken *very* quickly. Given the outcomes of several dozen climate "summits" during the last 30 years, thermal inertia will easily beat human diplomacy. (And while we are at it, please note that melting all of Earth's ice might raise sea levels by about 280 ft (also see ◘ Fig. 1.5).) Sea-level rise is only one way in which a heating planet will inflict misery. Diseases of plants and animals (including humans) will become worse, for example. Oceans will become more acidic.

Thus, given thermal inertia's time lag, delivering results a half-century or more after they take place, present-day global warming can appear deceptively less consequential than it really is. Debating these effects years after they are set in motion makes it more difficult to construct a new energy paradigm before the damage becomes irreparable.

Rising temperatures also affects oceanic acidity levels, significantly impacting plant and animal life (◘ Fig. 1.6). This change is particularly troubling because acidity kills plankton, the foundation of the entire oceanic food chain.

We cannot just say it; we must do it, with the sense of emergency reserved for threats that truly threaten our tenure upon this world as thinking, feeling human

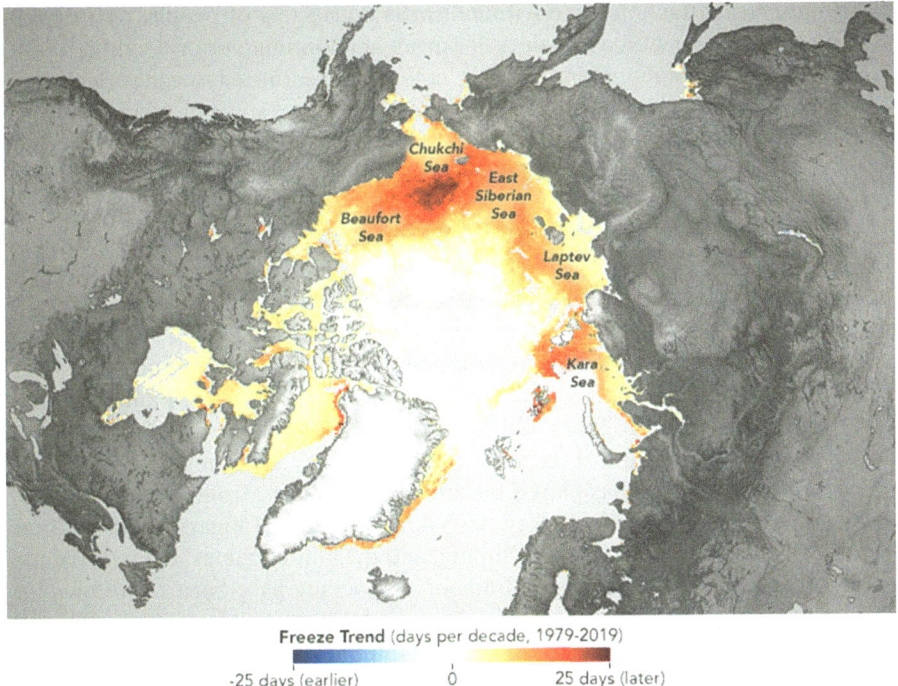

Freeze Trend (days per decade, 1979-2019)

-25 days (earlier)   0   25 days (later)

☐ **Fig. 1.5**   Freeze trend (days per decade, 1979–2019), Arctic Ocean. Source: NASA (2021). Public domain

☐ **Fig. 1.6**   World temperature rises, 2010–2019 average vs. 1951–1978 baseline. Source: NASA. Public domain

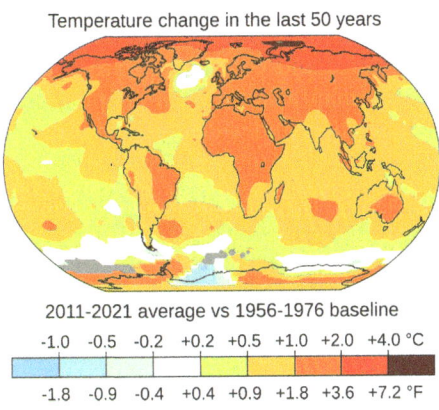

Temperature change in the last 50 years

2011-2021 average vs 1956-1976 baseline

-1.0	-0.5	-0.2	+0.2	+0.5	+1.0	+2.0	+4.0 °C

-1.8	-0.9	-0.4	+0.4	+0.9	+1.8	+3.6	+7.2 °F

beings. That is where we are today. This is not soft stuff. It is the toughest assignment that humanity has faced during its existence on Earth and one of humankind's own making. Without it, we fail our task to maintain a habitable Earth. As Benjamin Franklin once said, from a time before fossil fuels, we shall hang together, or we will most assuredly hang separately.

1

We must shed attachments to nationalism as well as fear of peoples outside our own cultures, religions, etc. This does *not* mean abandoning our own cultures, languages, etc. and becoming monochromic, colonized creatures. It means celebrating all of our cultures in a spirit of shared survival, fostering a spirit of cooperation, in which war and its progeny nationalism have become obsolete. Planetary survival allows no other path for humankind and the plants and animals that provide us sustenance.

1.2.6 Nationalism and War as Suicide Missions

Human nationalism that results in international conflict has become an assassin of democracy as well as an impediment to solving global problems. Paul Krugman wrote in the *New York Times* (2019, A-25) that "In their 2018 book, *How Democracies Die*, the political scientists Steven Levitsky and Daniel Ziblatt documented how this process has played out in many countries, from Vladimir Putin's Russia, to Recep Erdogan's Turkey, to Viktor Orban's Hungary. Add to these India's Narendra Modi, China's Xi Jinping, and the United States' Donald Trump, among others. Bit by bit, the guardrails of democracy have been torn down, as institutions meant to serve the public became tools of ruling parties and self-serving ideologies, weaponized to punish and intimidate opposition parties' opponents. On paper, these countries are still democracies; in practice, they have become one-party regimes….And it's happening here [the United States] as we speak. If you are not worried about the future of American democracy, you aren't paying attention". This narrowing idea of human purpose on this planet is occurring at the very time that circumstances require common purpose, beyond a history of conflict between nations. The battle against nationalism requires continual attention; peace and cooperation are not the natural state of human affairs, judging from human history given peoples' behavior after weapons of mass killing have been invented and widely used in war. Our weapons today are so dangerous that we dare not to use them. A change in human behavior to favor cooperation in the name of human survival vis-a-vis climate is going to require basic changes in human behavior *quickly.*

1.3 Summary

As we obliterated nationalism as a driving force of international relations, we also will need to wipe away war as a means of dealing with international disagreements, even acute ones, *especially* the tough ones. Given the history of humanity, with is full of hatred and wars, such an accomplishment will be an extremely tall order. No one has ever asserted that preserving the future for generations to come will be *easy.* However, delivering the Earth to a greenhouse gas-enhanced future will be suicidal. If working to eliminate nationalism is tough, think of living in the wasteland as implied by the *other* option.

We are reminded continuously that the late Carl Sagan, one of our most insightful scientific public intellectuals, had an interesting theory about highly developed civilizations. Given the number of stars and planets that must exist in the vast reaches of the universe, he said, there must be other highly developed and organized forms of life. Distance probably will keep us from making physical contact. Dr. Sagan said that another reason why we may never be on speaking terms with another intelligent race (judging from our own example) could be their penchant for destroying themselves in relatively short order after reaching technological complexity.

As we examine the worldwide rise of partisan nationalism and the damage it has wrought on the worldwide pursuit of issues requiring worldwide solutions, such as scientific cooperation, public health, and others, we mix analysis of both. We use both historical description and analysis. This analysis concludes with a description of why we must avoid the isolating nature of nationalism that separates people if we are to deal with issues of worldwide concern and to maintain a sustainable, survivable Earth, preserving individual and national freedom as we face the Earth's existential problems—at one point those that disincline us from cooperation, as well as solving the technological issues that make such solutions possible. *This question bears repeating in the name of planetary survival.* Do not say that such a thing is not possible or inconvenient. What will become impossible to accept becomes the destination to which we are headed if we do not open a new book in humanity's library, allowing us to cooperate internationally at a time when nations must find ways to solve common problems, most importantly the climate crisis. Inability to cooperate at this stage may doom everyone, eventually, to an overheated, stormy future plagued by droughts and deluges portending shortages of food and other essential commodities, meanwhile destroying large coastal urban areas because of rising sea levels. Future historians may look back at our time and wonder why as well as how we let our world succumb to isolating nationalism at a time when time was so short on cooperative intervention which is crucial for survival. That is the route to surrender, the easy way today, but the worst path long term.

1.4 Questions and Exercises

 1. What is the weight of human beings vis-a-vis all other living things on planet Earth? Is that ration going up, or down, our staying roughly the same? Why?

2. Define "Anthropocene." Please elaborate within the context of global climate change.

3. How does "climate change" differ from "global warming"?

4. Compare political and economic stances on global warming of Joe Biden and Donald Trump, Jr. (If the salient personnel change before this book goes out of print, please substitute.).

5. What is the primary difference between renewable and nonrenewable energies? Provide examples of each.

6. Please discuss the word "hoax" with reference to global warming.

1

7. Which type of energy provides more jobs in the United States—solar or coal mining?

8. Please discuss this statement: "The story in 2021 has been full of weather drama—record strength hurricanes, record size, heat, and sheer power of wildfires, flooding rains, droughts, some of the strongest and longest-lived tornadoes in history (in December, no less)." Are all of these occurrences a matter of chance or not?

9. What is the major difference between weather and climate? What exactly is weather, vis-a-vis climate?

10. What attributes are shared by the weather and climate of Afghanistan and Arizona?

11. Some people refer to global warming (climate change) as an "existential problem" for Earth and its inhabitants. What is meant by this? Do you agree or disagree? How and why?

12. Please define "nationalism." Could nationalism pose a problem for dealing with climate change? Why or why not?

13. What is the relationship of nationalism with international cooperation? Which is more amenable to dealing with climate change? Why?

14. What is the Jet Stream, and why is it important in weather forecasting (and sometimes) estimating long-term climate change? For example, what is the Arctic Oscillation, and why it is important to climate science?

15. What is "heat dome," and how does it set up some areas for extremely hot days, such as 116 °F in usually moss-draped western Oregon? How does Lytton, a very small town in British Columbia, reach 121 °F, followed by a forest fire that burns 90% of town?

16. How can salmon, a cold-water fish, be used as a climate indicator?

17. Please discuss "thermal inertia," and how it can postpone the effects of global warming about 50 years on land and about 150 years in the oceans. How does this phenomenon change awareness of how much warming is "in the pipeline"? Also discuss "global ocean heat uptake," for similar effects.

18. What did the late Dr. Carl Sagan theorize could be the penchant for civilizations destroying themselves in relatively short order after reaching a stage of technological complexity?

References

Diamond, J. (2022, January–February). The old man and the TEree. *Smithsonian*, pp. 21–33.

Hansen, J., & Sato, M. (2021). August temperature update & gas bag season approaches. *Red Green and Blue*. Retrieved May 12, 2022, from http://redgreenandblue.org/2021/09/22/dr-james-hansen-august-temperature-update-gas-bag-season-approaches/#.

NASA, Visible Earth. (2021). The long decline of arctic sea ice. Retrieved May 12, 2022, from https://visibleearth.nasa.gov/images/147746/the-long-decline-of-arctic-sea-ice/147746f.

Owour, Y. A. (2021, December). The spiritual voice of the forest. *National Geographic*, pp. 98–105.

Union of Concerned Scientists. (2021). Top annual CO_2 emitting countries, 2019. In: *Each country's share of CO_2 emissions.* Retrieved May 12, 2022, from https://www.ucsusa.org/resources/each-countrys-share-co2-emissions.

Further Reading

Cave, D., Bubola, E., & Sang-Hun, C. (2021, May 22). Long slide looms for world population, with sweeping ramifications. *New York Times*. Retrieved May 24, 2021, from https://www.nytimes.com/2021/05/22/world/global-population-shrinking.html.

Ducey, M. (2022, February 20). Sand hill cranes are flowing into central Nebraska. *Omaha World-Herald.* pp. B-1, B-2.

Science: Why So Urgent?

Saving Ourselves from Ourselves

Contents

© The Author(s), under exclusive license to Springer Nature
Switzerland AG 2023
B. E. Johansen, *Global Warming and the Climate Crisis*,
https://doi.org/10.1007/978-3-031-12354-2_2

⊜ Learning Objectives

1. Levels of carbon dioxide in Earth's atmosphere over time; relationship to temperature.
2. The key role of carbon dioxide in warming of the Earth over time.
3. Atmospheric gases and other components besides carbon dioxide that retain heat.
4. The role of thermal inertia in the delay of climate change's effects.
5. Fossil fuels: their origins and role in carbon dioxide emissions.
6. The role of carbon dioxide and other gases in atmospheric change during the distant paste (Note: the Pliocene).
7. The relationship between atmospheric carbon dioxide levels and sea levels and melting of ice.
8. The reaction of the human body when exposed to excess heat, especially at night.
9. Excess carbon dioxide plays a major role in acidification of the oceans, threatening any animal with a calcium shell, including phytoplankton, the basis of the oceanic food chain.
10. Students should learn how all of these factors (and others) are interrelated to pose a basic peril to life on Earth within the next century if major changes are not made to how we obtain and use energy.

Overview

Students should begin this course with a basic understanding of how carbon dioxide and other greenhouse gases retain heat in the atmosphere and why such retention will be dangerous (Actual effects of carbon dioxide and other greenhouse gases' retention are covered in ▶ Chap. 3). The effects of carbon dioxide, methane, etc. are not translated into heat instantly. Thermal inertia delays the effects, about 50 years in air and over land and about 150 years over oceans. Please note the Pliocene, 2–3 million years ago, when atmospheric carbon dioxide was about the same as today (about 420 parts per million), but temperatures were much higher. The difference is thermal inertia. The human body is acclimated to withstand a certain amount of heat level for a given amount of time, above such levels for a sustained period of time. Above that level (each organism has its own tolerance), sea life dies. The hydrological cycle also accelerates as temperatures rise, causing droughts and deluges to intensify, both of which pose problems for agriculture and human food supplies. For these and other reasons, rating the level of greenhouse gases in Earth's atmosphere lays a bet on a future Earth devoid of life as we and our ancestors have known it.

2

Four-inch-diameter hail; 50 inches of rain from *one* hurricane; 4 feet of snow from *one* blizzard; 250-mile-an-hour wind gusts from *one* tornado; a foot of snow in Texas; a thaw on the same day in Fairbanks, Alaska; 108 °F. June 28, 2021, in Seattle; a city in where air-conditioning was a rarity only a few years ago; 116 °F. in Portland, Oregon; 124 °F. in Lytton, British Columbia; meteorologists and fire-fighters observing a new type of disaster—a "firenado"—in Australia. These days, the news often looks like one very dangerous weather report.

2.1 Scientific Background

2.1.1 Carbon Dioxide Has No Motives. It Just Holds Heat

┌─ **Definition** ──────────────────────────────────

The trademark of global warming is heat, of course. In 2021 and 2022, around the world, we have had an abundance of that. Anyone who has not at least noticed that it is getting warmer and weirder has a serious case of carbon dioxide amnesia.

Carbon dioxide does not have a grudge against humans or other living things. It does not care if you drive an SUV or ride a bicycle. It does not care if you ignore its influence in atmospheric chemistry. All carbon dioxide does, in this instance, is to hold heat. The higher the CO_2 level, eventually, the higher the temperature. And at the current rate that we are generating it, a rising level of CO_2 will make our planet a very miserable (and given enough time), uninhabitable place.

2.1.2 A Variety of Malign Weathers

It is great to hear nearly everyone talking about climate change these days. Too bad, here in the shadow of very strong hurricanes and vast fires over large swaths of the US West, that it takes so much death and misery to get a thrust for change moving—all that malign weather on one continent, with much more elsewhere. On all continents, fires run rampant in some areas and deadly floods in others. The giant sequoia trees in California were threatened by fire, as is the resort town of South Lake Tahoe. Rains were on the highest point of Greenland. A three-digit afternoon greets residents in northern Siberia. Newfoundland experiences a hurricane. I hope people would not forget until the next huge hurricane makes a mess of almost half of the United States.

2.1.3 A Path That Stabilizes Climate

No bragging is intended here, but I started down this path in 1995. I read some of the scientific literature and began to ask why, in our daffy political environment, were not more people raising this issue, pounding the drum, regarding events that many scientists had come to regard as an existential threat for the entire Earth, along with all of its human, other animals, and plants? We can have a stable climate, clean air, and an unpolluted ocean and clean sources of energy that yield good-paying jobs. "It is up to the public to make sure that we get onto a path that stabilizes climate and allows all the creatures of Creation to continue to thrive on this planet," said Hansen et al. (1981).

2.1.4 Why Is Action so Urgent Now?

The effects of climate change are not theoretical, and they are not speculative problems that can be handed off to future generations. Economic activity around the world, as well as the lives of animals and plants, is being affected today by rising temperatures. This is not merely a matter of a few degrees on the thermometer but of alterations in human practices which shape the environment that sustains all of us.

Why is expeditious action on global warming so urgent? Why are so many scientists now sounding dire warnings that a decade or two of business-as-usual fossil fuel production and consumption will carry the Earth over various "tipping points" beyond which human ability to influence climate change may become irrelevant? Do not we have plenty of time for such a slow-motion crisis to unfold? No, we do not have. Two words that hardly ever come up in the public discourse over this issue demand attention as to why we do not have the luxury of time: thermal inertia.

2.1.5 Thermal Inertia and the Geophysical Facts of Global Warming

Knowledge of thermal inertia and feedbacks help to explain why an overwhelming consensus of climate scientists around the world has been ringing alarm bells for several years about Earth's changing climate—more specifically, the gradual, but accelerating rise in levels of carbon dioxide, methane, and other trace gases in the lower atmosphere because humans are combusting fossil fuels at a faster rate than nature has done at any point in the geophysical record. Carbon dioxide and methane levels in the atmosphere are more than 30% higher than at any time for which science has reliable proxy records—as of this writing, about 800,000 years, measured by Antarctic ice cores. We have begun to taste only the first fruits of this overload of greenhouse gases.

2

The amount of warming that we feel now is being restrained by the enormous thermal inertia of the oceans that cover two-thirds of Earth's surface. Once that warming is realized, however, it will endure for centuries, even if human consumption of fossil fuels stops completely. No additional forcing will be required to raise global temperature to that of the Pliocene, 2–3 million years ago, "a degree of warming that would surely yield dangerous climate impacts" said James Hansen et al. Equilibrium sea-level rise for today's 415 p.p.m. CO_2 is at least several meters, judging from paleoclimate history about 50 years ago (on land) and 100–150 years ago (in the oceans).

This is important and worth repeating: today's weather and climate (on land) reflect carbon dioxide levels, methane levels, etc. about 1970. If you are more than 60 years of age, perhaps you remember all those anti-Vietnam war rallies. In the ocean, recall when Warren Harding was the president and World War I was a recent memory. Roughly 150 years ago, the Wounded Knee massacre was in the future, and Thomas Edison was seeking the correct filament that would commercialize electric light.

I hope I have your attention, because taking our exercise in geophysical time travel into the future, if we quit using fuels that emit fossil fuels *today*, the proportion of greenhouse gases in the atmosphere and oceans will continue to rise for 50–150 years into the future. After that, assuming that we do not emit a single molecule of CO_2 or MH4, their levels will begin to fall. These are the geophysical facts. And that is why getting on top of this problem is so important. Today, most scientists know this. Odds are that most other people, including a number of rancid-mouthed political figures, do not. One may ask how and why such ignorant people are laying plans for future generations.

2.1.6 Carbon Dioxide's Natural Cycle

As part of Earth's natural cycle, the greenhouse effect is very necessary to life on Earth. Without it, the planet's average temperature would be minus 2 °F. Natural warming makes much of the world—outside of the Arctic, Antarctic, and deserts—habitable. It is an added warming provoked by human combustion of fossil fuels that causes a problem. Like chocolate, a little carbon dioxide and methane are a good thing; too much is toxic to the system. Fossil fuels provide us comfort and convenience, but raising their use in fundamental ways presents the challenge of the century and, most probably, for several centuries to come. Unless we wean ourselves from fossil fuels, and do so quickly, the *real* problems will begin after the middle of the twenty-first century. Sir John Houghton, one of the world's earliest experts on global warming, told the London *Independent*: "We are getting almost to the point of irreversible meltdown, and will pass it soon if we are not careful" (Lean, 2004, 8).

"Failure to act now is the most risky and most expensive thing we could do" warned a statement issued on June 2, 2008, by 1700 prominent scientists and economists under the aegis of the American Association for the Advancement of Science (AAAS). The strength of the science on climate change compelled the signers to warn of its growing risks, including "sea-level rise, heat waves, droughts, wildfires, snowmelt [that will exhaust the sources of glacier-fed drivers], floods and disease, as well as increased plant and animal species extinctions" said lead signatory James McCarthy, the president of the AAAS (Slash, 2008). The scientists and economists asserted that smart greenhouse gas reduction strategies will ignite economic growth, generate new domestic jobs, protect public health, and strengthen energy security.

The evidence accumulates relentlessly. In one week late in 2013, reports indicated that mass loss from the Greenland's ice sheet had quadrupled in 20 years and that warmer-than-average ocean waters had stoked super typhoon Haiyan's intensity before it became the strongest tropical cyclone to hit land in recorded history, in Asia. The same has been said of Hurricane Ida, which was a strong but not remarkable storm when it crossed into the Gulf of Mexico early in August 2021. Gaining strength from Gulf waters in the upper 80 °F to near 90 °F., it blew up to nearly a Category 5 before slamming the coast as one of the strongest and most damaging cyclonic storms in US history. While other factors including upper-level winds can affect cyclone intensity, water temperature is the most important.

Regulators cancelled the 2014 shrimping season in the Gulf of Maine because of overharvesting and warming waters. The value of the harvest had shrunk from more than $10 million in 2011 to $1.2 million during a shortened harvest in 2013. This drumbeat of news is not unusual (Straneo & Heimbach, 2013, 36; Normile, 2013, 1027; Maine: Shrimp, 2013, A-21).

A report released late in 2008 by the US Geological Survey suggested that climate change in the United States was accelerating much more rapidly than previous studies had estimated. The report by the US Climate Change Science Program utilized findings of the periodic reports by the United Nations Intergovernmental Panel on Climate Change (IPCC) with special attention to rapid loss of sea ice in the Arctic Ocean and its relationship to enduring drought in the US Southwest. The report also raised estimates of sea-level rise by the end of the twenty-first century, to a level of about 4 ft, up from a consensus figure of 1.5 feet advanced by the IPCC in its 2007 assessment. The Greenland and Antarctic ice sheets are losing an average of 48 cubic miles of ice a year at that time, a rapid rise from earlier years. Konrad Steffen, the director of the Cooperative Institute for Research in Environmental Sciences at the University of Colorado at Boulder and lead author of the section on ice sheets, said that "The models the IPCC used did not factor in some of the dynamics that scientists now understand about ice sheet melting. Among other things, Steffen and his collaborators identified a process of lubrication, in which warmer ocean water undercuts coastal ice sheets and accelerates melting" (Eilperin, 2008b).

The report did not affix human blame to the extensive drought in the US Southwest during the previous several years, but Richard Seager, a senior research scientist at the Columbia University's Lamont-Doherty Earth Observatory, said that nearly all of the 24 computer models that the group used indicate that human activity will contribute to a continuing drought. "If the models are correct, it will transition in the coming years and decades to a more arid climate, and that transition is already underway," Seager said, adding that such conditions would probably

2

include prolonged droughts lasting more than a decade (Eilperin, 2008b). As of this writing (late 2021), that drought had continued and accelerated at least 13 more years.

2.1.7 Thermal Inertia and Sea-Level Rise

James E. Hansen estimated that thermal inertia, given business-as-usual greenhouse gas emissions, would provoke a 25-m (more than 80-foot) sea-level rise within the subsequent two centuries (after the year 2100). Run your eyes across a map of the Earth and notice how many urban areas lie within 25 m of sea level: Shanghai, London, Mumbai, New York City, Miami, Jakarta, Kolkata, and many others. This is an estimate, so assume that Hansen is off by half—not likely, since he is a careful scientist, but possible. Half of 25 m is still quite a bit of sea-level rise, and astonishing as well, given the human affinity for acquiring homes and businesses within 12.5 m of any given high tide line.

If large-scale melting of the world's mountain glaciers, along with a large part of Greenland's and parts of the West Antarctic ice sheets, could add up to 25 m of sea-level rise worldwide, shall we lay bets on the date that first-floor toilets will back up from flooding at the White House, about 50 feet above Washington D.C.'s Tidal Basin? The British government is already discussing how much time will pass before its seat of government in London may have to abandon a city swallowed by rising waters.

According to James E. Hansen, "Global warming differs from previous pollution problems in two fundamental ways. With water pollution or common air pollution (smog), the problems occur immediately when the pollutant is emitted. If we decide there is a problem and stop emitting them, the problem goes away. However, global warming is caused by greenhouse gases that have a lifetime of hundreds of years. So we can't wait until we have a full-blown problem and then say "Oh, we better stop emitting these." It is too late. Carbon dioxide can hang around for centuries.

"The second major difference with the global warming problem is that the climate system responds slowly to the gases that we add to the air. Because of the great thermal inertia of the ocean, only about half of the eventual warming due to gases already in the air has been realized. The Earth has warmed 1.5 °F so far, but there is another one degree already in the pipeline. Moreover, there are surely more gases in the pipeline, because of power plants that we have in place and vehicles that we are not going to take off the road."

2.1.8 Destroying Creation for Future Generations

"One and one-half degrees!" said Hansen. "Who cares about that? Even with another degree or two in the pipeline, who cares about a few degrees? Well, we had better all care about it, because we have already brought the planet close to some tipping points. If we pass those tipping points there will be dramatic consequences.

We will leave an impoverished planet for our children, we will have been lousy stewards of creation, [and] we will have destroyed creation for future generations" (Johansen, 2007, 7).

In 50 more years, when our children are grandparents, the planetary emergency of which we are now tasting the first course will be a dominant theme in everyone's lives, unless we act now. Hansen and other scientists tell us that within a decade or two, thermal inertia will take off on its own, portending a hot, miserable future for coming generations.

Carbon dioxide represents about half of global warming. Methane, with residency in the air of a month or less, adds another significant fraction, about half that of carbon dioxide. A number of other trace chemicals, many of them created by human industry, contribute the rest of our atmospheric greenhouse gas overload. In 2020, atmospheric methane reached 1800 parts per billion, a new high for at least the previous 80,000 years, compared to about 700 parts per billion in pre-industrial times, 250 years ago, mainly because of increased emissions from human sources. The atmospheric concentration has been falling slowly since then, due mainly to increased efficiency in the oil and gas industries, which have reduced the amount of methane flared into the open air and wasted.

The Importance of Feedbacks

In addition to thermal inertia, a second important principle that influences climate change involves feedbacks, including albedo (reflectivity). Melting arctic ice exposes open water in summer, when the sun shines at the top of the world. Dark ocean water absorbs more heat than lighter-shaded ice and snow, causing even more heating and more melting. In the meantime, around the Arctic Circle, on land, permafrost melts, adding still more carbon dioxide and methane, accelerating the natural process that feeds upon itself. To these natural processes, add the trigger of increasing human emissions.

2.2 Impact of Global Warming Around the Globe

2.2.1 Permafrost Is Not Longer so Permanent

Across Alaska, Northern Canada, and Siberia, scientists have been finding signs that permafrost also is melting at an accelerating rate. As permafrost melts, additional carbon dioxide and methane convert from solid form, stored in the earth, to gas, releasing carbon dioxide and heat to the atmosphere. Once again, human contributions of greenhouse gases are provoking a natural process. Parts of the Trans-Siberian Railway's track have twisted and sunk due to the melting of permafrost, for example, causing delays of service of several days at a time. Scientists in Siberia report methane and carbon dioxide bubbling out of melting permafrost that now sometimes does not refreeze, even during winter. Greenland lost a record amount of ice during an extra-warm 2019, with the melt massive enough to cover California in more than 4 feet (1.25 m) of water.

"Not only is the Greenland ice sheet melting, but it's melting at a faster and faster pace," said Ingo Sasgen, a geoscientist at the Alfred Wegener Institute in Germany (Borenstein, 2020). In 2019, Greenland melt added 0.06 inches (1.5 mm) to global sea-level rise. That sounds like a tiny amount but "in our world it's huge, that's astounding," said study co-author Alex Gardner, a NASA ice scientist (Borenstein, 2020). Add more water from melting in other ice sheets and glaciers, along with an ocean that expands as it warms—and that translates into slowly rising sea levels, coastal flooding, and other problems, he said.

The world's capital of permafrost is Russia, given its band of frozen land that reaches almost halfway around the world in the Arctic. Years ago, when science was less sophisticated, ground that was frozen at the surface (or under frozen lakes) was assumed to be permanent. Not so, especially in our times, when "permafrost" is thawing wherever it exists.

Russians have become perceptive experts at the movements of permafrost as well and have learned that frozen ground or water on thawed surface may be deceptive. As Joshua Yaffa explained in *The New Yorker* (2022), "During frigid winters, temperatures under [lakes'] surface[s] remain above freezing. Unfrozen water allows microbes to keep digesting organic matter long after the surrounding landscape is covered in snow. Water also has a powerful erosion effect." Edges of the thawed water continue to thaw and collapse, taking pieces of not-so-permanent earth with it into lakes, releasing methane, a greenhouse gas.

Walter Anthony, a professor at the University of Alaska Fairbanks, said that "Once permafrost thaws to the point where it creates depressions filled with water, the thaw starts to go deep and fast and expands laterally—you can't really stop it" (Yaffa, 2022). These effects are compounded by the fact that snow acts like a blanket, preventing the full effect of cold air from reaching the "brew" under the ice. That plus the fact that air temperatures are increasing more rapidly in the Arctic and Antarctic than anyplace else on Earth.

"Much of the scientific community has come to see permafrost thaw…as a slow-motion disaster," wrote Yaffa. This naturally created methane is slowly released across the large expanse of Russia, Alaska, and Canada, sometimes slowly bubbling to the surface in tell-tale lumps, and mixes with other greenhouse gases, including those released by human combustion, all contributing to warming of Earth's lower atmosphere. For many years, projections of the United Nations Intergovernmental Panel on Climate Change (IPCC) had trouble factoring in the addition of Arctic methane, and even recently its estimates have been too vague to be useful, making it a "wild card" of climate science, according to Ted Schuur, who leads a project on permafrost thaw and climate change at the University of Northern Arizona (Yaffa, 2022).

2.2.2 Carbon Sinks to Sources

In the meantime, "Across the Arctic," Yaffa wrote (2022), "ecosystems are shifting from carbon sinks—which absorb more greenhouse gases than they release—to carbon sources," which are net sources of them. In addition, fires, which have become common around the Arctic in recent years, have been adding to the greenhouse gas overload—with eight million hectares burned during 2021 alone, an area about the size of Maine—releasing the equivalent of more than 500 megatons of carbon dioxide (Yaffa, 2022) (also see ◻ Fig. 2.1).

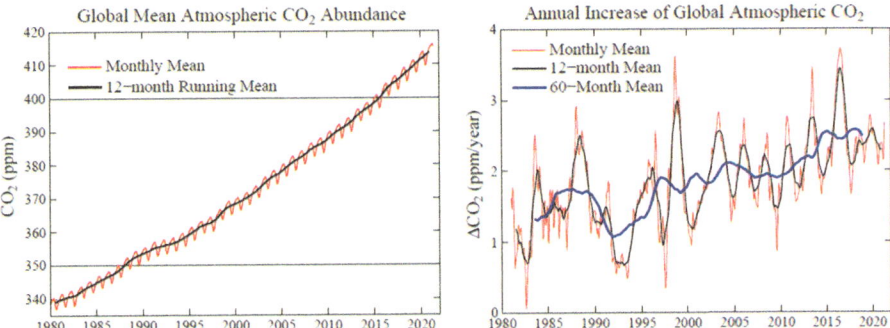

⊙ Fig. 2.1 (Left) Global mean atmospheric CO_2 abundance. (Right) Annual increase of global atmospheric CO_2. (Reproduced with permission from Hansen & Sato, 2021). Rise of carbon dioxide from 40 years ago to today. Note that the preindustrial cycle was between roughly 180 and about 280 parts per million. Today (e.g., 2022), it is about 420 p.p.m., having more than doubled in less than a century. Also note the steep trajectory of the curve

This environmental horror story has not been lost on Vladimir Putin, Russia's president, who "in 2003 had remarked that global warming simply meant "we'll spend less on fur coats," [who recently said] of the country's permafrost zone, "We have entire cities built on permafrost in the Arctic. If it all starts to thaw, what consequences will Russia face? Of course, we are concerned"" (Yaffa, 2022). The days of Putin's offhand remarks about fur coats have ended.

A study in the *Proceedings of the National Academy of Sciences* found that carbon dioxide levels during the last few years have increased at a faster rate than at any time since they have been measured. Human activity that produces carbon dioxide is increasing at an accelerating rate. In addition, the ability of natural "sinks" to absorb CO_2 on land and sea is declining. "All of these changes characterize a carbon cycle that is generating stronger-than-expected, and sooner-than-expected climate forcing," the authors of this study wrote (Canadella et al., 2007).

We have reached the point where global warming has been accepted by most thinking people as a threat involving every living being on our precious Earth. However, many of us have not a clue as to how to solve this problem. Such thinking (and doing) goes beyond solar and wind power, electric cars, and so forth. It has to do most basically with how we think and relate to each other as human beings not only on a personal level but also as world citizens. Thus, the first requirement necessary to solving this problem is acceptance of science-based truth.

2.2.3 The Perils of Warmer Nights

Nights are warming faster than days across most of the United States, with potentially deadly consequences. June 2021 was the hottest June on record in North America, with more than 1200 daily temperature records broken in the final week alone (Fountain, 2021a, 2021b, 2021c, 2021d, 2021e). But overlooked in much of the coverage were an even greater number of daily records set by a different—and

2

potentially more dangerous—measure of extreme heat: overnight temperatures (Bhatia & Choi-Schagrin, 2021). July 2021 was the hottest on record for any month in the recorded record (usually since about 1880 in the United States). Again, the trend was the same: night temperatures were above averages by a wider margin than daytime highs.

On average, nights are warming faster than days across most of the United States, according to the 2018 National Climate Assessment Report. It is part of a global trend that is being accelerated by climate change. Unusually hot summer nights may lead to an excessive number of deaths, according to climate scientists and environmental epidemiologists, because they diminish people's ability to cool down from the day's heat. "What's making the news is the highs, but nighttime minimums have an impact on mortality," said Lara Cushing, an environmental health scientist at the UCLA (University of California, Los Angeles) Fielding School of Public Health (Bhatia & Choi-Schagrin, 2021).

2.2.4 Why Warmer Nights Are Deadlier

Humans' bodies need time to cool after the hottest of days. Usually, cooling occurs, during sleep, when body temperature declines. After a hot day, "It's really important that people have an opportunity to bring their core body temperature down," said Kristie Ebi, an environmental health scientist at the University of Washington. "When it's really hot at night, you don't have that relief and it puts more physiological strain on your body" (Bhatia & Choi-Schagrin, 2021).

Heat waves that are both excessively hot and humid, in which sweat does not provide much cooling to dissipate body heat, also may compel consequences such as organ failure. The longer a heat wave persists, the more dangerous it may become for older people, young children, and pregnant women. Heat waves are also more likely to affect people whose wages depend on outdoor work, such as agriculture or construction, those who are homeless, and people with medical problems such as cardiovascular diseases and diabetes.

In 2006, a heat wave led to nearly 150 heat-related deaths in California, according to ▶ coroners' reports. (There were ▶ nearly 600 excess deaths during that period, suggesting an even greater effect.) What made that particular heat wave dangerous was its humidity, which traps heat at night, resulting in unusually high nighttime temperatures that caught Californians' off guard, said Tarik Benmarhnia, an environmental epidemiologist at the University of California, San Diego (Bhatia & Choi-Schagrin, 2021). When cities are affected by extreme heat, poorer communities tend to be most vulnerable, he said. Heat-related deaths and hospitalizations in the 2006 California heat wave were highest in ZIP (Zone Improvement Plan) codes with fewest air-conditioners. The highest-income homes were ▶ more than three times as likely to use central air-conditioning compared with the lowest-income ones.

Some cities seek to alleviate the effects of heat waves by opening cooling centers, checking in on vulnerable people, and providing bottled water. It is really the nighttime that is the problem," said Rupa Basu, the chief of air and climate epidemiology at the CalEPA's Office of Environmental Health Hazard Assessment (Bhatia & Choi-Schagrin, 2021).

2.2.5 Why Are Nights Getting Warmer?

In some parts of the Pacific Northwest during later June 2021, temperatures soared nearly 30 °F above their average, an extreme that "would have been virtually impossible without climate change," said Geert Jan van Oldenborgh of the Royal Netherlands Meteorological Institute (Choi-Schagrin, 2021).

As temperatures rise, the air can hold more moisture. Water vapor accounts for around 85% of the greenhouse effect, according to Alexander Gershunov, a research meteorologist at the Scripps Institution of Oceanography at the University of California, San Diego. The water vapor does not cause the initial warming, but there is a feedback loop: higher temperatures increase moisture in the air, and more moisture traps more heat close to the ground's surface, like a blanket, which leads to even more warming (Choi-Schagrin, 2021).

"Of all the extreme weather events, heat waves are the most directly related to climate change," he said. He thinks of climate change as the "steroids" behind heat waves. "In general, minimum temperatures are warming faster than maximum temperatures in the U.S.," said Claudia Tebaldi, an Earth scientist and climate modeler at the Pacific Northwest National Laboratory (Choi-Schagrin, 2021).

A few days after the heat wave from hell broke, hundreds were believed to have died in it: 20 in Washington State, 50 in Portland, 70 in Oregon, and hundreds in British Columbia. What is more, salmon (a cold-water fish) were dying en masse, and many beaches in and near Puget Sound were inundated with bacteria, not a pleasant place to spend a sizzling afternoon.

2.2.6 Patterns of Warming: Nighttime, Urban Hot Spots

Without fail, the most intense increases in global warming have been occurring at night, in large urban areas. During the day, urban landscapes suck up heat in large expanses of concrete, brick, and asphalt. A parking lot, for example, is notably hotter than a nearby urban park with grass and trees; older, bigger cities (e.g., New York City outside of Central Park, London, Shanghai, etc.) have higher heat loads due to concentration of buildings and streets. Even air-conditioning expels hotter air outside while providing cooler air inside buildings. This phenomenon has long been called the urban heat island effect. Heat held in urban areas also takes a longer time to dissipate than areas in the countryside, further inhibiting the cyclical low in the diurnal cycle.

The human body recharges during the cooler part of the diurnal (day/night) cycle; without it, the rate of death from heat-related maladies rises. During the moist intense heat waves, more people without air-conditioning who live concen-

2

trated in urban apartments with very little air circulation are the most likely to succumb to heat illness. "Not having that break from the heat is really hard on the human body—it builds up," said Jennifer Vanos, a professor in the School of Sustainability at Arizona State University.

These conditions intensify in thickly settled parts of urban areas where residents are more likely to be people of color, especially Blacks. Satellite imagery has been used to indicate that people of color who live in hotter parts of cities suffer from a higher death rate than usual in the worst heat waves. Neighborhoods with mainly Black residents average about 3 °F. higher than those usually populated by Whites with more trees and fewer heat-absorbing surfaces (Bhatia & Katz, 2021, A-12). Urban heat islands usually correlate not only with possession and use of air-conditioning but with income levels generally. This is not an isolated problem; this is not a single one-off anecdote, said Angel Hsu, a professor of public policy and environment at the University of North Carolina: "You are not going to see the impact on a linear basis," said Hsu. "It happens exponentially" (Bhatia & Katz, 2012).

Where Was Our Warming? It Was AWOL in the Arctic

Snow coverage in the Arctic continues to shrink. In Nebraska, in February, it was as cold as a _____ _____. (Fill in your own expletive.) A record cold and heavy snow in February did not mean that the whole Earth was cooling. It did mean that the upper air currents were twisting and turning in some weird ways.

February 16, 2021, brought Omaha a low of minus 23 °F., record daily cold. It was it a day when perhaps anyone in the area wanted to ask global warming specialists "Where is our global warming when we *really* need it?" Given a with-a-quick lesson in the geophysical facts, a.k.a. "The climate plays tricks on us." The carbon dioxide level is still with us, at about 420 parts per million in 2021. It is still holding more heat than it has in at least the last couple of million years. So what is going on? Climate involves changes over time. Weather is today's wind in our faces. Weather is the story; climate is the plot.

The coldest day I could recall in Omaha before February 16, 2021, was in December of 1983. It was a memorable day mainly for a low of about minus 22 °F. That was a temperature, not wind chill. I was walking along Dodge Street from Dundee, at about 52nd Street, westward during my second Omaha winter, to the University of Nebraska at Omaha, where I was beginning what would be a 38-year career, as an assistant professor of journalism (and, as turned out, other things as well) when a man I had never previously met stopped his car in the midst of a very busy street at rush hour, leaned over, shoved the front passenger door open, and commanded "GET IN!" "YES SIR!" I replied, escaping the coldest day of my life, until February 16, 2001.

On February 16–17, 2021, the forecast temperature bottomed at plus 7 in Fairbanks, Alaska; the average there is minus 13 °F. In Austin, Texas, the forecast low one day later was plus 7. The average there in February is 45. The fact that a coincidental low of 7 °F was reached 2 days apart in Fairbanks and Austin was an atmospheric prank played on us by the Arctic Oscillation, in which the jet stream (which steers storms and upper-air wind patterns at a height of jet aircraft), north to Alaska, and then southward and eastward, plunging to the Gulf of Mexico, and then north and slightly east up the US East Coast, sucking relatively warm air, which is loaded with moisture, out of the Gulf Stream. (Austin also had a quick 6.4 inches of snow.)

"Bomb Cyclones"

The above is one intriguing way in which a large, violent blizzard that meteorologists are fond of calling a "bomb cyclone" can enter the record books as a fruit of global warming. And how could that happen? During the winter of 2021–2022, for example, the Gulf Stream was averaging 10 °F above average, increasing the contrast with the cold air sweeping in off the continent.

As the contrast intensifies, the wind speeds as air circulates counterclockwise around the storm, colliding with cold air over the land, causing the storm to intensify, wringing out prodigious amounts of snow over the United States' northeast and Middle Atlantic states. The western side of the storm whipped cold air into Texas and nearby states (also into the southern United States), often causing deadly ice storms. This pattern (and others, such as the "ice-albedo feedback, mentioned above and below) considered as a whole may affect the entire Northern Hemisphere. Again, thanks to NASA's Earth Observatory, witness: "Throughout 2020, the Arctic Ocean and surrounding seas endured several notable weather and climate events. In spring, a persistent heat wave over Siberia provoked the rapid melting of sea ice in the East Siberian and Laptev Seas. By the end of summer, Arctic Ocean ice cover melted back to the second-lowest minimum on record. In autumn, the annual freeze-up of sea ice got off to a late and sluggish start."

2.2.7 A Long Decline in Ice Cover

Forty years of satellite data showed that 2020 was just the latest in a decade-long decline of Arctic sea ice. In a review of scientific literature, polar scientists Julienne Stroeve and Dirk Notz outlined some of these changes: in addition to shrinking ice cover, melting seasons are getting longer and sea ice is losing its longevity (also see ◘ Fig. 2.2).

The longer melting seasons are the result of increasingly earlier starts to spring melting and ever-later starts to freeze-up in autumn. Averaged across the entire

2

Not Just Declining, Sea Ice is Becoming Younger
Percent of Ice in the Arctic Ocean by Age (During the First Week of November)

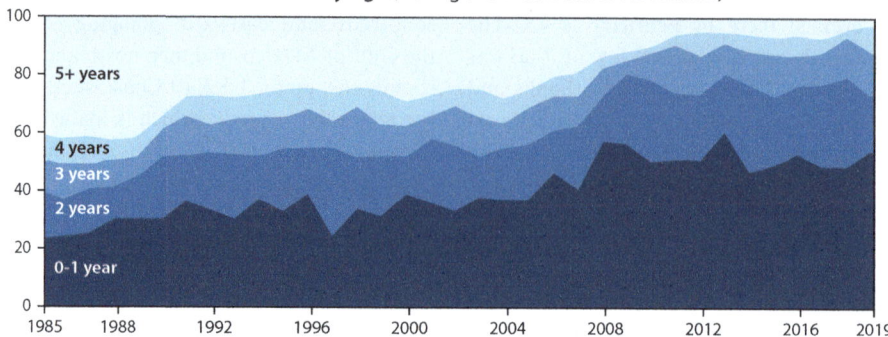

□ **Fig. 2.2**   Sea ice age: 1985–2019. Courtesy NASA Earth Observatory. Public Domain. The chart above demonstrates another way the Arctic is changing: the average age of sea ice is becoming younger. At the start of the satellite record, much of the ice covering the Arctic Ocean was greater than 4 years old. Today, most of the ice covering the Arctic Ocean is "first-year ice"—ice that forms in winter and does not survive a single summer melt season. (After sea ice reaches its minimum extent each September, the remaining ice graduates to second-year status)

Arctic Ocean, freeze-up is happening about a week later per decade. That equates to nearly one month later since the start of the satellite record in 1979.

The change is part of a cycle called the "ice-albedo feedback." Open ocean water absorbs 90% of the Sun's energy that falls on it; bright sea ice reflects 80% of it. With greater areas of the Arctic Ocean exposed to solar energy early in the season, more heat can be absorbed—a pattern that reinforces melting. As a result, the Arctic sea ice pack is becoming more fragile. In summer 2020, ships easily navigated the Northern Sea Route in ice-free waters and even made it to the North Pole without much resistance.

2.2.8 We Are Asphyxiating the Oceans

For anyone situated in the center of a large landmass, it is easy to forget that two-thirds of the Earth's surface is covered by water, which supports and sustains us. According to new studies in scientific journals that have not received much ink in the popular press, we are asphyxiating the oceans at a rapid rate as oxygen levels decline and corals die. The primary cause of this asphyxiation is the steady warming of the atmosphere that our ruling plutocracy insists does not exist. Surface warming can be even more dangerous in lakes, which by definition are surrounded by rapidly warming land. Even large bodies of inland water, such as the Great Lakes, warm much faster than open oceans.

The decline in oxygen levels is a worldwide trend, as described by Denise Breitburg and colleagues in *Science* (2018). "This oxygen loss, or deoxygenation, is one of the most important changes occurring in an ocean increasingly modified by human activities that have raised temperatures, carbon-dioxide levels, and nutrient inputs and have altered the abundances and distributions of marine species," they

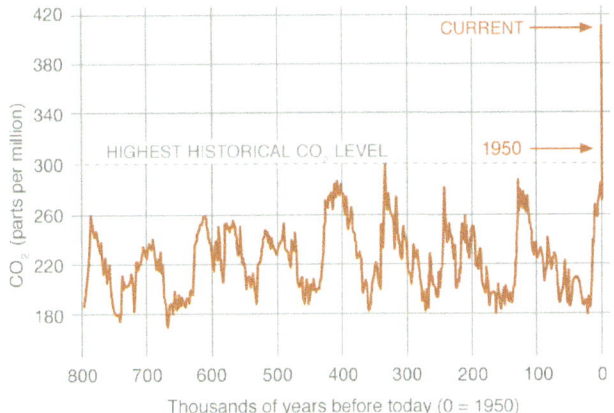

Fig. 2.3 World's historical carbon dioxide levels. Source: NOAA, NASA. Public Domain. Note the cycle at roughly 180 parts per million and 280 p.p.m. for 800 years until 1950, followed by a nearly vertical spike

wrote. In the open oceans, deoxygenation has been intensified by rising acidity provoked by carbon dioxide absorption, as well as the injection of nutrients from agriculture and sewage that are predominant in Lake Eire around Toledo, as well as other inland bodies of water.

The decline in oxygen levels has intensified since the 1950s, due nearly entirely to human activities, from increasing emissions of carbon dioxide and other greenhouse gases and overload of nitrogen-based fertilizers. All of these activities combine along coastlines at the mouths of major rivers (such as the Mississippi in the United States and the Ganges in India and Bangladesh to produce oxygen-starved "dead zones") (Fig. 2.3).

"Oxygen is fundamental to biological and biogeochemical processes in the ocean," wrote Breitburg and colleagues (2018). "Its decline can cause major changes in ocean productivity, biodiversity, and biogeochemical cycles. Analyses of direct measurements at sites around the world indicate that oxygen-minimum zones in the open ocean have expanded by several million square kilometers and that hundreds of coastal sites now have oxygen concentrations low enough to limit the distribution and abundance of animal populations and alter the cycling of important nutrients."

2.2.9 Corals: Death Due to Rising Temperatures

Coral reefs, the richest source of food in the oceans, also are steadily becoming more vulnerable due to rising temperatures, which bleach [kills] them. A study published in *Science* contained the work on ocean oxygen starvation described just how quickly the coral holocaust has developed. "Large-scale coral bleaching events, in which reefs become extremely fragile, were virtually unheard-of before the 1980s. But in the years since…the frequency of coral bleaching has

2

increased to the point that reefs no longer have sufficient recovery time between severe episodes," wrote Kendra Pierre-Louis and Brad Plummer in the *New York Times* (2018).

When coral bleaches, they wrote, "Overheated seawater causes corals to part ways with symbiotic plantlike organisms called zooxanthella that live inside of them. In addition to giving coral reefs their bright colors, zooxanthella also provide corals with oxygen, waste filtration, and up to 90 percent of their energy. Absent zooxanthella, corals not only take on a ghostly pallor, hence the term bleaching, but they are also more susceptible to death." Given enough time between bleaching events (10–15 years), corals can recover. The problem is that as ocean temperatures slowly rise, the intervals allowing recovery at sustainable temperatures are becoming shorter. Worldwide bleaching conditions are usually associated with El Niño conditions (the most severe of which have occurred in 1982–1983, 1998, and 2015–2016). These cause ocean temperatures to rise suddenly in tropical ocean waters that contain most coral reefs.

A study by Terry P. Hughes and colleagues (2018) examined 100 reefs worldwide and found that rising temperatures had reduced recovery time for reefs 50% in less than 40 years (1980–2016). Temperature peaks have risen higher over time, also increasing damage to reefs (Hughes, 2018). "As global warming has progressed, tropical sea surface temperatures are warmer now during current La Niña conditions than they were during El Niño events three decades ago," they wrote. "Consequently, as we transition to the Anthropocene, coral bleaching is occurring more frequently in all El Niño–Southern Oscillation phases, increasing the likelihood of annual bleaching in the coming decades." The year 1980 was chosen to begin the study because before the major El Niño of 1982–1983, mass bleaching was nearly unknown in the historical record (also see ◑ Fig. 2.4).

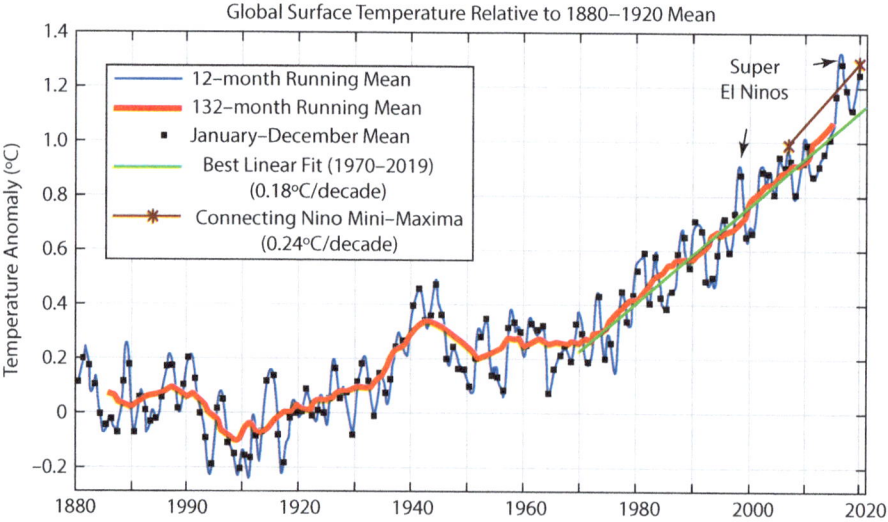

◑ **Fig. 2.4** World temperatures 1880–2020. Source: NASA. Public Domain. While the curve sometimes recedes, the rising trend over time is easy to see

"That year," according to Hughes et al., "Reefs across the tropical eastern Pacific exposed to warm El Niño year waters bleached. Coral reefs in Costa Rica, Panama and Columbia experienced 70 to 90 percent mortality. Most reefs in the Galápagos Islands, the cradle of Darwin's theory of evolution, experienced 95 percent mortality. While many mass bleaching were prompted by El Niño events, which tend to warm Pacific Ocean temperatures, the bleaching event that hit the Great Barrier Reef in 2017—the reef's first biennial (back-to-back bleaching, coverings 2 consecutive years—occurred at the beginning of a La Niña event, when ocean waters should have been cooler. It's a sign that global warming is steadily pushing up ocean temperatures even in cooler years," by mid-century, given present trends, provoking killing bleaching that will be the norm in large parts of the world's oceans which contain coral reefs.

2.2.10 Displaced Seasons in the Himalayas

During the winter of 2020–2021, Northern India suffered a gigantic dam breach that killed more than 250 people because of rapidly melting ice. Satellite photos from NASA indicated that ice can melt and now can occur at any altitude and at any season in the Himalayas. The cause here is climate change that can occur in a very irregular fashion, making protection for people who live there very difficult. There is a lesson here for all of us: as temperatures rise, changes often are not even. Sometimes, as NASA photos show, winter can alternate with spring or fall and then reverse itself.

More evidence of screwed-up seasons came from India's far north, where, in mid-winter, usually generous winter snows went AWOL. Instead, winter arrived early and laid down copious snow. By January, much of it had melted, in large part causing another dam break that unleashed a flood which killed several people.

As a bulletin from NASA's Earth Observatory said, "Each January, the Himalayas are typically blanketed with fresh snow. But unusual winter melting in recent months has left several glaciers and mountains without a new coating. Many glaciers are bare even at their crests. The high glacier passes of Nanpa La and Nup La, located around 30 miles (50 kilometers) northwest of Mount Everest…indicated that From October 2020 to January 2021, the average snow line—the boundary where snow-covered surfaces meet bare ground—on these glaciers rose around 100 meters (330 ft), indicating significant melt". "A substantial area of the glaciers are now probably experiencing melting year round," said Mauri Pelto, a glaciologist at Nichols College. "In years past, most melting stopped during winter and the snow line didn't move, but that's not really the case now" (Patel, 2021).

The world's most rapid period of melting ice began, slowly, a few decades after the start of the industrial age, with its appetite for coal and then oil and then natural gas. It has increased, drop by drop, speeding up, and then for 150 years, to our time, as walls of ice now crumble from the edges of the Greenland and Antarctic ice caps. The end of this ice world *is* being televised and observed by scientists peering from space satellites. It can be observed and measured but to date cannot be stopped, much less reversed. It takes more than scientists to stop this melting, which has reached the size of a swimming pool 10 feet deep the size of Ireland, every year, for the last 20 years (Greshko, 2021, 30). To stop "our reeling cryosphere….We need political will" (Greshko, 2021, 33). To reverse and then stop the combustion of fossil fuels—bearing in mind that because of thermal inertia (explained elsewhere in this chapter), the melting will continue (if any ice remains) for several decades hence.

Melting season in the region around Mount Everest is usually concentrated during the summer monsoon (April–September). In recent years, however, abnormally warm temperatures have extended the melting period, in some instances by as much as 4 months. As of January 22, 2021, weather stations at the Everest Base Camp reported maximum temperatures above freezing for 8 days that month. On January 13, temperatures peaked at 7 °C (45 °F).

Matthews and colleagues also observed less snow accumulation during the summer monsoon in recent years (Matthews et al., 2020). Generally, the summer monsoon delivers about 75% of the high peaks' annual snow accumulation. The team, however, noted an increase of rain and melting during the summer monsoon ▶ in 2019 and 2020 that reduced the amount of lasting snowfall near Mt. Everest. Due to less-than-normal seasonal snowfall, the warm temperatures are melting snow that was accumulated before the monsoon season. "I have seen snow-free situations last into January before, but usually the snow line isn't this high to start with," said Pelto (Patel, 2021).

"We have been basically seeing spring and summer-like conditions in the middle of winter," said Tom Matthews, a climate scientist at Loughborough University (United Kingdom) who helps manage weather stations at Mount Everest placed during the Rolex National Geographic Expedition. These melting events, said Matthews, are associated with pulses of warm air carried in by prevailing winds from the west (Patel, 2021).

Colder weather and new snowfall are needed to pause the melting, said Pelto. Since January 23, temperatures have dipped below freezing, but snowfall is still light and conditions are dry. Dry, colder weather drives sublimation, converting the solid snow directly to a gaseous state into the air and further reducing the snow cover.

"I don't think the ablation [melting] season is near ending. Until you get new snowfall, you will continue to experience melting and sublimation." said Pelto. "It is evident that the ablation season length and ablation area on these high elevation glaciers is expanding" (Patel, 2021).

Dominated by thin first-year ice, along with some older ice thinned by warm ocean water, the Arctic sea ice pack is becoming more fragile. In summer 2020, ships easily navigated the Northern Sea Route in ice-free waters and even made it to the North Pole without much resistance.

Fortunately, summers are still not entirely ice-free. "We've been hovering for some time around 4 million square kilometers of Arctic sea ice each summer," said Stroeve, a researcher at the University of Manitoba (Patel, 2021). She added that she intends to examine which conditions and processes could push sea ice to the next "precipitous drop"—when the extent of summer ice cover drops to a new benchmark of three million square kilometers.

Mar-a-Lago Is Due to a Dunking

The narrow nature of nationalistic thought is truly amazing. Witness former President Donald Trump and his often-repeated assumption that global warming is a hoax. Making such assumption, he probably is not concerned (or even aware) that his Mar-a-Lago estate on Florida's east coast is due to a dunking into rising seas, as the earth under it also subsides. Future histories may make of Trump and his estate an example of what ignorance of geophysical reality may cause.

Mar-a-Lago sits on a barrier island at Palm Beach, Florida, roughly 3–6 feet above the high tide line, which makes its owner's flat-out denial of global warming rather foolish. Denying the geophysical facts will not spare Mar-a-Lago a wet, salty demise. It is not simply rising seas that are going to put Trump's prized estate under water. Like most of that giant sandspit we call Florida (and, for that matter, most of the United States' Atlantic Ocean and Gulf of Mexico coastlines), Mar-a-Lago sits on land that is slowly sinking on seas that are rising, accelerating relative sea-level rise (Smith and Levin, 2016).

Thermal inertia—the amount of sea-level rise already "in the pipeline" due to fossil fuels already burned—guarantees a dunking for Mar-a-Lago. The amount of time all of this will require is subject to some debate. The best estimate may be 50–100 years—sooner, perhaps, if the area is struck by one or more major hurricanes during that period. Mar-a-Lago was built about 90 years ago. In 90 more years, it will very probably be fish food. Donald Trump, who likes games with odds, can bet on it.

Trump has no choice. Ignoring these facts will not postpone the reckoning. Mar-a-Lago is an ironic metaphor for a planet on suicide watch.

Even so, ignorance was in control of Trump's limited world. One will recall that Trump tweeted in 2012 that global warming is a hoax concocted by the Chinese to make US industry less competitive. In early December 2016, he told *Fox News* that "nobody really knows" (Johansen, 2015) whether climate change is real. He also chose Oklahoma Attorney General Scott Pruitt, also a staunch denier of climate change, to run the US Environmental Protection Agency, where ignorance of climatic reality had become a litmus test.

As Trump denied that climate change was a problem, officials in Palm Beach had overhauled 12 pumping stations to vastly increase the amount of seawater that they can pump into the Intracoastal Waterway with a 20-year, $120 million plan that eventually will be able to suck up almost one million gallons per minute. "I just deal with the reality that sea levels are rising," said Palm Beach Town Manager Thomas Bradford. "I don't want to rile people up about it" (Johansen, 2015). While the [now ex-president] president ignores climate change, brackish water is already bubbling out of the ground near the Trump estate during "king tides," when the sun and moon align, adding power (thus height) to movement of water adjacent to Trump's east-facing front yard and many other places. Low-lying cities along the US East Coast (one example being Norfolk, Virginia) already are afflicted with flooding from

2

"king tides," as the ocean slowly rises and the land slowly sinks. Millions of homes along the US Gulf and Atlantic coasts have been or will be subject to these flooding tides that can be partly attributed to climate change.

Please now return to Mar-a-Lago and Donald Trump's ignorance of the geophysical facts. According to Palm Beach County's online climate change mapping tool, roughly a quarter of Mar-a-Lago's verdant, palm tree-lined grounds will flood if sea levels rise 2–3 ft. The town of Palm Beach also has required higher seawalls around homes built on the adjacent seacoast. Suzanne Goldenberg wrote in *The Guardian* (2017) that "the water is already creeping up bridges and advancing on access roads, lawns and beaches because of sea-level rise Palm Beach County's own internal documents call for a 2-feet sea-level rise within the next 40 years. In 30 years, the grounds of Mar-a-Lago could be under at least a foot of water for 210 days a year because of tidal flooding along the intra-coastal waterway, with the water rising past some of the cottages and bungalows, [an] analysis by ► Coastal Risk Consulting found[that]....Parts of the estate are already at high risk of flooding under

heavy rains and storms, the analysis found. By 2045, the storm surge from even a category-two storm would bring waters crashing over the main swimming pool and up to the main building."

What is now Trump's estate was built during the mid-1920s by the Post family (the cereal company that became the General Foods). Trump acquired it in 1985 for less than $ten million, which was a relative bargain. The name "Mar-a-Lago" is a linguistic mishmash that can be translated as "sea to lake"), referencing the Atlantic Ocean and the Fort Worth lagoon, at the back (west) of the property. It is maintained as a private club, with a six-figure initiation fee and five-figure annual dues.

"Somewhat tantalizingly," wrote David Owen (2016) in *The New Yorker*, "it wouldn't take much of a sea-level rise or storm surge to inundate the entire place, since its sweeping lawns, like most of the rest of southern Florida, lie just a few feet above high tide. Perhaps a thorough soaking will convince Trump that global warming is more than a Chinese hoax and worth doing something about—unless he just builds a big seawall around the entire property, and makes the Chinese pay for it."

Warming and Birthing Complications

A study that examined more than 32 million births in the United States indicates that women exposed to higher-than-average temperatures or atmospheric pollution had an elevated chance of

having premature, stillborn, or underweight children. While it is too early to establish how strong the connection between the two may be, it is present and significant, according to Dr. Bruce

Bekkar, a gynecologist and obstetrician who took part in the study, which was concluded in 2020. He also chairs the Public Health Advisory Council of the Climate Action Campaign, of San Diego, California.

Bekkar said that early births had dropped about 20% across areas where fossil fuel plants had shut down in California. In addition to heat waves' role in early births, growing incidence of wildfires (and their role in air pollution) also increased the number of premature births. "It's going to get worse," said Bekkar of the relationship between warming temperatures and air pollution. ["It's] going to put increasing pressure on premature birthrates" (Cramer, 2021).

2.2.11 On a Beach in the Pliocene

Scientists have been looking for shells on beaches along the US East Coast, from near Washington, D.C., to Orlando, Florida. These are not ordinary shells on today's beaches but the surf line of the Pliocene, about 3 million years ago. The shells are fossilized, and they washed up the last time the carbon dioxide level in the atmosphere was 400 parts per million, *under* the level that we are now breathing. The project, funded by the National Science Foundation, is called "Pliomax."

The shoreline during the Pliocene between the present sites of Washington, D.C., and Orlando, Florida, roughly followed today's "fall line," where rivers become too shallow, narrow, and rocky to navigate most commercial ships, often between 60 and 100 miles inland. The question on the scientists' minds that has become especially salient following Hurricane Sandy, which pushed the New Jersey coast back about 35 ft, is this: how high will the ocean rise after the temperature rises the 4–5 °F anticipated by the end of this century by some climate modelers? Temperatures in the Pliocene were roughly equal to those projected for the end of the twenty-first century. A century or two after that, plus another century to melt ice as thermal inertia in the ocean catches up with surface temperatures, 60–100 miles of the US East Coast may be under water, east to west. This equation assumes that atmospheric carbon dioxide stops dead in its tracks at 400 p.p.m., a level that was surpassed soon Pliomax began. According to data collected from ice cores, at the start of the Industrial Revolution, the worldwide carbon dioxide level was about 285 parts per million. By 1959, it was 316; by 1970, 326; by 1980, 339; by 1990, 354; by 2000, 370; by 2010, 390; and by 2020; 414.

CO_2 levels have risen without pause since the ascendancy of fossil fuels began about 1850 as you can see in ◘ Fig. 2.5, and we are at about 420 parts per million now (2022, when this book probably will be published). Many scientists agree that any level above 350 is dangerous for the Earth and its inhabitants. Our atmosphere passed that level in 1989.

2

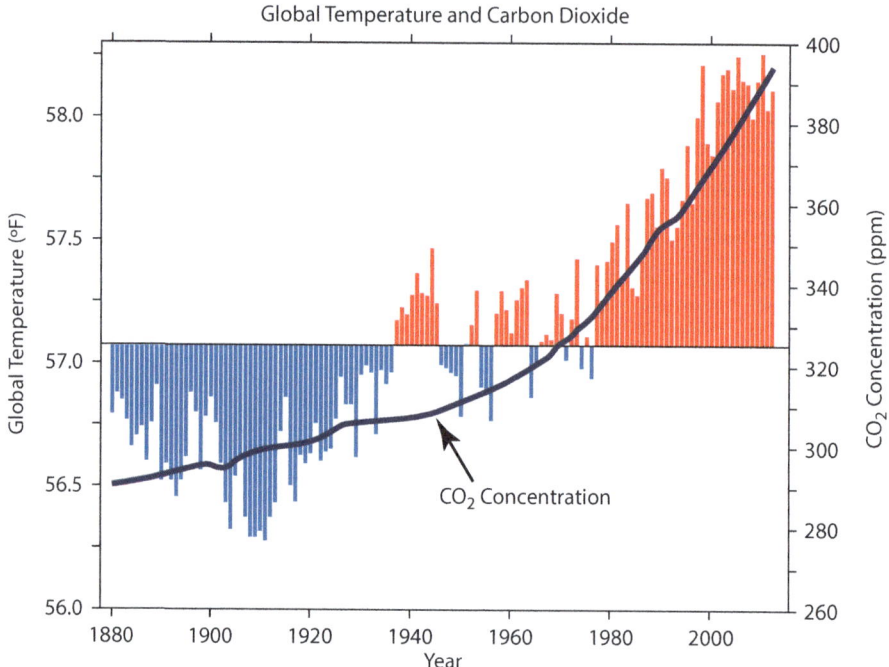

● **Fig. 2.5** This graph overlays increases in temperatures with the carbon dioxide curve, 1880–1920, showing the similarity of both, even during a relatively short period of time. Source: NASA/ GISS. Public domain

2.2.12 Greenhouse Gas Levels Correlate with Melting Ice

Within 300 years of "business as usual" (i.e., between 1900 and 2200), greenhouse gas levels (which very likely will be higher by then, accelerating this process) will have melted much of Greenland's ice, as well as the West Antarctic Ice Sheet. Portions of the larger East Antarctic Ice Sheet will be shedding glacial ice. Mountain glaciers and the Arctic Ocean's ice cap will exist only in history books and dusty, decade-old copies of the *National Geographic*.

Given today's CO_2 levels, and the speed with which they are rising, major sea-level rise is not a question of if, but when. Three hundred years may seem like a long time to you and me, but it is a blink of the eye in geologic time, much less than the time period required by the natural causes that raised the ocean to the levels that washed up shells near present-day Washington, D.C. Unless the levels of greenhouse gases (mainly CO_2 and methane) stop rising and then decline to no higher than 350 p.p.m., this script has already been written. This cake has been baked. Every pump of the gas pedal today is making it more likely. We do not know the exact timetable, because land itself also slowly rises and falls. We do know the seas will rise and that storms will cause short-term inundation above the average. We also know that most of the land that comprises the Eastern Seaboard is subsiding, that is, slowly sinking. The height at which some of the fossilized shells have been found as part of the Pliomax survey indicates that the level of sea-level rise is not a question with a single, linear, answer. In South Africa, some shells have been found 64 feet above modern shorelines. The most commonly cited figure for Pliocene sea level is about 80 feet above today's.

"I wish I could take people that question [about] the significance of sea-level rise out in the field with me," Maureen E. Raymo, who works at the ▶ Lamont-Doherty Earth Observatory, a unit of Columbia University, told the *New York Times*. "Because you just walk them up 30 or 40 feet in elevation above today's sea level and show them a fossil beach, with shells the size of a fist eroding out, and they can look at it with their own eyes and say, 'Wow, you didn't just make that up'" (Gillis, 2013).

"Absolutely, unequivocally, nature has changed before," Richard B. Alley, a leading climate scientist from Pennsylvania State University told the *New York Times*. "But it looks like we're going to do something bigger and faster than nature ever has." "I can merely tell you that every time in recent Earth history where we've had these kinds of temperatures for any protracted period of time, two polar ice sheets have catastrophically collapsed," said ▶ Jerry X. Mitrovica, an earth physicist at Harvard University (Gillis, 2013).

2.2.13 West Antarctica's Tipping Point

NASA satellites have detected a steadily widening rift in Antarctica's Pine Island glacier. This glacier acts as a plug, like a cork in a bottle, between the ocean and the West Antarctic Ice Sheet. Should this ice sheet melt into the sea, the world ocean would rise several feet. There are various estimates, and they indicate just how fragile such things can be. This "cork" is now breaking up. Greenland's ice is also melting at an accelerating rate. By 2020, more scientists were regarding the dissolution of the West Antarctic Ice Sheet as a high probability, possibly beginning in the twenty-second century. Two papers published during May 2014, in *Science* and *Geophysical Research Letters*, described upwelling of warm water from the deep ocean that is expected to provoke a slow-motion collapse of the West Antarctic Ice Sheet (Joughin et al., 2014; Rignot et al., 2014). This collapse was forecast in 1978 by glaciologist, John H. Mercer (who died in 1987) who was an outlier at that time. In the decades since, his ideas have gathered credibility. By 2014, scientists were setting a very rough timetable for the ice sheet's dissolution at 200–900 years.

A decrease in the melt rate back to earlier levels would be "too little, too late to stabilize the ice sheet," said Ian Joughin, a glaciologist at the University of Washington and lead author of a salient paper in *Science*, which concluded "There's no stabilization mechanism" (Gillis & Chang, 2014). "The basic problem is that much of the West Antarctic ice sheet sits below sea level in a kind of bowl-shaped depression in the earth," explained Justin Gillis and Kevin Chang in the

2

New York Times. "As Mercer outlined in 1978, once the part of the ice sheet sitting on the rim of the bowl melts and the ice retreats into deeper water, it becomes unstable and highly vulnerable to further melting" (Gillis & Chang, 2014).

2.2.14 Drought and Deluge

Global warming is not merely a matter of rising temperatures. Warming temperatures also change the behavior of the Earth's hydrological cycle, increasing the severity of storms, as well as the frequency and intensity of both droughts and deluges. Atmospheric moisture increases with temperature, and theory, as well as an increasing number of daily weather reports, strongly indicates that changes in precipitation patterns may vary wildly across time and space. The hydrological cycle seems to be changing more rapidly than temperatures. Such changes can be highly uneven, episodic, and often nasty. Both droughts and deluges are likely to become more severe as temperatures warm. They may even alternate in some regions. The US Southwest has been suffering a multi-decade drought by 2020 that had continued for several decades, for example, when the remains of Hurricane Norbert pumped up the monsoon and gave Phoenix its wettest single day on record, almost 3 inches. The evening national news suddenly was filled with images of cars drowning up to their rooftops there, as well as in Las Vegas, Nevada, and Riverside, California, east of Los Angeles. A few days later, the drought returned.

The summer of 2021 became a poster child for droughts and deluges across the United States, because of a stagnant oscillation of the jet stream, as explained in ▶ Chap. 2. The western half of the United States and adjacent Canada baked in record heat and drought that has been intensifying for at least 20 years. The part of the continent east of the Mississippi River was buried or drenched time and again by snow and then rain and then the sopping wet remains of severe hurricanes. This pattern can vary from time to time, as is illustrated by a NASA adaptation of satellite photos during a rare period when below-average temperatures dominated most of North America, Northern Europe, and Siberia (◘ Fig. 2.6). Most of the time, temperatures trend above average as they generally rise as in ◘ Fig. 2.7.

2.2.15 South Asian Downpours

India, with its annual monsoon dry season that usually alternates with heavy rains, has adapted to a drought-deluge cycle. About 90% of India's precipitation falls between June and September during an average year, so heavy rain in Mumbai (Bombay) during late July is hardly unusual. On July 26 and 27, 2005, however, 37.1 inches of rain fell in Mumbai within 48 hours, the heaviest such downpour on record for an Indian city (and, most probably, any sizable city) in two diurnal cycles. The deluge contributed to more than 1000 deaths in and near Mumbai and the surrounding Maharashtra State. The metropolitan area of 17 million people was largely closed down by the rain, as several people drowned in their cars. Mass

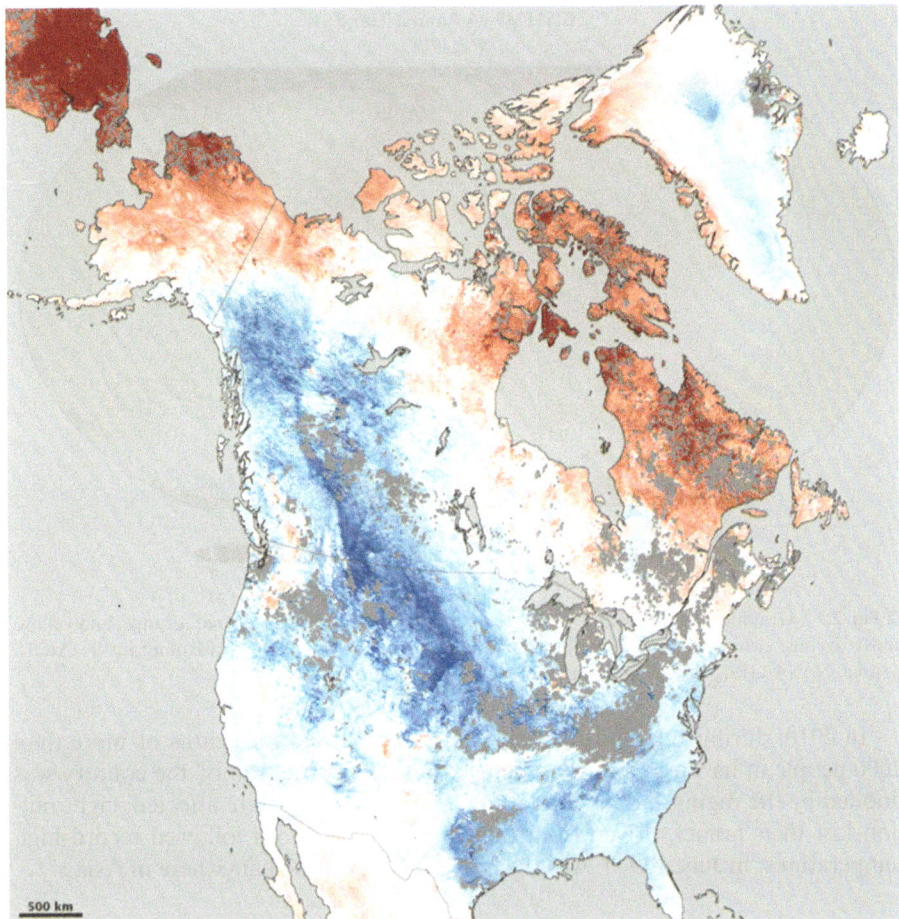

■ **Fig. 2.6** Northern hemisphere temperature variation, 1979–2019. Note that the jet stream usually flows between the blue (below) average and the red (above average) areas. Storms often roughly follow this line, with some variations. Source: NASA/NOAA/GISS (Public Domain)

transit and telephone services stopped. Other people were electrocuted by wires that fell onto flooded streets. Tens of thousands of animals also died (Record Rainfall 2005, A-12).

As floods devastated Pakistan in the summer of 2010, an extremely rare deluge inundated the town of Leh, in Ladakh, India, which sits in what is usually one of the driest deserts on the planet. The village sits in what is usually a high-altitude desert screened from most precipitation by surrounding mountains. The average rainfall there in August is a fraction of an inch. In the early morning of August 6, 2010, however, a half-hour deluge swept much of the village away, killing 150 people, with several hundred others missing. The storm was so intense, and isolated, that it missed a weather station in the valley and went unmeasured.

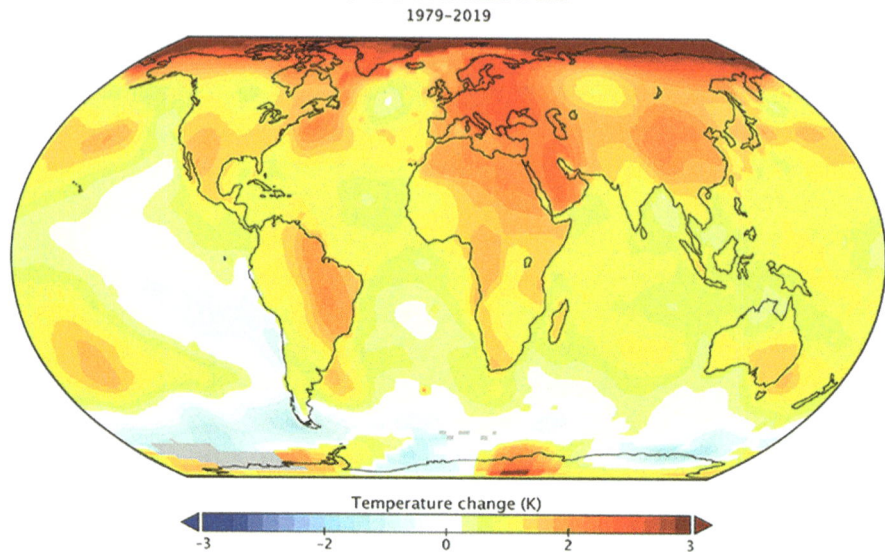

GISTEMP v4 Annual Trend
1979–2019

Temperature change (K)

-3   -2   0   2   3

☐ **Fig. 2.7** Graphic presentation of global temperatures vis-a-vis averages (red, orange, and yellow, above average; white, average; blue, below average). Note the intensity of warming in the Arctic. Source: NASA/GISS. Public Domain

In 2010, during July and August, Pakistan mourned the deaths of more than 2000 people in its worst monsoon deluge on record. One-fifth of the country was flooded by the raging Indus River, and 20 million people were affected; eight million lost their homes, driven from their homes. The floods followed record high temperatures; in July, 129 °F was the highest ever recorded anywhere in Asia.

2.2.16 Floods: Columbia to China

During the fall of 2010, Colombia experienced its worst rainy season in at least 30 years, leaving more than 130 people dead and a million homeless. Also during 2010, by the end of July, 1072 people were killed by floods in China, according to Shu Qingpeng, a deputy director of the Office of Flood Control and Drought Relief, with the central government. About 140 million people were affected by floods in China during those 7 months; a million homes had been destroyed, and economic damage was estimated at $31 billion.

The Rio Negro, a major Amazon tributary, went from a record high to record low levels between 2009 and 2010, a severe deluge-drought cycle. An Associated Press report said that "Floating homes along the Rio Negro now rest on muddy flats, and locals have had to modify boats to run in shallower waters in a region without roads. Some riverbanks have caved in, although no injuries have been reported. Enormous fields of trash and other debris have been revealed by the disappearing waters. The drought is hurting fishing, cattle, agriculture and other businesses, prompting authorities to declare a state of emergency in nearly 40 municipalities" (Azzioni, 2010).

2.2.17 Wildfires to Drowning, and Back Again, in Australia

Many areas that were beset by some of Australia's worst bush fires in 2019 and 2020, the most damaging in Australia's recoded history, in 2020 were inundated by extremely heavy rainfall that produced massive floods which forced many people from their homes. Australia has had more alternating floods and droughts than any other large land area on Earth—and some of the same extremes punished the same areas. An unusual convergence of climate factors leads to these alternating patterns, and temperature rises may be making them worse and doing more damage.

Damien Cave of the *New York Times* described one such area: "Two years ago [2019], the fields outside Christina Southwell's family home near Wee Waa, the cotton capital of Australia, looked like a dusty, brown desert as drought-fueled wildfires burned to the north and south. Last week [including December 5, 2021], after record-breaking rains, muddy floodwaters surrounded her, along with the stench of rotting crops. She had been trapped for days with just her cat, and still didn't know when the sludge would recede....All it leaves behind is this stink, and it's just going to get worse. Life on the land has always been hard in Australia, but the past few years have delivered one extreme after another, demanding new levels of resilience and pointing to the rising costs of a warming planet. For many Australians, moderate weather—a pleasant summer, a year without a state of emergency—increasingly feels like a luxury" (Cave, 2021, December 11).

In a second year of the weather phenomenon known as ▶ La Niña, meteorologists were predicting even more flooding for Australia's east coast, adding to the stress from the pandemic, not to mention from a recent *rural mouse plague of biblical proportions.*

Andy Pitman, a director of the ARC Center of Excellence for Climate Extremes at the University of New South Wales, said that "Australia is surrounded by powerful climate-driven oceans, from the tropical South Pacific to the colder Southern Ocean off Antarctica. The El Niño and La Niña patterns tend to hit Australia harder than they do in other places, with harsh droughts that end with major floods" (Cave, 2021).

2.2.18 Water Shortages and Floods Across India, Bangladesh, and Pakistan

People across India have been facing periodic critical drought conditions and water shortages. As of June 25, 2019, nearly 65% of the country's reservoirs were running dry. One of the worst affected areas has been the west-central state of Maharashtra, where 6 of the 17 reservoirs had dried out.

The water shortages have been only partially caused by lower-than-average rainfall. Chennai at one point had been ► devoid of rain for nearly 200 days. In 2018, the city experienced one of its ► weakest northeast monsoon seasons, on record in our memory, which usually runs from October to December and provides a large proportion of the area's annual rainfall. Additionally, the southwest summer monsoon season from around June through September had been delayed across India and unable to deliver substantial rainfall to Chennai.

Chennai has been ► struggling to meet water demands for several years. The National Institution for Transforming India (a government think tank) ► reported that Chennai, India's sixth largest city, was one of 21 cities that could run out of groundwater by 2020. Many ► news reports state*d* that the water shortages were due to low rainfall as well as poor water management.

On the other hand, India frequently has a staggering number of deaths and immense damages from flooding during its monsoons. Cyclones (hurricanes) can spin in off the Bay of Bengal and deliver rainfall and wind that in the past has taken the lives of hundreds to thousands of people. Mumbai once had 3 feet of rain (37 inches) in 2 days. Two feet from one storm has fallen on Chennai, on India's southwest coast (37 inches) fell once in Mumbai (Bombay) in 2 days. Snow melting off the Himalayas also can clause deadly flooding.

2.2.19 Extreme Precipitation and Scary Math

Warmer air generally holds more water vapor, and because of warming, the Earth's atmosphere is about 5% moister than 40 years ago, a factor in the increasing severity of deluges. Where it is not raining, more heat also provokes faster evaporation, increasing drought. Scientific studies are beginning to bear out what many incidents of extreme precipitation have been telling weather watchers who keep an eye on the meteorological sky: a warmer atmosphere produces more rain and snowfall and a greater risk of damaging floods.

These studies also lend credence to a scary prospect and one that popular media coverage largely misses: while increases in temperature are linear, intensity of precipitation has been said by some scientists to increase exponentially (Trenberth et al., 2003, 1205–1217). One must wonder what the atmosphere has in store for us once temperature increases get *really* serious.

"Climate models have improved a lot since ten years ago, when we basically couldn't say anything about rainfall," said Gabriele Hegerl, a climate researcher at the University of Edinburgh, Scotland (Schiermeier, 2011). Hegerl and colleagues compiled data from weather stations in the Northern Hemisphere and then compared it with precipitation models. The study covered the years 1951 through 1999.

A second study associated damaging floods in 2000 in England and Wales with temperature increases. Myles Allen of the University of Oxford and colleagues found that human-induced climate change "may have almost doubled the risk of the extremely wet weather that caused the floods (Pall, 2011 382-85)." While increases in extreme precipitation have been recognized on an incident-by-incident basis for more than ten years, these studies provide systematic evidence. "What has been considered a 1-in-100-years event in a stationary climate may actually occur twice as often in the future," said Allen (Schiermeier, 2011).

2.2.20 Rising Levels of Greenhouse Gases and Extreme Precipitation

Hegerl and colleagues wrote in *Nature*: "We show that human-induced increases in greenhouse gases have contributed to the observed intensification of heavy precipitation events found over approximately two-thirds of data-covered parts of Northern Hemisphere land areas....Changes in extreme precipitation projected by models, and thus the impacts of future changes in extreme precipitation, may be underestimated because models seem to underestimate the observed increase in heavy precipitation with warming" (Min et al., 2011, 382).

The study of flooding in the UK said "The U.K. floods of October and November 2000 occurred during the wettest autumn in England and Wales since records began in 1766. These floods damaged nearly 10,000 properties across that region, disrupted services severely, and caused insured losses estimated at £1.3 billion (c. $2 billion). Although the flooding was deemed a "wake-up call" regarding impacts of climate change at the time· Here the scientists presented "a multi-step, physically based "probabilistic event attribution" framework showing that it is very likely that global anthropogenic greenhouse gas emissions substantially increased the risk of flood occurrence in England and Wales in autumn 2000" (Pall, 2011, 382).

An organization including meteorologists from 189 countries during August 2010 said that summer's heat waves, droughts, and floods had displayed global warming's effects. "Several diverse extreme weather events are occurring concurrently around the world, giving rise to an unprecedented loss of human life and property. They include the record heat wave and wildfires in the Russian Federation, monsoonal flooding in Pakistan, rain-induced landslides in China, and calving of a large iceberg from the Greenland ice sheet," said the World Meteorological Organization (WMO).

The year 2010 was not unusual. In January 2011, near-record floods swept a large area of Northeastern Australia, including Brisbane, a city of two million people, forcing evacuation of 30,000 homes and businesses, killing at least 15 people. At the same time, floods rampaged through towns near Rio de Janeiro, Brazil, killing more than 600 people, most of them in landslides. Many were killed in their beds during the wee hours of the morning, including at least 250 in Teresopolis, a suburb of Rio de Janeiro.

2

2.2.21 Food Crises Add to Cross-Border Migrations

The Intergovernmental Panel on Climate Change (IPCC) issued a report in August 2019 describing a looming food crisis that among other things tied lack of food provoked by climate change to increasing migration across borders. The report said that food shortages are more likely to affect poor countries than rich ones, increasing the number of climate refugees.

This is already happening, in Syria, for example, and as the report points out, in Central America (Guatemala, Honduras, Nicaragua, Costa Rica, and El Salvador), a drought exceeding historical proportions played a major role in creating food shortages which have had a direct role increasing the number of people seeking asylum in the United States at the Mexican border that ex-President Donald Trump exploiting to increase political support for his cherished border wall. Between 2010 and 2015, the number of refugees seeking entry to the United States from these three countries increased about 500% (Flavelle, 2019).

Roughly 500 million people worldwide live on land that has recently turned to desert, and soil fertility has been depleted between 10 and 100 times faster than it is forming (Flavelle, 2019). These migrations are redefining politics around the world as nationalist regimes that oppose immigration rise to power. In the meantime, spreading agriculture has been draining wetlands, exposing stores of carbon dioxide which, once released, cause CO_2 levels to rise, prompting temperatures to rise even more quickly.

Rising levels of carbon dioxide and other greenhouse gases increase temperatures in the atmosphere and oceans, melting ice and reinforcing a process that increases the probability of extreme weather, including floods, droughts, and violent storms such as hurricanes, over time shrinking the amount of land available for agriculture. Rising temperatures also decrease the nutritional value of many foods. In combination, these factors, along with increasing populations, will increase famine. Cynthia Rosenzweig, a senior research scientist at the NASA Goddard Institute for Space Studies, said that "A particular danger is that food crises could develop on several continents at the same time" (Flavelle, 2019).

2.2.22 Long-Term Drought Patterns in the US West (and Other Places)

As carbon dioxide levels rise, atmospheric circulation patterns change. Spin the globe and you will notice that most of the world's deserts lie between 20 and 40° north and south latitude. Near the equator, warm, moist air rises, cools, and unleashes downpours. In the upper troposphere, the air spreads north and southward toward both poles, descending again at about 30° north and south latitude, drying the regions it reaches, creating deserts. For reasons that are not yet fully understood, as temperatures rise, these "Hadley Cells" reach further north and south, expanding arid areas on all continents not supplied directly with ocean-borne moisture. One such area is the US Southwest; this dry area expands from time to time over large parts of the US Midwest. "Data reveal a 4-degree latitudinal shift already," James Hansen and colleagues have written. "[This is] larger than model predictions, yielding increased aridity in the southern United States, the Mediterranean region, Australia and parts of Africa" (Hansen et al., 2008).

The US Southwest drought that began in the US Southwest about 2000 has been the longest there in 1200 years, according to a study issued in 2022. Unlike this drought, the most recent one was caused mainly by the greenhouse gas overload in the atmosphere, and it may not be over. The newest drought also was longer than several Maya droughts that preceded the present episode, between about 1000 C.E. and 1400 C.E. Before the latest data, about the 1200-year-old drought is the most intense in the present US Southwest. Had been estimated at 500 years.

"Exceptional conditions in the summer of 2021, when about two-thirds of the West was in extreme drought, 'really pushed it over the top,'" said A. Park Williams, a climate scientist at the University of California, Los Angeles, who led an analysis using tree ring data to gauge drought. As a result, 2000–2021 is the driest 22-year period since 800 A.D., which is as far back as the data goes," wrote Henry Fountain in the *New York Times* on February 14, 2022. Julie Cole, a climate scientist at the University of Michigan who was not involved in the research, said that while the findings were not surprising, "the study just makes clear how unusual the current conditions are" (Fountain, 2022). The study was published in *Nature Climate Change's* February 14, 2022, issue under the title "Rapid Intensification of the Emerging Southwestern North American Megadrought in 2020–2021."

An analogue for present-day drought in the Western US West may be the multi-episode "mega-drought" that hit the Mayas, in Central America between roughly 800 and 1300 CE. This drought, which was caused by natural cycles, not by human-induced greenhouse warming (The Mayas' and others' emissions of carbon were limited to wood burning), also destroyed some Native American cultures, such as the Anasazi (c. 1130–1160) and Mesa Verde (1250–1300). Temperatures warmed, and the atmospheric circulation (Hadley Cells) that helps to determine precipitation patterns moved northward. Today, with rapid warming, these conditions may be recreated but with turbocharged speed due to the worldwide emissions of fossil fuels.

The early and mid-twentieth century was the wettest in 1000 years in most of the American West. After that, a much drier spell set in, and global warming is probably only one of several reasons. A La Niña weather pattern in the Pacific sometimes (but not always) shifts storm tracks northward in the US West.

When rains fail, wildfires shift northward to places such as Canada's prairie provinces, ravaging large areas.

Diminishing snows melt earlier and give the fire season a boost (Kunzig, 2008). Drought has continued to intensify. By 2014, California and nearby states were experiencing the worst drought since detailed precipitation reports have been kept. By 2021, the drought cycle and historic heat had become even worse, as wildfires larger than any in the region's recorded history scorched large parts of the West. Historians of wildfires will remember the Dixie and Caldor and the intense efforts of more than 4000 firefighters to stop the blazes from scorching South Lake Tahoe (developed in detail in the next chapter). The drought included a 108 °F afternoon

2

in Seattle and 116 °F in Portland, Oregon. The fire season also has expanded, with fires lasting into October in some areas. During the last half of September 2021, the Sequoia National Park was closed due to the approach of massive wildfires that were threatening the continent's largest and oldest trees, providing an idea of these fires' historical impact.

The seasonal snow pack is vital for generation of electricity, as well as for agriculture, recreation, and drinking water. It also provides water for salmon to reach their breeding grounds. As winter weather warms, more abundant (and often flooding) winter rains run off too quickly to be useful, especially during dry Western summers. "We have known for decades that the hydrology of the West is changing, but for much of that time people said it was because of Mother Nature and that she would return to the old patterns in the future," said Tim Barnett of the Scripps Institution of Oceanography at the University of California at San Diego. "But we have found very clearly that global warming has done it, that it is the mechanism that explains the change and that things will be getting worse" (Kaufman, 2008).

2.2.23 Ocean Acidity and Coming Extinctions

Carbon dioxide overload in the atmosphere also acidifies the oceans, endangering marine animals that live in shells, from corals to sea urchins. Acidity also shrinks the number and size of phytoplankton, the basis of the oceanic food chain. By the early years of the twenty-first century, carbon dioxide levels in the oceans were rising more rapidly than at any time since the age of the dinosaurs, according to work published by Ken Caldeira and Michael E. Wickett. They wrote (2003) "We find that oceanic absorption of CO_2 from fossil fuels may result in larger pH changes over the next several centuries than any inferred in the geological record of the possible 300 million years, with the possible exception of those resulting from rare, extreme events such as bolide impacts or catastrophic methane hydrate degassing."

A "bolide" is a large extraterrestrial body (usually at least a half mile in diameter) that impacts the Earth at a speed roughly equal to that of a bullet in flight. A "bolide" could be a large asteroid. "Methane hydrate degassing" involves the rapid conversion of solid methane deposits on ocean floors to gaseous form in the atmosphere by warming temperatures.

Carbon dioxide is being injected into the oceans far faster than nature can neutralize it. Seawater is usually slightly alkaline, at about pH 8.2. The pH of the oceans has fallen 0.1 during industrial times. The scale is logarithmic, so a 0.1 change indicates a 30% increase in the concentration of hydrogen ions. Under a business-as-usual scenario, the pH may fall by roughly 0.5 by the year 2100.

Since the industrial revolution began, human beings have infused roughly 120 billion tons of carbon dioxide into the oceans. By 2006, the seas were absorbing an additional two billion tons of CO_2 per year. Every day, each citizen of the United States adds, on average, 40 pounds of carbon dioxide into the world's oceans. Continued carbon dioxide overload is acidifying the oceans at a rate of 100 times greater than at any other time in the geophysical record (Riebesell et al., 2007). One may surmise what this is doing, long term, to the acidity of the world's oceans and seas. It is rather analogous to the dangerous rise in carbon dioxide, methane, etc. overload in the atmosphere.

2.2.24 "Doomsday Scenario" for the Oceanic Food Chain

Levels of ocean surface acidity are now more than 30% above pre-industrial levels (about 150 years). This danger is most notable in the colder waters of the Arctic and Antarctic, which hold relatively more carbon dioxide than warmer ocean waters. A report on ocean acidification by Britain's Royal Society said that "without significant action to reduce CO_2 emissions," there may be "no place in the future oceans for many of the species and ecosystems we know today" (Kolbert, 2006). Stated more simply, increasing acidification of the oceans due to rising levels of carbon dioxide may threaten a large number of ocean species with extinction, including pteropod snails that feed salmon and other fish that human beings eat (Holland, 2001).

Once acidity reaches levels that dissolve calcium shells in the tropics, "It's a doomsday scenario for coral reefs," said Caldeira—that is, for corals not already killed by rising water temperatures. He anticipates that coral reefs will survive only in walled-off enclosures where acidity has been controlled by humankind, in open-ocean zoos. "Our emissions are huge compared with natural fluxes," said Caldeira. "If you could stop emissions and wait 10,000 years, natural processes would probably take care of most of it" (Holland, 2001, 110–111). Emissions, however, are not being curtailed. The result is a doomsday scenario for the oceanic food chain and major damage to people and other animals that depend on sea life in their diets.

Scientists have been investigating what such acidification may do to other animals with calcium shells that are exposed to increasing acidity. Gretchen Hofmann of the University of California, Santa Barbara, reported that rising ocean temperatures and acidification could be fatal to the purple sea urchin (*Strongylocentrotus purports*). At a pH of 7.8, the larvae of the purple sea urchin built skeletons with great difficulty. Warming the water in which the sea urchins lived compounded the effect (Kintisch & Stokstad, 2008).

2.2.25 Accelerating Ocean Acidity and Coral Reefs

Concern about rising acidity of the oceans was reinforced in January 2009 as 155 scientists from 26 countries organized by several international groups under the aegis of the United Nations released a report. "Severe damages are imminent," the group said in a summary of its deliberations (called the Monaco Declaration) at a symposium in Monaco during October 2008. "The chemistry is so fundamental and changes [are] so rapid and severe that impacts on organisms appear unavoidable," according to James Orr, a chemical oceanographer at the Marine Environmental Laboratory in Monaco who headed the symposium's scientific committee.

According to the declaration, "Ocean acidification may render most regions chemically inhospitable to coral reefs by 2050." The group said that ocean acidification will continue to increase unless atmospheric levels of carbon dioxide in the air stop rising. Referring to ocean acidity, Jason Hall-Spencer, a British marine biologist and reader at Plymouth University, said "Unfortunately, the biggest tipping point, the one at which at which the ecosystem starts to crash, is mean pH 7.8, which is what we are expecting to happen by 2100. So that is rather alarming" (Kolbert, 2014, 118).

2.2.26 Drought in the Amazon Valley

In 2010, the Amazon Valley, which has been notable for its heavy rain and lush forests, experienced its second drought in 5 years, with more in following years. The 2005 drought had been called a 100-year event by some scientists. "This is what's quite alarming—that we've seen these two very unusual events," said Simon Lewis, a University of Leeds (Great Britain) forest ecologist. "And those two unusual events are consistent with those predictions that suggest that the Amazon may be severely impacted over the next few decades by these droughts" (Joyce, 2011). Writing in *Science*, Lewis and colleagues said the droughts have been aggravated by northward movement of warm water in the Atlantic Ocean. That shift is carrying moisture north, away from the Amazon basin.

Longer term, while increases in temperature are linear, intensity of precipitation has been said by some scientists to increase exponentially (Trenberth et al., 2003). One must wonder what the atmosphere has in store for us once temperature increases get *really* serious and thermal inertia catches up with rising temperatures (see ▶ Chap. 2).

The Amazon Valley, an area that was a net source of oxygen, has, in recent years, frequently become a net source of carbon dioxide. Forests that once dominated the region have been replaced by large, scorched savannahs left by fires, prairies eaten down by cattle and other farm animals, gold (and other metals') strip mines, roads, and urban areas, as well as large swaths of bare mud that once was forest, now being repurposed as roof tiles for spreading suburban housing developments.

In large parts of Northeastern Brazil, forests and jungle have been replaced by savannah and then by farms and finally assembly grounds for these roof tiles. "On a recent day," wrote Jack Nicas in the *New York Times*, "With temperatures approaching 100 degrees, [a] river had run dry, the crops would not grow, and [a] family's 30 remaining cattle were quickly consuming the last pool of water" (Nicas, 2021, A-1). The entire scene was backed up by brown, nearly stripped dry hills.

This is desertification, a byproduct of human mismanagement of formerly verdant, green, oxygen-bearing land, accelerated by warming temperatures. It is spreading around the Earth by money-making capitalists who believe that this is the normal course of human endeavor. These include powerful people in high places, such as Brazil's President Bolsonaro.

Worldwide, half a billion human beings in 2021 lived on land that is becoming too dry to sustain them much longer (Nicas, 2021, A-1). This type of scenery is now commonplace on Australia's growing deserts (also ravaged by fire), parts of northern and northeastern China, the Southwestern quadrant of the United States, large parts of Southern Russia, Africa's growing Sahara, and other places. It is part of a chain of circumstances that robs surface soil of its ability to grow anything except, occasionally, a few weeds. Brazil's northeast experienced its worst drought on record between 2012 and 2017. In addition, each time that workers strip 3–5 feet of soil not get at the clay that may cover houses in Sao Paulo, the soil has required several million years to form. The ceramic company pays landowners about $10 for 30 tons of clay (Nicas, 2021, A-10).

The Senator from Coal Country

The science is evident on this matter, but a devotion to fossil-fueled thinking still enjoys something of a lock on the political rhetoric of the United States' Republican Party and even some Democrats. Witness ex-President Donald Trump, dismissing the entire body of evidence in one word ("Hoax!"), as he argued for increased mining of coal to restore jobs in an industry that even the capitalistic marketplace has decided is obsolete. Another example has been Joe Manchin III, a Democrat of West Virginia, where coal has a very important influence on politics, with many strategic campaign contributions. Manchin accentuates the coal industry's role in providing jobs for his constituents as record floods destroy some of their homes.

Manchin earned a considerable amount of publicity in 2021 as the only Democratic US senator to oppose President Joe Biden's $2 trillion "Build Back Better" plan, which included about $558 billion to combat climate change. Manchin was the only Democratic vote in an evenly split Senate, which meant that his one vote could kill the bill. Allowing it to pass, on the other hand, would have pinched fossil fuels, most notably oil, coal, and natural gas, while favoring renewable energy sources, such as wind and solar. Senator Manchin asserted that such a bill would harm his state's constituents, which would be correct. Such a bill also would place a major drain on Senator Manchin's personal bottom line. For example, Manchin's family owns a business (part of Enersystems) that sells subgrade coal from abandoned coal mines to a heavily polluting coal-burning power plant in his state.

What does Senator Manchin get out of this? He founded the aforementioned family coal brokerage that paid him between $500,000 and $1 million in 2020. (These numbers come from governmental disclosure forms that require declaration of income from various activities within a range, as reported in the *New York Times*.) Enersystems as a whole was worth between $1 and $5 million. Manchin in 2020 disclosed that this business paid him dividends, interest, and business income worth $492,000 in 2020. Manchin routinely receives more money from coal, oil, and gas interests than any other US senator (Weisman & Friedman, 2021).

As we say in basic journalism, follow the money.

2

Mosquitos and Earthworms: Prosaic but Potent Climate Changers

The standard image that most of us conjure up when we think of climate change is oil wells, smokestacks, melting ice, floods, and fires. Few of us consider the climate-changing role of earthworms and mosquitoes, which are spreading northward at a quickening rate as temperatures warm. Climate scientists have been surprised as well, as worms wriggle and mosquitoes buzz into their climate models.

Both earthworms and mosquitoes are spreading northward in the Northern Hemisphere as temperatures warm, especially in the boreal forests that encircle the Arctic from Canada and Alaska to Northern Europe and Siberia. When I visited the Arctic almost 20 years ago to write about the effects of warming there, I was surprised, on landing at the airport on Iqaluit, on Baffin Island, to be greeted by small swarms of very large, slow mosquitoes that were so easy and satisfying to swat.

Much closer to the equator, I recall, visiting India, that one had to be aware of which kind of mosquito came out at what time of the day (usually morning or evening) and what they could spear you with (malaria or dengue fever, usually). India has some of the best mosquito repellent on the planet, creams with pleasant odors, much nicer than our industrial-smelling raid. In India, concerning mosquitoes, 1.3 billion potential targets know what they are doing.

In most of the United States, outside of the very rare instance of West Nile virus, mosquitos are an itchy pest but otherwise relatively harmless. However, in the event that one becomes a target for a mosquito that does carry West Nile, it can kill you (after several months of agony) just as assuredly as untreated dengue or malaria. In a warmer world, mosquitoes are spreading northward, along with all the maladies that they carry.

"Climate change is poised to increase the spread of dengue fever, which is common in parts of the world with warmer climates like Brazil and India, a new study warns, wrote Kendra Pierre-Louis and Nadja Popovich in the *New York Times* (2019)." "Worldwide each year, there are about 100 million cases of dengue infections severe enough to cause symptoms, which may include fever, debilitating joint pain and internal bleeding. There are an estimated 10,000 deaths from dengue—also nicknamed breakbone fever—which is transmitted by *Aedes* mosquitoes that also spread Zika and chikungunya."

A study published in *Nature Microbiology* during June 2019 forecast expansion of dengue-transmitting mosquitoes into the mid-latitudes of the southeastern United States, coastal China, and parts of Australia's Outback. The warmer it gets, the larger the mosquitoes' habitation ranges. Warmer weather also encourages them to reproduce faster, as well as earlier in spring and later in the fall, as periods of weather without a killing freeze expands.

2.2.27 **Earthworms Spread Northward, Too**

Cindy Shaw, a carbon-research scientist with the Canadian Forest Service, was studying an area in northern Alberta that was recovering from oil and gas exploitation, when she saw evidence of earthworm activity for the first time. Until then, the textbook on earthworm activity in that area was that they had disappeared in an ice age 10,000 years ago. "I was amazed," she said. "At the very first plot, there was a lot of evidence of earthworm activity" (Mitchell, 2019). "Now," wrote Alanna Mitchell in the *New York Times*, "invasive earthworm species from southern Europe—survivors of that frozen epoch, and introduced to this continent by European immigrants centuries ago—are making their way through northern forests, their spread hastened by roads, timber and petroleum activity, tire treads, boats, anglers and even gardeners" (2019).

The climatic bottom line is this: as the worms eat, they release carbon dioxide that had been stored in the forest floor. With the help of the worms, the boreal forests are turning from carbon "sinks" to carbon sources, and this is going on in many of the boreal forests that ring the Arctic. The incursion of earthworms is so new to climate science that it has not been factored into models that anticipate future carbon levels and temperatures. "It is a significant change to the carbon dynamic and how we understand it works," Shaw said. "We don't truly understand the rate or the magnitude of that change" (Pierre-Louis & Popovich, 2019).

The worms in question here feed on the surface of the forest floor. Others burrow into the soil and break it down, freeing nutrients. They break down the soil's organic matter, acting as a fertilizing agent, helping trees and other plants to grow at a more rapid pace, and sequestering carbon. However, as this type of worm increases decomposition of soil, it also adds carbon dioxide to the atmosphere. Scientists have yet to even estimate what the net balance may be. As worms spread northward, this question is becoming more compelling.

2.2.28 **Earthworms' Eating Habits and Rising CO_2 Levels**

Alanna Mitchell, of the *New York Times*, wrote that in the boreal forest, "This spongy layer of leaf litter contains most of the carbon stored in the boreal soil. As it turns out, most of the invading earthworms in the North American boreal appear to be the type that love to devour leaf litter and stay above ground, releasing carbon. Erin K. Cameron, an environmental scientist at Saint Mary's University in Halifax, Nova Scotia, who studies the boreal incursion of earthworms, found that 99.8% of the earthworms in her study area in Alberta belonged to *Dendrobaena octaedra*, an invasive species that eats leaf litter but does not burrow into the soil. In 2015, Dr. Cameron discovered that forest floor carbon is reduced by between 50% and 94%, mostly in the first 40 years. That carbon, no longer sequestered, goes

2

into the atmosphere. By 2009, earthworms had worked their way into 9% of Northeastern Alberta's boreal forests. By 2049, she projected, that figure would be half, portending more carbon dioxide from what used to be an inconsequential animal.

Kyungsoo Yoo, a soil scientist at the University of Minnesota, has found earthworms that have wriggled as far north as the edge of permafrost in Alaska, north of the boreal forests, reinforcing melting there and increasing the amount of carbon dioxide being released. One can see where this is headed. A few billion hungry earthworms will be a potent climatic force.

2.2.29 California Fires: Why There, Why Now?

Enormous wildfires have been constantly raging in California in recent summers, at a size and intensity heretofore unknown, at least in human history. Eight of the state's ten largest fires on record—and 12 of the top 20—have scorched areas of the state within the past 5 years (as of the end of the 2021 fire season), according to the ▶ California Department of Forestry and Fire Protection (Cal Fire). Together, those 12 fires have burned about 4% of California's total area—a Connecticut-sized amount of land.

The Dixie fire and the August fire complex (2020) stand out for their size. Each of these burned nearly ▶ one million acres—an area larger than Rhode Island—as they burned for several months in Northern California. Several other large fires, as well as many smaller ones in densely populated areas, have been catastrophic vis-a-vis the number of structures destroyed and ▶ lives lost. These fires together have destroyed 40,000 homes, businesses, and pieces of infrastructure. The Dixie fire destroyed 1329 structures and cost hundreds of millions of dollars to fight.

"The numbers are really worrisome, but they are not at all surprising to fire scientists," said Jon Keeley, a US Geological Survey scientist based in Sequoia National Park (Voiland, 2021). He is among several experts who say that a confluence of factors has driven the surge of large, destructive fires in California: unusual drought and heat exacerbated by climate change, overgrown forests caused by decades of fire suppression, and rapid population growth along the edges of forests.

"The current drought is unprecedented," said Keeley in 2021. "Each of the past three decades has had substantially worse drought than any decade over the last 150 years" (Voiland, 2021). In the short term, drought exacerbates fires by sapping trees and plants of moisture and making them easier to burn. Over the long term, it adds ▶ vast amounts of dead wood to the landscape and makes intense fires ˉ more likely.

The 2020–2021 and 2021–2022 droughts have been especially extreme. "The last two years in California have brought compound drought conditions—effectively, very dry winters followed by relentless summer heat and atmospheric aridity," explained John Abatzoglou, a climate scientist at the University of California, Merced. "This has left soil and vegetation parched across much of California, so the landscape is capable of carrying fire that resists suppression" (Voiland, 2021).

Data from the ▶ Western Regional Climate Center indicated that the northern two-thirds of the state has received only half of usual rainfall over the past several years. The US Drought Monitor has categorized about 85–90% of California as experiencing "exceptional" or "extreme" drought for all of summer 2021. And the period between September 2019 and August 2021 ranked as the second driest on record for the state, according to data from the National Centers for Environmental Information.

Daniel Swain, a climatologist at the University of California, Los Angeles, added that one of the most direct ways that climate change is influencing California fires is by raising temperatures. "Heat essentially turns the atmosphere into a giant sponge that draws moisture from plants and makes it possible for fires to burn hotter and longer," he said. Meteorological data shows that the 2-year period from September 2019 through August 2021 ranks as the third warmest on record in California, with temperatures that were roughly 2.9° (1.6 °C) degrees warmer than average. Air can absorb about 7% more water for every degree Celsius it warms (Voiland, 2021).

Abatzoglou noted that some of the harrowing scenes across Northern California during 2020 and 2021 were due to an extreme and unusual ▶ dry lightning siege in mid-August that ignited thousands of fires in one night. "But in 2021 I am less convinced of bad luck," he said. "Climate change is aiding in the warming and the more rapid drying of fuels that predispose the land to large fires" (Voiland, 2021).

2.2.30 Drought and Syria's Civil War

Syria's civil war and the rise of the Islamic State in Iraq and Syria (ISIS) have been associated by scientists with prolonged drought in the region beginning in 2007 which has roots in global warming. According to Elizabeth Kolbert, writing in *The New Yorker* (2015, 23–24), "The country [Syria, in 2008] experienced its driest winter on record. Wheat production failed, many small farmers lost their herds, and prices of basic commodities more than doubled." Within months, as the drought continued, hundreds of thousands of people abandoned their homes and farms in the countryside and moved to Damascus, Homs, and other cities, crowding them with more than a million refugees from war in Iraq. By 2015, more than four million Syrian refugees had moved to Turkey, Lebanon, and Jordan, as well as several European countries. By 2016, NASA had issued a report using tree rings to determine that the drought in this region from 1998 to 2012 was the worst in at least 900 years.

Researchers from the University of California—Santa Barbara—and Columbia University described civil unrest linked to collapse of farming in Syria and the migration of 1.5 million farmers to cities, with related poverty that provoked civil unrest after 2007. Because of the civil war, weather records after that time are scarce. Droughts have become more frequent and intense in Syria; three of Syria's longest droughts have occurred during the last 30 years, as temperatures have risen and winter precipitation has declined.

2

"There are various things going on, but you're talking about 1.5 million people migrating from the rural north to the cities," said climate scientist Richard Seager at Columbia, co-author of a study in the *Proceedings of the National Academy of Sciences.* "It was a contributing factor to the social unraveling that occurred that eventually led to the civil war," wrote Seth Borenstein for the *Associated Press* (2015). The study's lead author, Colin Kelley, said that climatic change combined with the oppression by the Assad regime, immigration of at least one million refugees from Iraq, and political instability across the region to cause the civil war. However, said Seager, this is the "single clearest case" ever presented by scientists of climate change playing a part in conflict because "you can really draw a blow-by-blow account with the numbers" (2015).

2.2.31 Heat, Drought, and the Rise of Boko Haram

As with the civil war in Syria and social and economic breakdown fostering the rise of the terrorist group, Boko Haram has roots in spreading deserts, with accompanying heat and drought. At the height of the crisis, several million people in northeastern Nigeria fled the violence sparked by Boko Haram, many of whom became sick and starved. "More than 8 million people are in need of humanitarian assistance, 5.1 million are severely malnourished, most of them children," according to a report in *Science.* "The displaced have crowded into squalid camps and towns too destitute to deal with the influx," wrote Leslie Roberts in *Science* (2017). "Food, clean water, and sanitation are scarce or nonexistent, and these conditions create a perfect breeding ground for disease. In a deadly cycle, malnutrition renders children more susceptible to infection and less able to fight it. Epidemics of malaria and measles rage, polio has resurfaced, and child mortality is off the charts" (Roberts, 2017).

The unrest in Nigeria is a major contributor to migration across the Mediterranean Sea into Europe, as thousands of desperate people fleeing drought, famine, and jihadi violence cross the Sahara desert into Libya. In some northern Nigerian cities, the smuggling of human beings has become the main source of economic activity. Nigeria is the Africa's richest country, but distribution of wealth is wildly uneven. "As Nigeria's economy has grown – spurred by oil extraction, agriculture, and foreign investment – so has the percentage of its citizens who live in total poverty," wrote Ben Taub in *The New Yorker.* "Some wealthy businessmen travel with paramilitary escorts; police officers demand bribes at gunpoint, and crippled beggars crawl through traffic ...tapping on car windows and pleading for leftover food" (2016, 36–49).

Boko Haram (whose name loosely translated means "Western education is a sin") exceeds even other terroristic groups in cruelty. It has been notable for an especially brutal kidnapping more than 200 Nigerian schoolgirls and the massacre of several hundred civilians in the town of Gamboru Ngala. The brutal nature of the group's actions springs at least in part from instability spurred by a warming, drying climate that has deprived many farmers and herders of their homes and livelihoods. Changes in nature, according to a report in *The Guardian* (published in the UK), have played a role in societal breakdown:

"Instability in Nigeria…has been growing steadily over the last decade - and one reason is climate change." In 2009, a UK Department for International Development did a study warning that climate change could contribute to increasing resource shortages in the country due to land scarcity from desertification, water shortages, and mounting crop failures.

A more recent study by the Congressionally funded US Institute for Peace confirmed a "basic causal mechanism" that "links climate change with violence in Nigeria." The report concludes "…poor responses to climatic shifts create shortages of resources such as land and water. Shortages are followed by negative secondary impacts, such as more sickness, hunger, and joblessness. Poor responses to these, in turn, open the door to conflict." Unfortunately, a business-as-usual scenario sees Nigeria's climate undergoing "growing shifts in temperature, rainfall, storms, and sea levels throughout the twenty-first century. Poor adaptive responses to these shifts could help fuel violent conflict in some areas of the country" (Ahmed, 2014).

According to Prof. Sabo Bako of Ahmadu Bello University, during the 1980s, Boko Haram was preceded by the Maitatsine sect of Northern Nigeria, which emerged during the 1980s as victims of several ecological disasters found themselves destitute, leaving them in a chaotic state of absolute poverty and social dislocation in search of food, water, shelter, jobs, and means of livelihood.

Many Boko Haram recruits also have been rendered homeless, having lost all that they owned to severe drought and that has inflicted food shortages on Chad Niger, along Nigeria's northern borders. Boko Haram's recruits are a minority of about 200,000 farmers and herders who have been forced off their farms by intensifying heat and drought. "While a good number of these men were found in major cities like Lagos, pushing water carts and repatriating their earnings to the families they left behind," said Africa Review, "others were believed to have been lured by the Boko Haram."

2.2.32 Most of Boko Haram Are Not Muslim Fanatics

In Northern Nigeria, where activity by Boko Haram has been most intense, 70% of the population lived on less than the equivalent of one US dollar per day per person. David Francis, one of the first reporters to cover Boko Haram from outside of Africa, said that "Most of the foot soldiers of Boko Haram aren't Muslim fanatics; they're poor kids who were turned against their corrupt country by a charismatic leader."

Drought and heat have reduced arable land in Northern Nigeria and neighboring states and intensified conflict between farmers, who are mostly Christian, and cattle herders, who are overwhelmingly Muslim. This conflict, which also has roots in climate change, has paralleled Boko Haram's terror campaign but appears unrelated. Both are related to ecological breakdown, however. Nigeria also has added 125 million people in 50 years, adding to pressure on the land and demand for food and energy. Lake Chad, on Nigeria's northwestern border with the nation of the same name, has lost 90% of its surface area in roughly the same period. Its fishing industry has collapsed, and surrounding farms and fields have been routinely starved for water.

Erika Eichelberger reported in *Mother Jones* (June 2014) that climate change has also had contributed to the devastation by heat and drought. Nigeria as a whole has warmed by about 1.4 °F since the mid-twentieth century, with larger increases in arid northern areas. Eichelberger wrote that "as is the case with other insurgent movements around the world, economic hardship also helps drive recruitment. Poverty and unemployment in the north have reinforced the Boko Haram narrative that says the government has been corrupted by Western values, and thus cares more about enriching itself than helping Nigerians," according to a report issued in 2011 by US Institute for Peace, which is funded by the US Congress.

2.3 Outlook and Solutions

2.3.1 How Much Heat and Drought Will Destroy Agriculture?

The travails of a warming climate may be as personal and intimate as one person's suffering heat illness during one afternoon or as universal as tens of millions sharing misery. Witness June of 2021 (not yet even summer on the calendar) when high record high temperatures reached 118 °F in Phoenix, Arizona, and 124 °F in Baghdad, Iraq. One can only guess how long ago such places lost their suitability for practicing the business of agriculture. Yet, all of us must think of certain things because survival requires food, and since all of us need food, we must engage in commercial transactions to get it. This is absurdly simple and profoundly complex at the same time. For most of the Earth's more affluent citizens, access to a varied, tasty diet is so easy that we forget a life on a planet where the basic business of acquiring enough to eat may become impossible because most of the planet is simply too hot and dry or, ironically, too wet to allow the growing, commercial distribution and preparation of food for a majority of people at any price. In the past, entire civilizations have grown and then nearly starved to death due to changes in temperature and precipitation patterns.

Witness, please, the Maya, in what is today Central America, where civilization thousands of years old was crushed by a four-century-long drought and series of heat waves that began about a thousand years ago. Before these conditions set in, food was grown, people were born; received educations; matured; held occupations as diverse as farmers, priests, and sports stars; and with their surplus energy-built pyramids then died. All of them spent their entire lives engaged in many complex commercial transactions, many of which involved the growth, sale, and preparation of food, without buying, selling, or otherwise using the energy of fossil fuels— no oil, no coal, and no natural gas. All of this occurred without producing a single whiff of what we take to be the commercial transactions that are at the root of today's climate crisis.

2.3.2 Zero Greenhouse Gas Emissions by 2050?

Every day, month, and year of business-as-usual fossil fuel consumption increases the probability of a climatic emergency within the twenty-first century, which will become worse after that. A study published on February 27, 2008, in *Geophysical Research Letters* by H. Damon Matthews of Montreal's Concordia University, Montreal, and Carnegie Institution senior scientist Ken Caldeira asserted that dangerous consequences (droughts, deluges, and coastal floods, among others) may be avoided only if humankind stops adding fossil fuel emissions to the atmosphere *completely* by the year 2050.

Caldeira and co-author H. Damon Matthews of the Department of Geography, Planning and Environment, Concordia University, Montreal, Quebec, Canada, wrote "To hold climate constant at a given global temperature requires near-zero future carbon emissions. Our results suggest that future anthropogenic emissions would need to be eliminated in order to stabilize global-mean temperatures. As a consequence, any future anthropogenic [human-created] emissions will commit the climate system to warming that is essentially irreversible on a centennial timescale" (Matthews & Caldeira, 2008).

The zero-emissions by 2050 target is meant to prevent a temperature rise of more than 2 °C (3.6 °F), at which many scientists agree feedbacks that will cause warming to accelerate on its own, causing widespread distortions in the hydrological cycle, eventual sea-level rises that will destroy coastal cities, and extinctions of flora and fauna. Oregon State University Professor Andreas Schmittner said that "The warming continues much longer even after emissions have declined...Our actions right now will have consequences for many, many generations. Not just for a hundred years, but thousands of years" (Eilperin, 2008a).

2.3.3 Toward an Ice-Free Arctic?

In the Arctic, September is the peak ice-melt season. By August 2007, polar ice had retreated to near-record low levels, exposing dark ocean water to soak up even more of the sun's warmth. During the summer of 2007, the Arctic Ocean lost almost one quarter of its sea ice *in one year.* Ice loss continued in an uneven fashion through at least 2021. As is often the case, provocations of this record ice loss were both natural and human. Greenhouse gases emitted by human transport and industry were part of the story. The rest involved persistent high pressure over the Arctic that allowed more sunshine than usual, along with winds from the south that summoned relatively warm, ice-eroding air. The startling Arctic ice melt of 2007 sent scientists scrambling to rewrite their projections for an ice-free Arctic summer. Before the summer of 2007, projections commonly gave the Arctic ice cap until the end of the twenty-first century to melt; after the loss of so much ice in one year in 2007, informed observers gave the summer ice cap a decade or two. Artic ice has continued to decline since 2007, in an irregular fashion. Since 1980, Arctic

2

Ocean has lost an average of 230 tons of ice each year (the entire Earth has lost an estimated 28 trillion tons of ice since the middle 1990s).

At the same time, Greenland, which contains most of the Northern Hemisphere's ice, had been melting more quickly. Even the island's northeastern quadrant, which had been stable, has been losing ice—10 billon tons a year between April 2003 and April 2012 (Kahn, et al., 2014; Koch, 2014, 4-A). The melting has continued since then but sometimes at a slightly lower rate. As of this writing, we are broaching 425 p.p.m., which is above the 400 p.p.m., the level of the Pliocene, 2–3 million years ago. This implies enough warming to drown some low-lying cities once thermal inertia catches up.

2.3.4 A Dramatic Acceleration of Warming at Mid-Century?

Many climate scientists believe that the middle of the twenty-first century will witness dramatic acceleration of global warming due in part to exhaustion of natural carbon sinks. At about the same time, various feedback loops also are expected to accelerate increases in atmospheric greenhouse gas levels and, consequently, worldwide temperatures. These include several natural processes that add greenhouse gases to the atmosphere, such as melting permafrost in the Arctic, as well as an increasingly dark Arctic Ocean that will absorb more heat as the ice cap melts. Please see two illustrations above (◘ Figs. 2.6 and 2.7) that show global methane levels and their increase over time.

In each of these cases, human-provoked warming caused by an overload of greenhouse gases in the atmosphere is expected to aggravate existing problems like a bank account drawing an environmentally dangerous form of compound interest. Evidence is accumulating that these processes already are underway. The danger, according to many people who are familiar with the paleoclimatic record, is that once this journey has begun in earnest, any return trip will become a matter of many centuries as well as copious pain and suffering for humans, plants, and animals.

Frozen bogs in Siberia contain an estimated 70 billion tons of methane. If the bogs become drier as they warm, according to one observer, "The methane will oxidize and the emissions will be primarily CO2. But if the bogs stay wet, as they have been recently, the methane will escape directly into the atmosphere...with twenty times the heat-trapping power of carbon dioxide" (Romm, 2007). About 600 million tons of methane are emitted each year from human and natural sources, so release of even a small fraction of the 70 billion tons of methane in the Siberian bogs will accelerate warming dramatically.

Another example of feedback loops that will reinforce warming temperatures (and probably drought) occurs in the tropics. A rainforest or jungle creates its own weather, as the foliage recycles moisture. In our time, however, human beings seeking food and shelter compete with nature. Roughly 20% of the Amazon rainforest has been destroyed, and approximately another 20% has been damaged by logging, which allows drying sunlight to reach the forest floor. Some models suggest that after 50% of the Amazon's rainforests have been destroyed, increasing heat and drought will reinforce each other in a feedback loop to further reduce local rainfall, accelerating drought and local temperatures, and increasing release of carbon currently locked in Amazon soils and vegetation in a vicious cycle (Romm, 2007). During the decade and a half after this was stated as a theory, droughts in tropical and subtropical areas have occurred, with two examples being the Amazon Valley and Southern Africa.

When studying climate change, we must remember that (as the often-remembered environmentalists' saying goes) "All things are connected." For example (offshore of Eastern Africa), warming air and sea currents have created perfect conditions for stronger cyclones, which move onshore and dump warm torrential rains, creating excellent conditions for breeding of locust swarms. The locusts obliterate crops planted by millions of people, threatening at least half of them with starvation. The first wave of locusts (in 2019) bred even more for repeated cycles in 2020 and 2021. Five million acres of pesticides were dumped on the devastated fields in 2020 and 2021, an effect toxifying the fields for years to come. "I think we can assume that there will be more locust outbreaks and upsurges in the Horn of Africa," said Keith Cressman, a desert locust expert with the United Nations Food and Agriculture Organization (Cyclone of Locusts, 2022, 59). In other parts of East Africa, drought has been killing crops.

Another example of a killing chain of circumstances involves meerkats which live in Africa's Kalahari desert. They thrive in heat, to a point, but global warming is heating their habitat beyond tolerance for them, but also for their food sources, such as termites.

2.3.5 The Endurance of Feedbacks

Feedbacks are dangerous because they compound themselves. Ongoing warming liberates carbon dioxide and methane from Arctic and tropical soils and, ultimately, with enough atmospheric warming, even from the oceans. These feedbacks become self-reinforcing and feed upon themselves. The atmosphere is only one of many places where the Earth's system stores carbon dioxide. Roughly one-third of the Earth's carbon is stored in far northern latitudes (mainly in tundra and boreal forests), so the speed with which warming of the ecosystem releases this carbon dioxide into the atmosphere is vitally important to forecasts of global warming's speed and effects.

The amount of carbon stored in Arctic ecosystems equals two-thirds of the amount presently found in the atmosphere. The Arctic has been the most rapidly warming region on Earth. The possibility of a runaway greenhouse effect has provoked a lively scientific debate over the "tipping point" at which such an effect, having been triggered by human consumption of fossil fuels, will develop a life of its own, increasing the level of greenhouse gases in the atmosphere (and consequent warming) beyond any possibility of human mediation. At that point, we will find ourselves riding a runaway train that will continue to accelerate as long as the levels of greenhouse gases in the atmosphere continue to rise, *and* for several decades after they stabilize or fall, because of thermal inertia.

2

The amount of time that temperatures and sea levels may continue to rise even after greenhouse gas levels stabilize is not yet known with any degree of certainty, however. One scientific observer believes that "The rise in mean global SAT [surface average temperature] and the world ocean level (due to thermal expansion) may…*continue for several centuries after the stabilization* of the carbon-dioxide concentration [in the atmosphere], due to the gigantic thermal inertia of the oceans." The same observer also asserts that "The response of ice sheets to earlier climate changes may continue for several centuries after the climate stabilizes" (Kondratyev et al., 2004, 45).

2.3.6 The "Methane Burp" Hypothesis

The ultimate climatic nightmare becomes possible after the oceans warm enough to convert presently solid methane deposits in the oceans to atmospheric gas, further accelerating warming due to melting permafrost and ice sheets, and desiccation of rainforests. Geophysical evidence suggests that the Earth has suffered bouts of severe warming in the distant past from natural causes that were intensified by release of the planet's stores of greenhouse gases that usually reside in solid form, beginning with peat and permafrost on land, followed by methane hydrates in the oceans.

In cold water, methane clathrates form crystal structures which are somewhat similar to water ice. Warming temperatures of seawater could destabilize the clathrates and release some of their stored methane. Roughly ten trillion tons of methane is trapped under pressure in crystal structures with permafrost or on the edges of the oceans' continental shelves, constituting "the Earth's largest fossil-fuel reservoir," according to Gerald Dickens, a geologist at James Cook University in Townsville, Australia (Pearce, 2021). The greenhouse gas potential of all the methane stored in clathrates on the continental shelves and in permafrost worldwide is roughly equal to that of all the world's coal reserves.

Such events are not common, but they have been devastating, involving mass extinctions of flora and fauna. Considerable scientific inquiry is now aimed at estimating just how much human-provoked warming might initiate such a chain of disastrous feedbacks. During past periods of rapid warming, methane in gaseous form has been released from the seafloor in intense eruptions. An explosive rise in temperatures on the order of about 8 °C during a few thousand years accompanied a methane release 55 million years ago, called the Paleocene/Eocene Thermal Maximum.

The "methane burp" will not be tomorrow's news, but climate scientists pay attention to such things because rising greenhouse gas emissions could be taking us down a similar path eventually, a matter of many centuries. James Hansen believes that even a 2 °C worldwide temperature increase could "whipsaw" the Earth into a new, malign climate regime, via feedbacks, that would constitute "a different planet." At that point, increases in greenhouse gases would engage a self-perpetuating feedback loop that could not be stopped. The final destination of this climatic journey, as well, as the time of arrival is not now known.

A group of British and German research scientists using sonar has found that at least 250 sources of methane bubbles have been rising from methane hydrates in the seabed off Norway, an early indication, perhaps, of the "methane gun" that could add to the atmosphere's overload of greenhouse gases. The group reported in *Geophysical Research Letters* that the methane was rising near West Spitsbergen. Temperature records indicate that this area has warmed by about 1 °C during the last 30 years, destabilizing some of the hydrates. Professor Tim Minshull of the National Oceanography Centre at Southampton told the British Broadcasting Corporation (BBC) News: "We already knew there was some methane hydrate in the ocean off Spitsbergen and that's an area where climate change is happening rather faster than just about anywhere else in the world," he said. "Our survey was designed to work out how much methane might be released by future ocean warming; we did not expect to discover such strong evidence that this process has already started" (Westbrook et al., 2009).

This research indicates that the methane release may be part of a long-term pattern that reaches, in some cases, to the last years of the most recent ice age. Most of the methane ejected from the hydrates reacts with oxygen in the ocean, forming carbon dioxide as carbonic acid, contributing to ocean acidification. "If this process becomes widespread along Arctic continental margins, tens of teragrams of methane per year could be released into the ocean," the scientists wrote (Westbrook et al., 2009).

2.3.7 Methane Burp as an Energy Source?

How powerful are the forces of climate change denial? To sample contrarians' sheer tone deafness to the music of geophysics, witness plans to *harvest* methane hydrates, the most abundant unconventional fossil fuel of them all, and *burn* it. Serious money is now being spent on test wells to access what its proponents believe will be an eternal source of energy (and profit). The bad news is that burning even a small fraction of Earth's methane hydrates will fry the planet and turn the trick faster than the natural "methane burp" that human combustion of other fossil fuels may someday provoke.

A successful test of technology to extract methane hydrates took place during 2012 at a site called Ignik Sikumi on Alaska's North Slope, financed by the US Department of Energy, the ConocoPhillips, and the Japan Oil, Gas and Metals National Corporation. The hype was astounding in a tragicomic sort of way. Company press releases sang of a "global energy source…buried beneath the world's oceans and the Arctic permafrost…that many think might dwarf today's fracking revolution: huge reservoirs of natural gas trapped in ice crystals" (Cockerham, 2013). Fossil fuel mavens call it "flammable ice," which could make fossil fuel-poor nations such as Japan fuel asset-rich.

The US Department of Energy has said that "methane hydrates may exceed the energy content of all other fossil fuels combined" and "could ensure decades of affordable natural gas and cut America's foreign oil dependence." National Energy Technology Laboratory Director Anthony Cugini said "These methane hydrates really create an opportunity that can move shale gas to the back page if you believe that. The resource is so large" (Cockerham, 2013).

Then there is the matter of carbon dioxide overload in the atmosphere. "You get into the question, are we going to stick with fossil fuels until we completely fry the atmosphere?" said Richard Charter, a senior fellow at the Ocean Foundation and a member of the Energy Department's methane hydrates advisory committee. Blowouts from hydrate wells could trigger subsea landslides and release huge eruptions methane into the atmosphere, adding to global warming. This would be the last thing the Earth needs at a time when our major problem is fossil fuel overload.

Carbon Dioxide Levels: How Much Is Too Much?

Regarding the maximum level of carbon dioxide required to prevent serious damage from global warming, Rajendra Pachauri, a former head of the Intergovernmental Panel on Climate Change (IPCC), has expressed support for a level of up to 350 parts per million. This is 100 p.p.m. under the IPCC's officially stated position, in its 2007 assessment, of 450 p.p.m. This change brings the IPCC into line with a position that James Hansen has been asserting for many years—a level that is now *below* atmospheric levels of slightly more than 420 p.p.m.

Bill McKibben, an author and climate activist who founded the ▶ 350.org after consulting with James Hansen, conveyed Pachauri's change of mind in an email (borrowing from an Agence France-Presse report) to people with whom he works: "As chairman of the Intergovernmental Panel on Climate Change, I cannot take a position because we do not make recommendations," said Rajendra Pachauri when asked if he supported calls to keep atmospheric carbon dioxide concentrations below 350 p.p.m. "But as a human being I am fully supportive of that goal. What is happening, and what is likely to happen, convinces me that the world must be really ambitious and very determined at moving toward a 350 target" (McKibben, 2009).

For activists, this was a good news, an indication that the sometimes-ponderous IPCC may listen to its own chief officer and "smell the coffee." For the Earth, it demonstrates just how close to a climatic train wreck we really are. In the meantime, the world's governments seem to be in no hurry to factor a 350 p.p.m. target into their actions. To do so would require putting the Earth on a crash carbon dioxide diet, a mass mobilization along the lines of World War II, or more, to change the ways in which we all live, work, generate, and consume energy. Hansen also has advocated this approach for years. As politicians fiddle, however, evidence mounts that the climatic stakes are rising.

2.3.8 "Natural Gas" Is a Fossil Fuel, Too

While carbon dioxide contributes more to global warming than any other element, the second most plentiful heat-trapping gas is methane. The focus remains on carbon dioxide because it can persist in the atmosphere for several centuries, while methane (CH4) dissipates after a much shorter period. Methane has 80 times the heat-trapping potential as carbon dioxide, so that consistent sources of it can be as dangerous as CO_2 before it dissipates. The use of natural gas (methane) as a "green" alternative to oil or coal (a frequent argument in its favor) is little more than advertising drivel. In addition, natural gas often is transported in pipelines, which are prone to leaking.

The only real alternative fuels are those that produce no greenhouse gasses—wind, sun, waves, and so forth.

2.3.9 Global Warming and Natural Limits

Understanding how natural systems operate is at the center of ecological thinking. What follows (on thermal inertia and climate science) may seem "off topic" to some readers at first glance, because it crosses disciplinary boundaries. It is really at the heart of the matter. To understand the importance of Earth-centered thinking, we must realize how natural systems operate. Such understanding is vital to a realization of how vital a paradigm change in our belief system contributes to a sustainable future. Do not let categorical thinking obscure a realization that, in this case (as in many others), everything is connected.

Today, the most pervasive and urgent illustration of fossil fuels' unsustainable legacy is global warming, as levels of greenhouse gases rise to dangerous levels. The fossil fuel age dawned just as the United States became the Earth's most powerful economy, built across an expanding territory with surging immigration (mainly, but not entirely, from Europe). Exploitation of coal, followed by oil and methane (natural gas) between the mid-nineteenth and early twentieth centuries, introduced machine labor representing the equivalent of a billion horses (or three billion human slaves). Not coincidently, perhaps, human slavery became economically as well as politically obsolete. As an illustration of just how much human labor was necessary to propel fossil-fueled machines between 1800 and 1970, consider that the number of human hours of labor going into an acre of wheat declined from 56 to 29. For an acre of cotton, the same figure declined from 185 to 24. The raising of food has become as mechanized as the manufacture of anything else: seven calories of energy (mainly fossil fuels) by 2014 was required to produce one calorie of food (Johnson, 2014, 14, 19, 39). This revolution in energy generation also increased production of heat-retaining greenhouse gases in Earth's atmosphere.

2.3.10 **The Due Bills for Our Use of Fossil Fuels Are Being Served**

The due bill for our use of fossil fuels is now being served. By 2015, scientists had figured that "burning the currently attainable fossil fuel resources is sufficient to eliminate the [Antarctic] ice sheet" (Winkelmann et al., 2015). This study was directed at Antarctica only, but all other ice on Earth would melt at the same time. How much time may be required to produce an ice-free planet? No one really knows. At present rates of increase in the burning of fossil fuels, the last chunk of ice may melt within several thousand years. This statement is so broad that it might as well be "Who knows"? Complete melting of the ice, factoring in delays due to thermal inertia, along with human efforts to cool the Earth may postpone the date of final ice melt. The momentum of thermal inertia would have made melting all but irreversible, however.

To reiterate, global warming is dangerous because it is a sneaky, silent, slow-motion emergency, demanding that we acknowledge a reality centuries in the future with a system of individual, legal, and diplomatic reaction that reacts in the past tense. Ken Caldeira, a researcher at Stanford University's Carnegie Institute of Science, told Chelsea Harvey of *The Washington Post* that "The legacy of what we're doing over the next decades and the next centuries is really going to have a dramatic influence on this planet for many tens of thousands of years" (Harvey, 2014).

In 2016, the atmospheric level of carbon dioxide in Earth's atmosphere breached 400 parts per million in all areas, at all seasons. Levels of methane and nitrous oxides, the two other principal sources of greenhouse gases, also reached record levels by substantial margins. During 2015 and 2016, world's temperatures, stoked by El Niño conditions, surged to a new record as well, above 2014's previous high. "We're moving into uncharted territory at a frightening speed," said World Meteorological Organization Secretary General Michel Jarraud (Warrick, 2014). By 2020, the radiative forcing of these gases had increased by about 40% since 1990.

As large rises in worldwide temperatures startled scientists early in 2016, James Hansen and 18 co-authors published a study in the open-access journal *Atmospheric Chemistry and Physics* (from the European Geophysical Union), making a case that several meters in sea-level rise could take place within a century, not the several hundred years projected by many other scientists. "My interpretation is that this is the beginning," Hansen said. "And it's one or two decades sooner than in our model" (Gillis, 2016). "I think almost everybody who's really familiar with both paleo and modern is now very concerned that we are approaching, if we have not passed, the points at which we have locked in really big changes for young people and future generations," Hansen said (Mooney, 2016). Limiting global temperature rise to 2 °C (3.6 °F) over preindustrial levels, as recommended by recent diplomatic efforts such as the 2015 Paris accords, will not prevent climate-driven changes that will force evacuation of many coastal cities, Hansen and colleagues warned.

2.3.11 Dawn of the Anthropocene

Increasing levels of greenhouse gases in Earth's atmosphere are part of a broader, intensifying, trend in the birth of a geological epoch now widely called the Anthropocene, during which human activities have become the primary force altering the planet, "sufficiently to produce a stratigraphic signature in sediments and ice that is distinct from that of the Holocene epoch" (Waters, et al., 2016). The Anthropocene actually began tens of thousands of years ago according to some, with the first use of fire and deforestation, but intensified with the advent of fossil fuels about 1800 CE. By 1950, humanity's role in shaping the Earth's system was dominant, as population and industrialization exploded and so did the carbon content of the atmosphere.

The controlling role of humanity includes not only infusion of carbon dioxide, methane, and other greenhouse gases into the atmosphere at levels previously not experienced since the Pliocene, 2–4 million years ago (with consequent increases in temperatures), but also rising levels of numerous artificial pesticides and herbicides, lead, fly ash, and other forms of air pollution, fertilizers, plastics, and radioactivity resulting from nuclear weapons testing. Nuclear radioactivity from past tests will be detectable in ice and sediments for at least 100,000 years. The amount of plastic manufactured each year equals the weight of the entire human race, and most of it ends up in landfills or the oceans. As human population has increased, extinction rates of other species excepting human beings have risen at an accelerating rate. "Unlike with prior subdivisions of geological time, the potential utility of a formal Anthropocene reaches well beyond the geological community," commented Waters and colleagues in *Science* (2016). "It also expresses the extent to which humanity is driving rapid and widespread changes to the Earth system that will variously persist and potentially intensify into the future."

2.3.12 A Narrow Window to Avert Catastrophe

Several well-known climate scientists wrote in *Nature Climate Change* said in 2016 that "The next few decades offer a brief window of opportunity to minimize large-scale and potentially catastrophic climate change that will extend longer than the entire history of human civilization thus far," wrote the 22 climate researchers, led by Peter Clark, from Oregon State University (Mooney, 2016).

The scientists said "Most of the policy debate surrounding the actions needed to mitigate and adapt to anthropogenic climate change has been framed by observations of the past 150 years as well as climate and sea-level projections for the twenty-first century. The focus on this 250-year window, however, obscures some of the most profound problems associated with climate change. Here, we argue that the twentieth and twenty-first centuries, a period during which the overwhelming majority of human-caused carbon emissions are likely to occur, need to be placed into a long-term context that includes the past 20 millennia, when the last Ice Age ended and human civilization developed, and the next ten millennia, over which time the projected impacts of anthropogenic climate change will grow and persist. *This long-term perspective illustrates that policy decisions made in the next few years to decades will have profound impacts on global climate, ecosystems and human societies—not just for this century, but for the next ten millennia and beyond"* (emphasis added) (Clark et al., 2016).

As temperatures rise, the combination of ocean water's expansion as it warms and the melting of ice found on land is raising sea levels. Human beings have an affinity for the oceans, and a large proportion of us live on or near coastlines. Glance at a map of the world and point out the large cities that will be in peril as sea levels rise a few feet—Shanghai, Kolkata, London, New York City, Miami, and many others. Increase the proportion of greenhouse gases in the atmosphere and change its circulation patterns, expanding convection patterns that meteorologists call "Hadley Cells," which causes declines in rainfall over some areas, with expanding deserts. Harvests fail, and people go hungry. We are adapted to the climate, as are all of Earth's flora and fauna. When the climate changes, everything changes.

A change in the balance of trace gases in the atmosphere requires a fundamental change in the ways that energy is obtained and used by everyone on Earth—the ways in which we transport ourselves, heat and cool our homes, and manufacture nearly everything we use in daily life. Such a change creates debate and dissension as scientists' warnings become more ominous, as the Intergovernmental Panel on Climate Change (IPCC) warns that burning even a fraction of proven fossil fuel reserves will damage the global ecosystem, possibly beyond repair. Scientists now anticipate that three quarters of known fossil fuel reserves must remain in the ground if warming is to be kept at a tolerable level—requiring companies that exploit these reserves to eventually write down their asset values to zero. The mother-lode model is failing in the marketplace as Peabody Coal, the biggest miner of that dirty fuel on Earth, went bankrupt as wind power flourished—because it costs less! In one way, a turning point has been reached, and it is a good one.

2.3.13 Building a Sustainable Future Is Not a Luxury

Even as wind and solar energy advance, however, politicians flush with oil company cash deny that global warming exists. Many other people have come to realize that building a sustainable future is not a luxury. The mother-lode mentality does not anticipate conditions of seven generations hence, as many Native American paradigms instruct. The Native paradigm is more in line with science's understanding of thermal inertia, by which the atmosphere reacts to today's greenhouse gas emissions a half-century from now on land and more than that in the oceans. The behavior of thermal inertia requires that we anticipate the future and not merely react to present conditions.

More people are realizing geophysical limits. Today, we are witnessing an energy system paradigm shift. For example (one of many), in 2020 for the first time, more than 60% of Iowa's electrical generation came from wind. That proportion has continued to rise. Plans call for 100% wind generation by about 2030. At the same time, technological change, as always, generates fear of unemployment. Paradoxically, such changes also always generate economic activity. A change in our basic energy paradigm during the twenty-first century will not cause the ruination of our economic base, as some deniers of climate change believe. Appreciation of sustainable models will enhance the economy.

The proportion of carbon dioxide in the atmosphere continues to rise worldwide, however, a trend that has not changed since the beginning of the industrial revolution. That level reached about 425 parts per million in 2022, as high as it was in the Pliocene, 2–4 million years ago, when sea levels were 100 feet higher and temperatures 4–6 °F warmer. This is a key figure, and one that indicates how much change has yet to be experienced because of thermal inertia, within the next few centuries. This cake is already being baked. In terms of geologic time, the change is coming about remarkably quickly. Carbon dioxide is a trace gas, a tiny proportion of the atmosphere; at 400 p.p.m., it comprises only one-tenth of one-third of 1% of the air. It is, however, a remarkably efficient retainer of heat, and it is now increasing at a rate more quickly than at any time in the geophysical record, hundreds of millions of years.

2.3.14 Sea Levels Are Still Rising

The price of overloading the air with carbon dioxide and methane continues to rise. The IPCC in 2014 projected that by 2100 rising sea levels and storm surges probably will swamp some of Asia's largest cities, among them Mumbai, Bangkok, Kolkata, Dhaka, Shanghai, Ho Chi Minh City, and Rangoon; in Europe, London will be at risk, and in the United States, New York City, New Orleans, Miami, and others will succumb to seas that rise in storm surges. The IPCC also projects that a rise in atmospheric carbon dioxide level to 430–480 parts per million will raise acidity levels in the world's oceans to a level by 2100 that will imperil nearly everything that produces a calcium carbonate shell, from some phytoplankton (the basis of the maritime food web) to corals, many of which arc also in trouble due to rising water temperatures. Carbon dioxide is being injected into the oceans far faster than nature can neutralize it. Seawater is usually slightly alkaline, at about pH 8.2. The pH of the oceans has fallen 0.1 during industrial times. The scale is logarithmic, so a 0.1 change means a 30% increase in the concentration of ions. Under a business-as-usual scenario, the pH may fall by roughly 0.5 by the year 2100.

2.3.15 Greenhouse Gas Emissions Accelerate

Reports from the front lines in the war on greenhouse gases are grim. For carbon dioxide, the quick reference is the Keeling Curve, now at about 425 parts per million. According to a report from the United Nations, the Keeling Curve is rising more quickly than ever (since measurements began in the 1950s and, by proxy, many years before that). We are failing this test. Methane's proportion is also rising at record rates. As greenhouse gas emissions rise, new sources of oil and natural gas are cropping up all over North America, including Alberta's tar sands, and "fracking" from North Dakota to Pennsylvania. The *New York Times* tells us that manufacturing for the great oil boom is bringing parts of the rust belt back to life (Schwartz, 2014).

In 2014, the then-Canadian Prime Minister Stephen Harper was aggressively promoting tar sand development with plans for 10,000 miles of pipelines across that country, as well as the better-known Keystone XL southward from the tar sand fields into the US Midwest. The Canadian Parliament with Harper's backing by 2014 also had revoked or annulled 70 environmental laws. During ensuing years, Justin Trudeau replaced Harper, the Keystone XL pipeline was canceled, and many other projects were downsized because of copious public protests.

However, as of 2016, Canadian oil producers expected their web of new pipeline capacity to provide a doubling of the country's production by 2025, from 3.5 million to 6 million barrels a day, mostly for export. This estimate includes a doubling of tar sand output. The road to climate apocalypse is *still* paved with dollar bills and corporate profits. The corporate builders still want to have their geophysical cake and eat it too, in direct contradiction to the geophysical facts.

According to NASA, the Alberta oil sand fields, which were first exploited in 1967, are the world's largest oil sand deposit, with a capacity to produce 174.5 billion barrels of oil—2.5 million barrels of oil per day for 186 years. By 2013, one-third of Alberta's economy was tied in some way to the tar sands, including royalties worth more than $4 billion during its 2011–2012 fiscal year.

As Canada pumps thirsty India and China full of oil and gas (and do not forget coal exports from Wyoming's strip mines), carbon dioxide levels rose not only to higher levels but at a faster rate than ever before, as more people with more money used more fossil fuels. In the oceans, which are already so acidic that some shellfish are going sterile, the water is now less able to absorb carbon dioxide.

The World Meteorological Organization of the United Nations put all of this into stark numbers. "The changes we're seeing are really drastic," said Oksana Tarasova, chief of the WMO's Global Atmospheric Watch told the *Washington Post*. The carbon dioxide level rose just short of 3 parts per million in 2013, much more than the 1–2 p.p.m. annual average of the past few decades. "We are seeing the growth rate rising exponentially," Tarasova said. Carbon dioxide is now 40% higher than the usual cyclical peak before the advent of the industrial age. Methane, the second most prevalent greenhouse gas, is 150% higher.

We are now "at the level that climate scientists have identified as the beginning of the danger zone," said Michael Oppenheimer, a professor of geosciences at Princeton University (Johansen, 2014). "It means we're probably getting to the point where we're looking at the 'safe zone' in the rearview mirror, even as we're stepping on the gas" (Warrick, 2014).

In the land where "stepping on the gas" is considered an unmitigated blessing, in 2000, North Dakota was 39th among the 50 states in per capita income. By 2014, it was sixth. Why? New oil and gas from the Bakken formation. The *New York Times* describes "a transformation spreading across the heartland of the nation, driven by a surge in domestic oil and gas production that is changing the economic calculus for old industries and downtrodden cities alike" (Schwartz, 2014).

In Ohio, "in an arc stretching south from Youngstown past Canton and into the rural parts of the state where much of the natural gas is being drawn from shale deep underground, entire sectors like manufacturing, hotels, real estate and even law are being reshaped. A series of recent economic indicators, including factory hiring, shows momentum building nationally in the manufacturing sector" (Schwartz, 2014).

New energy production is "a real game-changer in terms of the U.S. economy," said Katy George, who leads the global manufacturing practice at the McKinsey & Company, a consulting firm. "It also creates an opportunity for regions of the country to renew themselves" (Warrick, 2014). And so it is, until the boom dissolves into a hot, miserable bust in a climate so hot that the US Midwest's iconic corn goes sterile.

2.3.16 We Have Seen the Last Ice Age

Temperatures are now warmer in the Arctic than during at least the last 2000 years, at a time when the Earth is receiving less heat from the sun due to changes in its orbit. The difference is greenhouse gas emissions due to human activity, Darrell S. Kaufman and colleagues wrote in *Science* (2009, 1236) that a cooling trend until the twentieth century "was caused by the steady orbitally driven reduction in summer insolation.... [which] was reversed during the 20th century, with four of the five warmest decades of our 2000-year-long reconstruction occurring between 1950 and 2000" (2009, 1236).

This study also provides fresh evidence for indications that human-generated warming likely will interrupt the natural ice-age cycle. "The slow cooling trend is trivial compared to the warming that's been happening and that's in the pipeline," Kaufman told Andrew C. Revkin of the *New York Times*. After a cooling less than 0.5 °F per millennium, the Arctic has warmed 2.2 degrees since 1900. "The fast rate of recent warming is the scary part," said Jonathan T. Overpeck. "It means that major impacts on Arctic ecosystems and global sea level might not be that far off unless we act fast to slow global warming." "It's basically saying that greenhouse-gas emissions are overwhelming the system," said co-author David Schneider told Revkin (2009).

Thirty scientists from several countries collected lake sediments and glacier ice samples, as well as tree ring records and other proxy data to trace the slow cooling and then sudden warming, in the Arctic. Without increasing greenhouse gases, they assert that the slow cooling would have continued because the Earth's orbit, which "wobbles" over time, is about 620,000 miles further from the sun than 2000 years ago. With the accelerating pace of warming from human combustion of fossil fuels, it is very likely that the planet has seen its last ice age until that provocation is removed.

2

2.3.17 When Will Feedbacks Take Control?

How much time remains before critical feedbacks lock into place? A world's task force of senior politicians, business leaders, and academics has warned that the point of no return will be reached in a report, *Meeting The Climate Challenge*, which describes how cascading feedbacks due to human-provoked climate change will irretrievably commit the Earth to disastrous changes, including widespread agricultural failure, water shortages and major droughts, increased levels of disease, sea-level rise, and the death of forests (Byers, et al., 2005, 1).

The report asserted that the tipping point will occur as the average world's temperature increases 2 °C above the average prevailing in 1750, before the industrial revolution began. By 2005, temperatures already had risen an average of 0.8 °C, with at least another half-degree Celsius "in the pipeline," due to thermal inertia. The report also asserted that the tipping point will occur as the atmospheric concentration of carbon dioxide passed 400 parts per million. It was nearing 420 p.p.m. by 2022. The report was issued in 2005, with the level at 380 p.p.m. In 2014, the group's chief scientific adviser was Rajendra Pachauri, the chairman of the United Nations' Intergovernmental Panel on Climate Change (McCarthy, 2005, 1).

The report concluded: "Above the 2-degree [C.] level, the risks of abrupt, accelerated, or runaway climate change also increase. The possibilities include reaching climatic tipping points leading, for example, to the loss of the West Antarctic and Greenland ice sheets, which, between them, could raise sea level more than 10 meters over the space of a few centuries…and the transformation of the planet's forests and soils from a net sink of carbon to a net source of carbon" (McCarthy, 2005, 1). By 2020, these words of foreboding have become closer to reality.

2.3.18 Human Political Inertia: Are We in Trouble Yet?

"The biggest lag is in the political system," said geoscientist Michael Oppenheimer of Princeton University. At least two decades have passed discussing the seriousness of the threat, he said, and, as he sees it, another 20 years may pass before a worldwide diplomatic program that is up to the task is in place. In the meantime, the window of time before feedbacks take control narrows. "We can't really afford to do a 'wait and learn' policy," Oppenheimer said. "The most important question is: when do we commit to [contain global warming] to 2°C.? Really, there isn't a lot of headroom left. We better get cracking." The current pace, said Roger Pielke, Jr., of the University of Colorado at Boulder, "Isn't going to do it (Kerr, 2007, 1231).

Very little has changed, geophysically, in the intervening 15 years except that the carbon curve is higher and sharper. James Hansen's and colleagues' evaluation of global warming has continued to close any remaining window of opportunity. Hansen himself maintained optimism in public, but by 2014, he was telling e-mail correspondents that the lack of political progress was costing him sleep. According to a scientific article that Hansen and several co-authors prepared late in 2008, the atmosphere's carbon dioxide level not only must stop rising from 385 parts per million at that time but must be cut at least 10% to below 350 p.p.m. in order to avoid sustained, major damage to the Earth's flora and fauna, when factored into the world's system by thermal inertia. Failing such a reduction, humankind will have placed enough warming "in the pipeline" during years to come to provoke not only uncomfortably warmer temperatures but also significantly eroding Arctic and Antarctic ice, major sea-level rise, intensifying droughts and deluges, and animal extinctions on the land and in the oceans (including many of the world's coral reefs). The oceans may acidify to the point that anything in a shell will be in peril. Of the 2 °C rise required to provoke unstoppable feedbacks, about half is now "in the pipeline," according to the calculations of Hansen and his colleagues in 2008. By 2022, the situation will have become even more dire (Hansen et al., 2008). For anyone who still needs to cultivate long-range fright, bear in mind that by 2022, the world's level of carbon dioxide was flirting with 425 p.p.m., increasing by about 3 p.p.m. per year, going nowhere but up, as it has after oil, coal, and natural gas began to dominate world's energy sources about 170 years ago. Need I ask: "Are we in trouble yet?

2.3.19 The Limits of Climate Diplomacy

Even if all countries meet their promised emissions cuts, a rise in atmospheric greenhouse gases is likely to worsen extreme wildfires, droughts, and floods, the United Nations said in a report issued on September 16, 2021. Even that is unlikely because as of 2021's end, meeting these goals seems beyond humankind's reach. Thus, by the end of the present century, continued emission of greenhouse gases, even at reduced levels, "is likely to increase the frequency of deadly heat waves and threaten coastal cities with rising sea levels, the country-by-country analysis" concluded (Sengupta, 2021). United Nations Secretary General António Guterres said that this prospective situation indicates that it shows "the world is on a catastrophic pathway" (Sengupta, 2021).

Somini Sengupta et al. (2021) in a report for the *New York Times* said "Perhaps most starkly, the new report displayed the large gap between what the scientific consensus urges world leaders to do and what those leaders have been willing to do so far." Emissions of planet-warming gases are poised to grow by 16% during this decade compared with 2010 levels, even as the latest scientific research indicates that they need to decrease by at least a quarter by 2030 to avert the worst impacts of global warming. "Now science is shouting from the rooftops that it's time to level up actions in an order of magnitude sufficient to the challenge," Christiana Figueres, a former head of the United Nations climate agency, said in a statement. "All other geopolitical issues will fade into irrelevance if we fail to rise to the existential challenge that climate change presents."

Altogether, nearly 200 countries (i.e., nearly all the nations on Earth) have made voluntary pledges to reduce or slow down emissions of planet's warming gases under the Paris agreement, reached in 2015 with the aim of averting the worst climate impacts. Some countries have since strengthened their pledges, including some of the world's biggest emitters, including the United States, Britain, and the

2

European Union. However, pledges are just that. Nothing will change at ground level until these pledges, and many more, are met or exceeded.

A recent analysis by the Climate Action Tracker found that no major emitters have a climate pledge in keeping with the 1.5° target. Several countries, including Britain and the European Union, are close. The United States is not. "Governments are letting vested interests call the climate shots, rather than serving the global community," Jennifer Morgan, an executive director of the Greenpeace International, told the *New York Times* (Sengupta, 2021).

Jim Hansen's Climate Forecast: Hot and Hotter

James Hansen and colleague, Makiko Sato, have taken the measure of several factors that influence the direction and intensity of climate change.

Having "crunched the numbers," they anticipate that temperatures will rise moderately until about 2024 or 2025, and then (if major action is not forthcoming to severely cut the rise of greenhouse gases), it is off to the races, with all of the influences coming into line to reinforce each other. This is a major warning from a scientist who has a record of being able to spot climate changes and has been doing so for 40 years.

Hansen's and Sato's paper opens with this abstract: "Record global temperature in 2020, despite a strong La Niña in recent months, reaffirms a global warming acceleration that is too large to be unforced noise – it implies an increased growth rate of the total global climate forcing and Earth's energy imbalance. Growth of measured forcings (greenhouse gases plus solar irradiance) decreased during the period of increased warming, implying that atmospheric aerosols probably decreased in the past decade. There is a need for accurate aerosol measurements and improved monitoring of Earth's energy imbalance."

La Niña periods usually occur when global temperatures are stable or falling. The coincidence of La Niña with record high temperatures is a red flag to climate scientists, which indicates that everything else being equal (which, of course, is not usually how the atmosphere usually works) temperatures would be even higher during an El Niño year. Given the usual cycle of El Niño and La Niña, the warmer cycle should return in a few years.

2.3.20 A Changing Atmosphere and Spreading Drought

What do Los Angeles, California, Tehran, Iran, Perth, Australia, and Cape Town, South Africa, have in common, debilitating drought and latitude? Therein also lies a major, worldwide, and dangerous facet of a warming climate that is being reported only in pieces.

Changes in worldwide atmospheric circulation are a major factor in worldwide climate change. Even though warmer air generally holds more moisture, not everyone will see more precipitation in a globally warmed world. Many deserts already are expanding, in a worldwide pattern influenced by atmospheric circulation patterns that meteorologists call "Hadley Cells." For reasons that are not yet fully understood, as temperatures rise, the Hadley Cells reach further north and south of the equator. Most deserts around the world range between 20 and 40 degrees north and south latitude, and they have been expanding.

Near the equator, warm, moist air rises, cools, and unleashes downpours. In the upper troposphere, the air spreads north and southward toward both poles, descending at about 30° North and South latitude and creating deserts. The situation is similar in Syria, with its unrelenting civil war, provoked by an extended drought. Cape Town is running out of water, and Sao Paulo is experiencing a multiyear drought.

2.3.21 Local Droughts in a Worldwide Pattern

While the overall pattern is dry, the new reality also includes some brief, dramatic reversals. Southern California, for example, has had some dramatic flooding rains which promote growth of vegetation, providing fuel for fires during the next dry, hot cycle. Tehran, very similarly, has been in the midst of a drought when a huge blizzard struck during January of 2018, just as drought was provoking social unrest across Iran. A foot and a half of snow paralyzed the Tehran airport, even as mountain water reserves for Tehran, a city of 12 million, were at record lows.

"We were afraid we would never see snow again in this city." "When I was a child, schools used to be closed because of heavy snow, but these days they are closed because of air pollution," said Kaveh Madani, a deputy chief of Iran's Department of Environment. "This shows how our environment has changed, and what we have done to the city" (Erdbrink, 2018). Water levels were falling below levels sufficient to generate hydroelectric power.

Gholamreza Khoshkholq, the head of Tehran's electricity company, said that dams will not be able to generate enough power when demand peaks as temperatures rise. "There will be blackouts," Gholamreza told Thomas Erdbrink of the *New York Times* (2018). The ground under Tehran has been sinking as aquifers dry up. "Sinkholes are a constant presence across the country," wrote Erdbrink, "and parts of highways and overpasses in the capital have collapsed into the earth, in some cases dragging cars and people down with them."

2.3.22 More Human Contributions to a Steadily Warming Atmosphere and Oceans

As scientists study the interplay of effects that add up to humankind's contribution to a steadily warming atmosphere and oceans, find more contributing factors in this interplay of elements. It was not until a few years ago, for example, that wildfires, which have become hotter and more numerous, have been detected to be as large a contributor to ozone pollution as many large urban areas. By 2021, as wildfires spread, scientists found that all of the atmosphere, from the troposphere to the stratosphere, 40,000 feet vertically, was polluted to some degree by ozone, as well as wildfire smoke, which also contributes to warming.

Additionally, the oceans have been incrementally warming. Until recent years, oceanographers could count on El Niño conditions to warm the oceans on average, and La Niña to cool them, again on an average. The year 2021 was a La Niña year, and the oceans' temperature-sensitive areas warmed instead of cooled (Cheng et al., 2022).

The oceans take up about 93% of the excess heat resulting from greenhouse gas emissions. "Here," wrote Lijing Cheng et al. in *Advances in Atmospheric Sciences*, "We provide the first analysis of recent [Ocean Heat] changes through 2021 from two international groups. The world ocean, in 2021, was the hottest ever recorded by humans…" This analysis all linked the temperature rises to an increase in anthropogenic greenhouse gas concentrations [i.e., to the El Niño-Southern Oscillation (ENSO)] (Cheng et al., 2022).

Ice Free, Human Free, Eventually?

Most scientific projections of global warming's effects on our atmosphere, our climate, and our lives—recent reports of the IPCC provide an example—do not extend past the end of the present century. The projections do not stop there because global warming will end. It is not a matter of humanity, having had enough, being able to set a thermostat back a few degrees and reverse things. In fact, thermal inertia and various other feedbacks virtually guarantee a rapid amplification of warming into the next century and beyond if fossil fuels continue to be burned at anything like today's rate. Consensus science does not have the tools to forecast such things, however.

James Hansen is still intruding on the consensus. He has lately been asking what could happen to climate (as well as Earth's flora and fauna) if the human race burns every single ounce of available fossil fuels, that is, all of the Earth's coal, oil, tar sands, oil shale, fracked oil, and natural gas. The fossil fuel industry in the last few decades has vastly improved its technology and has increased its access to new reserves.

What will be the eventual outcome of "drill, baby, drill?" A recent examination by Hansen concludes that the Earth eventually could be ice-free, and all but bereft of living animal life, including human beings. A few acres of very high-priced real estate may remain on the highest mountains and Antarctica. Perhaps, as in the recent film *Elysium*, a fraction of the "one per cent" may escape to space stations. In the real world, however, they would not have to worry about invasion by Earth-bound rabble. The scalding surface would be too hot for that (Hansen, 2013b, September 16).

In the Long Range: Earth and Venus

James Hansen's consideration of Earth's climatic future takes him to the climate of Venus, which was the subject of his Ph.D. dissertation. It was the greenhouse warming of Venus, in fact, that led Hansen, during the 1970s, to devote his scientific career to the study of global warming on Earth.

Venus millions of years ago may have been more like Earth—a little warmer, perhaps, owing to its position closer to the Sun, but not too warm for liquid oceans. The geophysics of Venus differ from Earth's in one important way, however: it has no plate tectonics, which allow subsurface pressures on Earth to discharge such things as lava piecemeal. Such discharges express themselves in volcanic eruptions and earthquakes that can be deadly close at hand but across the entire Earth maintain a rough equilibrium.

On Venus, subsurface pressures are said to have built up to enormous levels and then exploded in a single, spectacular blast that provoked a runaway greenhouse effect. Temperatures rose, and the oceans boiled away. As a result, today's Venus has a surface atmosphere hot enough to melt lead. Most of Venus' carbon is now in the atmosphere, not in its crust (as on Earth), observes Hansen (Hansen, 2013a, April 15).

So that is Venus—no place for a picnic. What about Earth? Here, with technological ingenuity and a desire for profit, comfort, and convenience, human fossil fuel corporations are supplying us with products manufactured from carbon extracted from the crust, combusted as energy, and released to the atmosphere as carbon dioxide and methane. We are, in other words, mimicking natural processes on Venus. In geological time, this is happening remarkably quickly, although thermal inertia delays effects by about 50 years in the air (and on land) and a century or two in the oceans.

Thermal inertia and natural variability allow the climate change deniers among us to argue that geophysical facts do not matter. If the carbon dioxide level is now at 425 parts per million (at or higher as it was during the Pliocene, when oceans were about 100 feet higher), so what?

This is what: geophysics do matter. At some point—exactly *when* we do not yet really know, although the IPCC now says roughly 2040 on our present emissions path—feedbacks may take off on their own and accelerate the changes (in a "hyperthermal") (Johansen, 2014).

By that time, the climate deniers will have been discredited, but it will be too late. We will be on our way (*again, with a qualifier that no one now knows exactly when*) to a world where the air in most places will eventually become simply too hot to sustain human (or most other) life. "It is not an exaggeration to suggest," wrote Hansen in an e-mail post September 26, 2013, "based on best available scientific evidence, that burning all fossil fuels could result in the planet being not only ice-free but human-free."

2.4 Ten Reasons for Concern About Global Climate Change

2

1. **Thermal inertia:** Today's fossil fuel consumption does not turn to heat tomorrow; through thermal inertia, it takes about 50 years in the atmosphere and 150–200 years in the oceans. Thus, we, today, are facing the results of fossil fuel emissions from roughly 1960. The amount of fossil fuel burned per year has risen substantially since then.

2. **Climate change is cumulative.** Many of the feedbacks that provoke rising temperatures tend to accelerate over time, compounding each other. For example, rising emissions from human sources provokes melting permafrost, which adds even more carbon dioxide and methane to the atmosphere. Melting ice makes surfaces darker, which absorbs more heat.

3. **Global warming is most intense where most people do not see it,** such as in the Arctic and on the rims of the Greenland and Antarctic ice sheets, which also house much of the world's frozen water. Warmer winters in the Arctic provide less cold weather at lower latitudes. As much as we may rue cold snaps in winter, they serve a natural purpose. In the US West, Alaska, and British Columbia, warmer winters have played a role in insect infestations that are killing large areas of evergreen trees.

4. **Some skeptics argue that rising levels of carbon dioxide and other greenhouse gases do not matter.** At 180 parts per million CO_2, the Earth has had periodic ice ages for at least the last 800,000 years; at 280 p.p.m., it has had an interglacial, such as today. The carbon dioxide level is now about 425 p.p.m. and rising. Appreciating the effects of thermal inertia (see #1), one can see why average temperatures probably will be much higher 50–100 years given that the level of carbon dioxide continues to raise, as itches for the last 170 years, since fossil fuels became a dominant force in world's energy use.

5. **Rising carbon dioxide levels cause problems other than warmth,** most of them out of human sight. For example, rising CO_2 makes the oceans more acidic, imperilling anything in the ocean with a calcium shell. Rising acidity in the oceans also play a role in declining fertility of plankton, the base of the aquatic food chain.

6. **Beware "clean coal,"** which is an oxymoron, at least in our time. All schemes to capture or sequester greenhouse gases from coal are very expensive in energy and monetary expenditure, so much so that they make alternative fuels, such as wind and solar power, a bargain.

7. **Human short-sightedness:** Snow today, ice age tomorrow, sing some of the skeptics, ignoring the difference between weather and climate. *Weather is the story; climate is the plot.* While weather varies, there is an underlying trend in temperature, and it is rising.

8. **Mountain glaciers are the "water towers" of many cities** in South America (Peru, Bolivia, Chile) and Asia (large parts of India and China). When mountain glaciers melt, drought intensifies.

9. **Polar bears feed from the ice, so no ice means no bears,** at least not as we have known them. While some skeptics have said that polar bears will adapt to climate change and get with the program, do we really want them raiding garbage dumps and mugging tourists because they cannot eat ringed seals?

10. **The hydrological cycle speeds up as temperatures rise,** causing droughts and deluges to intensify, both of which pose problems for agriculture and our food supply.

2.5 Questions and Exercises

1. Discuss the "character" of CO_2 and methane. What does it "want"?
2. What will it take to get the human race unified and active to restore balance in the ecosystem vis-a-vis climate change?
3. Discuss James E. Hansen's and others' research into effects on climate of greenhouse gases (1980). What in Hansen's background set him on this path?
4. Why is expeditious action on global warming so urgent?
5. As under "Introduction," describe "thermal inertia." See above.
6. Discuss the rise in carbon dioxide, methane, and other greenhouse gases' rise in the atmosphere since the beginning of the industrial age.
7. If large-scale melting of the world's mountain glaciers, along with a large part of Greenland's and parts of the West Antarctic ice sheets, could add up to 25 m of sea-level rise worldwide, what cities (or parts thereof) would be in peril of sinking?
8. According to James E. Hansen, global warming differs from previous pollution problems in two fundamental ways. Name the ways, discuss, and if you wish, debate.
9. Carbon dioxide represents about half of global warming in the atmosphere. What comprises the other half?
10. On average, ▶ nights are warming faster than days across most of the United States, according to the 2018 ▶ National Climate Assessment Report. Why is nighttime warming dangerous to human beings?
11. In winter, Fairbanks, Alaska, may occasionally warm to 35 °F while snow falls in Houston, Texas. Discuss the atmospheric (especially Jet Stream) contortions that may lead to such differences between the Arctic and subtropical areas.
12. The decline in oxygen levels has intensified since the 1950s, due nearly entirely to human activities, from increasing emissions of carbon dioxide and other greenhouse gases and overload of nitrogen-based fertilizers. Discuss how to alleviate these problems.
13. "Large-scale coral bleaching events, in which reefs become extremely fragile, were virtually unheard of before the 1980s. But in the years since…the frequency of coral bleaching has increased to the point that reefs no longer have sufficient recovery time between severe episodes."

    What effects do warming waters have on coral reefs?
14. Melting season in the region around Mount Everest is usually concentrated during the summer monsoon (April–September). In recent years, however, abnor-

mally warm temperatures have extended the melting period, in some instances by as much as 4 months. *What effects do glacial melting have on surrounding ecosystems?*

15. "To hold climate constant at a given global temperature requires near-zero future carbon emissions. Future anthropogenic emissions would need to be eliminated in order to stabilize global-mean temperatures.... As a consequence, any future anthropogenic [human-created] emissions will commit the climate system to warming that is essentially irreversible on a centennial timescale"

    What are the implications for a world in which carbon emissions continue to rise?

16. Greenland, which contains most of the Northern Hemisphere's ice, had been melting more quickly. Even the island's northeastern quadrant, which had been stable, has been losing ice—10 billion tons a year between April 2003 and April 2012. The melting has continued since then. "Not only is the Greenland ice sheet melting, but it's melting at a faster and faster pace." *How much might sea levels rise world-wide at this pace, and what might her the implications for sea-level cities?*

17. Many climate scientists believe that the middle of the twenty-first century will witness dramatic acceleration of global warming due in part to exhaustion of natural carbon sinks. At about the same time, various feedback loops also are expected to accelerate increases in atmospheric greenhouse gas levels and, consequently, worldwide temperatures. *Based on your studies to this point, what are the implications for rising seas and other parts of a "dramatic acceleration of global warming" by "the middle of the twenty-first century"? Bear in mind that this will be less than half a human lifetime from when this book is published.*

18. "Frozen bogs in Siberia contain an estimated 70 billion tons of methane. If the bogs become drier as they warm, according to one observer, "The methane will oxidize and the emissions will be primarily CO_2. But if the bogs stay wet, as they have been recently, the methane will escape directly into the atmosphere...with twenty times the heat-trapping power of carbon dioxide." *With reference to #16 and #17, what are the global implications of this?*

19. "Feedbacks are dangerous because they compound themselves. Ongoing warming liberates carbon dioxide and methane from Arctic and tropical soils and, ultimately, with enough atmospheric warming, even from the oceans. These feedbacks become self-reinforcing, and feed upon themselves." *Please provide examples of this geophysical fact.*

20. "Without fail, the most intense increases in global warming have been occurring at night, in large urban areas." *Referring to the text, why is this, and what implications does it have for plants and animals, including humans? What are the implications for human birthing?*

21. "Scientists have been looking for shells on beaches along the US East Coast, from near Washington, D.C. to Orlando, Florida. These are not ordinary shells on today's beaches, but the surf line of the Pliocene, about 3 million years ago." *Why are scientists interested in the Pliocene? What might events 2–3 million years ago have for today's world?*

22. *With reference to the book, what are the figures for carbon dioxide levels in the atmosphere during the last 60 years? What are the implications of a similar rise within your lifetime?*

23. NASA satellites have detected a widening rift in Antarctica's Pine Island glacier. This glacier acts as a plug, like a cork in a bottle, between the ocean and the West Antarctic Ice Sheet.

    Why are students of Antarctic Ice's prospective melting keeping an eye on the Pine Island Glacier?

24. "Global warming is not merely a matter of rising temperatures. Warming temperatures also change the behavior of the Earth's hydrological cycle, increasing the severity of storms, as well as the frequency and intensity of droughts and deluges."

    Why and how is it that both drought and deluge can alternate with each other as the levels of greenhouse gases increase in Earth's atmosphere? Please ponder this apparent paradox.

25. "Two years ago, the fields outside Christina Southwell's family home near Wee Waa, the cotton capital of Australia looked like a dusty, brown desert as drought-fueled wildfires burned to the north and south. Last week [including December 5, 2021], after record-breaking rains, muddy floodwaters surrounded her, along with the stench of rotting crops." The same question may be asked about India.

    What special meteorological circumstances make Australia and India especially prone to alternating droughts and deluges?

26. While increases in temperature are linear, intensity of precipitation has been said by some scientists to increase exponentially (Trenberth et al., 2003, 1205–1217). One must wonder what the atmosphere has in store for us once temperature increases get *really* serious.

    Please reflect upon the above observation.

27. The Amazon Valley, an area that was a net source of oxygen, has, in recent years, frequently become a net source of carbon dioxide. Forests that once dominated the region have been replaced by large, scorched savannahs left by fires, prairies eaten down by cattle and other farm animals, gold (and other metals') strip mines, roads and urban areas, and acres and acres of bare mud that once was forest, now being cast as roof tiles for spreading suburban housing developments.

    Why have portions of the Amazon Valley now become savannah?
    Consider both human developments and changes in atmospheric patterns.

28. The Intergovernmental Panel on Climate Change (IPCC) issued a report in August 2019 describing a looming food crisis that among other things tied lack of food provoked by climate change to increasing migration across borders.

    How does prolonged drought provoke and prolong migrations around the world? (Hint: Roughly 500 million people worldwide live on land that has recently turned to desert, and soil fertility has been depleted between 10 and 100 times faster than it is forming.)

29. As carbon dioxide levels rise, atmospheric circulation patterns change. Spin the globe and you will notice that most of the world's deserts lie between 20 and

40° north and south latitude. Near the equator, warm, moist air rises, cools, and unleashes downpours. In the upper troposphere, the air spreads north and southward toward both poles, descending again at about 30° north and south latitude, drying the regions it reaches, creating deserts.

Please examine the role of Hadley Cells in the atmosphere and how they contribute to distribution of deserts and wet areas on Earth.

30. Carbon dioxide overload in the atmosphere also acidifies the oceans, endangering marine animals that live in shells, from corals to sea urchins. Acidity also shrinks the number and size of phytoplankton, the basis of the oceanic food chain. By the early years of the twenty-first century, carbon dioxide levels in the oceans were rising more rapidly than at any time since the age of the dinosaurs, according to work published by Ken Caldeira and Michael E. Wickett.

Examine how rising ocean temperatures (and other factors) increase acidity in the oceans. Please then reflect how such a rise in acidity is affecting the ocean food chain.

31. Referring to ocean acidity, Jason Hall-Spencer, a British marine biologist and reader at Plymouth University, said "Unfortunately, the biggest tipping point, the one at which at which the ecosystem starts to crash, is mean pH 7.8, which is what we are expecting to happen by 2100. So that is rather alarming" (Kolbert, 2014, 118). *Please interpret these phrases in light of the fact that for many millions of sea and land creatures, including human beings, the oceans are a major source of protein.*

32. The "Methane Burp" Hypothesis: The ultimate climatic nightmare becomes possible after the oceans warm enough to convert presently solid methane deposits in the oceans to atmospheric gas, further accelerating warming due to melting permafrost and ice sheets, and desiccation of rainforests.

Discuss the possibility of such a disaster, and do not just say "impossible!" because it has occurred in the distant past.

33. Here is a very important question: with regard to carbon dioxide, how much is too much? That is, too much to keep the planet from descending into feedbacks cycles that will destroy it? The consensus among scientists and activists seems to be 350 parts per million, a matter of no small importance, because the air we breathe is already pushing 420 p.p.m. with no sign of dropping.

So, what do we do? *Please be specific. What sources of CO_2, methane, etc. can we banish? This, students, is the existential question, a.k.a. damned serious business.*

34. *Please double down on this one. Global warming is dangerous because it is a sneaky, silent, slow-motion emergency, demanding that we acknowledge a reality centuries in the future with a system of individual, legal, and diplomatic reaction that reacts in the past tense.*

35. Increasing levels of greenhouse gases in Earth's atmosphere are part of a broader, intensifying, trend in the birth of a geological epoch now widely called the Anthropocene, in which human activities have become the primary force altering the planet.

So, can we do anything about this? Do we want to? Would anyone support trading nuclear weapons for sticks and stones, or automobiles for bicycles and running shoes? Is this even worth broaching? (Some airline engineers have flown a few jets on fuel partially comprised of waste frying oil.)

36. The IPCC also projects that a rise in atmospheric carbon dioxide level to 430–480 parts per million will raise acidity levels in the world's oceans to a level by 2100 that will imperil nearly everything that produces a calcium carbonate shell, from some phytoplankton (the basis of the maritime food web) to corals, many of which are also in trouble due to rising water temperatures. Carbon dioxide is being injected into the oceans far faster than nature can neutralize it. Seawater is usually slightly alkaline, at about pH 8.2. The pH of the oceans has fallen 0.1 during industrial times. The scale is logarithmic, so a 0.1 change means a 30% increase in the concentration of ions. Under a business-as-usual scenario, the pH may fall by roughly 0.5 by the year 2100.

    Is this a question, or a rant? With all due consideration, it is both, with a hint that flying aircraft on used frying oil is only part of a decently adequate answer. It is an all-hands-on-deck question, one that cannot be stalled off for future generations.

37. One more point: "We are now at the level that climate scientists have identified as the beginning of the danger zone," said Michael Oppenheimer, a professor of geosciences at Princeton University (Johansen, 2014). "It means we're probably getting to the point where we're looking at the 'safe zone' in the rearview mirror, even as we're stepping on the gas" (Warrick, 2014).

38. *The Arctic Oscillation, in which a semi-stationary jet stream, blue, colder than average, and red, warmer than average. This is one intriguing way in which a large, violent blizzard that meteorologists are fond of calling a "bomb cyclone" can enter the record books as a fruit of global warming. With this knowledge, explain how such a cyclone can lay down 2 feet of snow with 60-mile-an-hour winds on January 29, 2022 (the deepest snowfall in that city's history), and global warming can be held partially responsible. HINT: The water temperature in the Gulf Stream was 10 °F above normal.)*

References

A Cyclone of Locusts. (2022 January). *National geographic*. 59.

Adam, K. (2021, September 23). Boris Johnson Tells World Leaders to 'Grow Up' on Climate Change, Takes Aim at Kermit the Frog. *Washington Post in Seattle Times*. Retrieved January 24, 2021, from https://www.seattletimes.com/nation-world/boris-johnson-tells-world-leaders-to-grow-up-on-climate-change-takes-aim-at-kermit-the-frog/?utm_source=marketingcloud&utm_medium=email&utm_campaign=Morning+Brief+9-23-21_9_23_2021&utm_term=Registered%20User.

Ahmed, N. (2014, May 9). *Behind the Rise of Boko Haram—Ecological Disaster, Oil Crisis, Spy Games; Islamist Militancy in Nigeria is being Strengthened by Western and Regional Fossil Fuel Interests*. The Guardian, UK.

Albright, R., Caldeira, L., Hosfelt, J., Kwiatkowski, L., Maclaren, J. K., Mason, B. M., Nebuchina, Y., Ninokawa, A., Pongratz, J., Ricke, K. L., Rivlin, T., Schneider, K., Sesboüé, M., Shamberger, K., Silverman, J., Wolfe, K., & Zhu, K. (2021, August 31). Arc of Fires in the U.S. West. NASA Earth Observatory. Retrieved September 5, 2021, from https://earthobservatory.nasa.gov/images/148772/arc-of-fires-in-the-us-west/?src=eoa-iotd.

Allan, R. P. (2011). Climate change: Human influence on rainfall. *Nature, 470*, 344–445.

Andreoni, M. (2021, November 3). Brazil, facing skeptics, seeks role, as climate leader. *New York Times*, A-8.

Anonymous. (2021, August 16). Flooding in Turkey has killed 59, with dozens still unaccounted for. *New York Times*, A-4.

Archambault, D. (2016, August 24). Taking a Stand at Standing Rock. *New York Times*. Retrieved August 26, 2016, from http://www.nytimes.com/2016/08/25/opinion/taking-a-stand-at-standing-rock.html.

Azzioni, T. (2010, October 25). Drought has Amazon Tributary at Record Low Levels. Associated Press in Washington Post. Retrieved October 29, 2010, from http://www.washingtonpost.com/wp-dyn/content/article/2010/10/25/AR2010102502661_pf.html.

Barry, C. (2021, October 21). *Venice copes with higher tides and climate change.* Associated Press Morning Wire. Retrieved October 29, 2021, from https://apnews.com/article/floods-climate-science-lifestyle-business-7e87c573c7ad9129a8d1a4fcaaa3f9e1?user_email_address=ff355df48f6 07b11ff93cb7e60dbb14d&utm_source=Sailthru&utm_medium=email&utm_campaign=MorningWire_Oct20&utm_term=Morning%20Wire%20Subscribers.

Borenstein, S. (2020, August 10). *Record melt: Greenland lost 586 billion tons of ice in 2019.* Associated Press. Retrieved August 12, 2020, from https://apnews.com/6fcaab97241d34c83f448f019179ca6b.

Borenstein, S. (2021a, July 11). *Summer trend: West gets hotter days, east hotter nights.* Associated Press in Omaha World-Herald. p. A-5.

Borenstein, S. (2021b, July 25). *Numbers show climate change impact.* Associated Press in Omaha World-Herald. p. A-15.

Borenstein, S., & Jordans, F. (2021, November 13). *Compromise seals climate deal.* Associated Press in Omaha World-Herald. p. A-12.

Brown, M. (2021, December 8). *Wildlife smoke carries risks.* Associated Press in Omaha World-Herald. p. A-15.

Burns, J. (2009, August 18). Methane seeps from arctic sea-bed. *B.B.C. News*. Retrieved August 25, 2009, from http://news.bbc.co.uk/go/pr/fr/-/2/hi/science/nature/8205864.stm.

Byers, S., Snowe, O., & International Climate Change Task Force. (2005). *Meeting the climate challenge: Recommendations of the International Climate Change Task Force.* Institute for Public Policy Research.

Caldwell, C. (2021). Bankers took over the climate change summit. *New York Times*, A-21.

Carlowicz, M. (2021, August 23). [Hurricane] Henri Soaks the Northeast. *NASA Earth Observatory*. Retrieved September 1, 2021, from https://earthobservatory.nasa.gov/images/148742/henri-soaks-the-northeast?src=eoa-iotd.

Cave, D. (2021, December 11). First Fires, Then Floods: Climate Extremes Batter Australia. *New York Times*. Retrieved January 22, 2021, from https://www.nytimes.com/2021/12/11/world/australia/flooding-fire-climate-australia.html.

Chea, T. (2021, August 29). *California drought devastates almond producers* (p. D-7). Associated Press in Omaha World-Herald.

Cheng, L., Abraham, J., Trenberth, K. E., Fasullo, J., Boyer, T., Mann, M. E., Zhu, J., Wang, F., Locarnini, R., Li, Y., Zhang, B., Tan, Z., Fujiang, Y., Wan, L., Chen, X., Song, X., Liu, Y., Reseghetti, F., Simoncelli, S., Gouretski, V., Chen, G., Mishonov, A., & Reagan, J. (2022). Another record: ocean warming continues through 2021 Despite La Niña conditions. *Advances in Atmospheric. Sciences.* https://doi.org/10.1007/s00376-022-1461-3

Cockerham, S. (2013, November 28). *Will natural gas eventually come from sea ice? Prospect thrills some, dismays others.* McClatchy [Newspapers] Washington Bureau. Retrieved December 3, 2013, from http://www.mcclatchydc.com/2013/11/28/209032/will-natural-gas-eventually-come.html.

Cramer, M. (2021, November 14). Baby born 19 weeks early defies long odds and astonishes doctors. *New York Times*. Retrieved November 30, 2021, from https://flipboard.com/@newyorktimes/most-emailed-9b8fb5gsz/baby-born-19-weeks-early-defies-long-odds-and-astonishes-doctors/a-68cCytaCQ2Kmc1RbFKM4gw%3Aa%3A3195393-6092a3f538%2Fnytimes.com.

Cvijanovic, I., Santer, B. D., Bonfils, C., Lucas, D. D., Chiang, J. C. H., & Ziummerman, S. (2017). Future loss of arctic sea-ice cover could drive a substantial decrease in California's rainfall.

Nature Communications, 8, 1947. Retrieved July 2, 2017, from https://www.nature.com/articles/s41467-017-01907-4.

Davenport, C. (2021, October 14). U.S. Plans to Build Wind Farms along most of coastline in climate push. *New York Times*, A-18.

De'ath, G., Louygh, J. M., & Fabricius, K. E. (2009). Declining coral calcification on the great barrier reef. *Science, 323*, 116–119.

Eichelberger, E. (2014, June). How environmental disaster is making boko haram violence worse. *Mother Jones*. n.p.

Eilperin, J. (2008a, March 10). Carbon output must near zero to avert danger, new studies say. *Washington Post*, A-1. Retrieved March 21, 2008, from http://www.washingtonpost.com/wp-dyn/content/article/2008/03/09/AR2008030901867_pf.html.

Eilperin, J. (2008b, December 25). Faster climate change feared; New report points to accelerated melting, longer drought. *Washington Post*. Retrieved January 13, 2009, from http://www.washingtonpost.com/wp-dyn/content/article/2008/12/24/AR2008122402174_pf.html.

Einhorn, C. (2021, October 6). Coral reefs in decline worldwide, study finds. *New York Times*, A-11.

Erdbrink, T. (2018, January 28). Iranians prayed for rain, but were covered in snow. *New York Times*. Retrieved February 10, 2018, from https://www.nytimes.com/2018/01/28/world/middleeast/iran-tehran-snow-drought.html.

Fassihi, F. (2021, July 22). Severe water shortages add a volatile element to challenges in Iran. *New York Times*, A-10.

Flash Floods from Ida Swamp the Northeast. (2021, September 6). NASA earth observatory. Retrieved October 8, 2021, from https://earthobservatory.nasa.gov/images/148792/flash-floods-from-ida-swamp-the-northeast?src=eoa-iotd.

Flavelle, C. (2019, August 8). The food supply is at dire risk, U.N. experts say. *New York Times*, A-1, A-8.

Flavelle, C. (2021a, August 5). In bill, parties recognize climate change crisis. *New York Times*, A-5.

Flavelle, C. (2021b, September 24). The cost of insuring expensive waterfront homes is about to sky-rocket. *New York Times*. Retrieved October 1, 2021, from https://www.nytimes.com/2021/09/24/climate/federal-flood-insurance-cost.html.

Flavelle, C. (2021c, November 26). Saving history with wet-vacs in Washington. *New York Times*, A-1, A-14.

Flavelle, C., Barnes, J. E., Sullivan, E., & Steinhauer, J. (2021, October 21). Reports Lay Out Climate's Threat to U.S. Security. *New York Times*, A-1, A-13.

Fountain, H. (2017, December 7). In a warming California, a future of more fire. *New York Times*. Retrieved December 9, 2017, from https://www.nytimes.com/2017/12/07/climate/california-fires-warming.html.

Fountain, H. (2021a, July 20). A wildfire so overwhelming that it controls the weather. *New York Times*, A-1, A-11.

Fountain, H.. (2021b, July 7). North America Has Its Hottest June on Record. *The New York Times*. Retrieved from https://www.nytimes.com/2021/07/07/climate/climate-change-temperatures-june.html.

Fountain, H. (2021c, August 13). It rained at the summit of Greenland; The showers are another troubling sign of a changing arctic, which is warming faster than any other region on earth; That's never happened before. *New York Times*, A-8.

Fountain, H. (2021d, August 17). Dwindling water levels in lake mead force cuts in supply in the west. *New York Times*, A-13.

Fountain, H. (2021e, December 15). How bad is the western drought? Worst in 12 centuries, study finds. *New York Times*. Retrieved December 18, 2021, from https://www.nytimes.comoughgty/2022/02/14/climate/western-drought-megadrought.html?campaign_id=2&emc=edit_th_20220215&instance_id=53189&nl=todaysheadlines&regi_id=35795487&segment_id=82763&user_id=8953ac8150496623ee2c782e2065b2e1.

Fountain, H., & White, J. (2021, December 18). Rising from the Antarctic, A climate alarm: Wilder winds are altering currents. The sea is altering carbon dioxide. Ice is melting from below. *New York Times*. Retrieved December 18, 2021, from https://www.nytimes.com/interactive/2021/12/13/climate/antarctic-climate-change.html.

2

Friedman, L., Tabuchi, H., & Choi-Schagrin, W. (2021, August 10). Have-nots tell richer nations to fix climate. *New York Times*, A-1, A-5.

Friedman, T. (2018, January 23). The tweet trump could never send Tehran. *New York Times*. Retrieved January 24, 2018, from https://www.nytimes.com/2018/01/23/opinion/trump-iran-climate-change.html.

Fuller, T., & Hubler, S. (2021, August 27). Wildlife Smoke Chokes Lake Tahoe, Once an Oasis of Fresh Air. *New York Times*, A-1, A-13.

Gaarder, N. (2013, March 30). Arctic warming blamed for last year's heat and this year's chill. *Omaha World-Herald*. Retrieved April 1, 2013, from http://www.omaha.com/article/20130330/NEWS/130339998/1685#artic-warming-blamed-for-last-year-s-heat-and-this-year-s-chill.

Gaarder, N. (2021, December 17). Surreal storms raise questions on weather. *Omaha World-Herald*, A-1, A-2.

Gall, C. (2021, August 5). Turkey's Forest Fires Fuel a Political Inferno Online. *New York Times*, A-6.

Gates, D. (2021, October 24). As climate concerns threaten air travel, aviation industry banks on technology solutions. *Seattle Times*, A-1.

Gillis, J. (2013, January 21). How high could the tide go? *New York Times*. Retrieved from http://www.nytimes.com/2013/01/22/science/earth/seeking-clues-about-sea-level-from-fossil-beaches.htm.

Gillis, J., & Chang, K. (2014, May 12). Scientists warn of rising oceans as Antarctic ice melts. *New York Times*. Retrieved May 12, 2014, from http://www.nytimes.com/2014/05/13/science/earth/collapse-of-parts-of-west-antarctica-ice-sheet-has-begun-scientists-say.html.

Goss, M., et al. (2020). Climate change is increasing the likelihood of extreme autumn wildfire conditions across California. *Environmental Research Letters, 15*, 094016.

Green, J. (2021). *The apocalypse reviewed: Essays on a human-centered planet* (p. 17). Dutton.

Greshko, M. (2021, December). Speed. *The National Geographic*, 30–33.

Hansen, J. (2013a, April 15). *Making things clearer: Exaggeration, jumping the gun, and the Venus syndrome*. Retrieved April 25, 2013, from http://www.columbia.edu/~jeh1/mailings/2013/20130415_Exaggerations.pdf.

Hansen, J. (2013b, September 26). *An old story, but useful lessons*. Retrieved October 1, 2013, from http://www.columbia.edu/~jeh1/mailings/2013/20130926_PTRSpaperDiscussion.pdf.

Hansen, J., Johnson, D., Lacis, A., Lebendeff, S., Rind, D., & Russell, G. (1981). Climate impact of increasing atmospheric carbon dioxide. *Science, 213*, 957–966.

Hansen, J., & Sato, M. (2021) August temperature update & Gas Bag Season Approaches. *Red Green and Blue*. Retrieved May 12, 2022, from http://redgreenandblue.org/2021/09/22/dr-james-hansen-august-temperature-update-gas-bag-season-approaches/#.

Hansen, J., Sato, M., Kharecha, P., Beerling, D., Masson-Delmotte, V., Pagani, M., Raymo, M., Royer, D., Hansen, J., Sato, M., Russell, G., & Kharecha, P. (2013). Climate sensitivity, sea level, and atmospheric carbon dioxide. *Philosophical Transactions of the Royal Society A: Mathematical, Physical, and Engineering Sciences., 371*. https://doi.org/10.1098/rsta.2012.0294

Hauser, C. (2021, December 16). 'Off the charts' storm rolls across Midwest. *New York Times*, A-21.

Healy, J. (2021, September 7). In fight against Caldor fire, hotel workers are a backbone. *New York Times*, A-12.

Heating Up in Tokyo. NASA World Observatory. (2021, July 27). Retrieved August 2, 2021, from https://earthobservatory.nasa.gov/images/148650/fires-rage-in-turkey?src=eoa-iotd;   https://earthobservatory.nasa.gov/images/148616/heating-up-in-tokyo?src=eoa-iotd.

Holland, J. S. (2001, November). The acid threat: as CO_2 rises, shelled animals may perish. *National Geographic*, 110–111.

Horowitz, J. (2021, December 12). Greek Island burns in a sign of crises to come. *New York Times*, A-1, A-6.

Hubbard, B. (2021, October 21). 24 in Syria are executed over series of wildfires. *New York Times*, A-9.

Hubler, S. (2021, July 30). The year summer came with dread. *New York Times*, A-10.

Hurricane Ida Batters Louisiana. (2021, August 30). *NASA earth observatory*. NASA. Retrieved September 5, 2021, from https://earthobservatory.nasa.gov/images/148767/hurricane-ida-batters-louisiana.

Hurricane Larry Hits Canada's East Coast. (2021, September 11). *Omaha World-Herald* (p. A-9).

Innis, M. (2016, April 9). Climate-related death of coral around world alarms scientists. *New York Times*. Retrieved April 15, 2016, from http://www.nytimes.com/2016/04/10/world/asia/climate-related-death-of-coral-around-world-alarms-scientists.html.

Isai, V. (2021, August 11). Heat wave could worsen a devastating wildfire season in Canada. *New York Times*, A-6.

Isai, V., & Gundlock, B. (2021, August 5). Canadian farmers race to save cattle from drought. *New York Times*, A-4.

Jimenez, J. (2021, August 16). Was Earth's hottest month on record, NOAA says. *New York Times*, A-15.

Johansen, B. E. (2007, October 7). Global Warming, 'Thermal Inertia,' and Tomorrow's News. *Nebraska Report*.

Johansen, B. E. (2015, June 14). Trump's Mar-a-Lago is due for a dunking. Danbury (CT) *News-Times, Connecticut Post, and The Hour*. Retrieved June 20, 2015, from http://www.newstimes.com/opinion/article/Bruce-E-Johansen-Trump-s-Mar-a-Lago-is-due-11217139.php.

Joughin, I., Smith, B. E., & Medley, B. (2014). Marine ice sheet collapse potentially underway for the thwaites Glacier Basin, West Antarctica. *Science, 344*, 683.

Jourbert, L. (2021, August). The edge of survival. *National Geographic*, 110–137.

Joyce, C. (2011, February 7). *'Alarming' Amazon droughts may have global fallout*. National Public Radio in NASA Earth Observatory.

Kasakove, S. (2011, December 4). Experts say extreme weather will continue across the U.S. *New York Times*, A-10.

Kaufman, D. S., Schneider, D. P., McKay, N. P., Ammann, C. M., Bradley, R. S., Briffa, K. R., Miller, G. H., Otto-Bliesner, B. L., Overpeck, J. T., Vinther, B. M., & Arctic Lakes 2k Project Members. (2009). Recent warming reverses long-term arctic cooling. *Science, 325*, 1236–1239.

Kaufman, M. (2008, February 1). Decline in snowpack is blamed on warming water supplies in west affected. *Washington Post*, A-1. Retrieved February 15, 2008, from http://www.washingtonpost.com/wp-dyn/content/article/2008/01/31/AR2008013101868_pf.html.

Keeley, J., & Syphard, A. (2021). Large California Wildfires: 2020 Fires in Historical Context. *Fire Ecology, 17*, 22.

Kerr, R. (2007). How urgent is climate change? *Science, 318*, 1231.

Kerr, R. A. (2009). The many dangers of greenhouse acid. *Science, 323*, 459.

Khan, S. A., Kjær, K. H., Bevis, M., Bamber, J. L., Wahr, J., Kjeldsen, K. K., Bjørk, A. A., Korsgaard, N. J., Stearns, L. A., van den Broeke, M. R., Liu, L., Larsen, N. K., & Muresan, I. S. (2014). Sustained mass loss of the northeast greenland ice sheet triggered by regional warming. *Nature Climate Change*. https://doi.org/10.1038/nclimate2161

King, R. (2021, September). Harmful algal blooms are impacting lake Eire. *Spectrum News 1*. Quoting NASA. Retrieved September, 2021, from https://spectrumnews1.com/oh/columbus/weather/2021/09/11/harmful-algal-blooms-impact-lake-erie.

Kintisch, E., & Stokstad, E. (2008). Ocean CO2 studies look beyond coral. *Science, 319*(1), 029.

Kitsantonis, Niki. (2021, August11). Fires ravage Greek towns as countries send support. *New York Times*, A-6.

Koch, W. (2014, March 17). Greenland's Ice Decline Accelerates. *USA Today*, 4-A.

Kolbert, E. (2006, November 20). The darkening sea: What carbon emissions are doing to the oceans. *The New Yorker*, 66–75.

Kolbert, E. (2021, November 15). Running out of Time at the U.N. climate conference. *The New Yorker*, 21–22.

Kondratyev, K., Krapivin, V. F., & Varotsos, C. A. (2004). *Global carbon cycle and climate change*. Springer/Praxis.

Kunzig, R. (2008, February). Drying of the west. *National Geographic*, 102.

Kwai, I. (2021, July 27). Between heat and floods, England endures extremes. *New York Times*, A-10.

Lean, G. (2004, November 7). Global warming will redraw map of the world. *London Independent*, 8.

Maine: Shrimp Season is Called Off. (2013, December 4). *New York Times*, A-21.

Matthews, H. D., & Caldeira, K. (2008). Stabilising climate requires near-zero emissions. *Geophysical Research Letters, 35*, L04705. https://doi.org/10.1029/2007GL032388

Matthews, T., Perry, L. B., Koch, I., Aryal, D., Khadka, A., Shrestha, D., Abernathy, K., Elmore, A. C., Seimon, A., Tait, A., Elvin, S., Tuladhar, S., Baidya, S. K., Potocki, M., Birkel, S. D., Kang, S., Sherpa, T. C., Gajurel, A., & Mayewski, P. A. (2020). Going to extremes. *American Meteorological Society, 101*, 11. https://doi.org/10.1175/BAMS-D-19-0198.1

McCarthy, M. (2005, January 24). Countdown to global catastrophe. *London Independent*, 1.

McDonald, B. (2011, October 12). The large and costly battle to contain a megafire. *New York Times*, A-10.

McKibben, B. (2021, August 4). It's not the heat. It's the damage. *The New Yorker*.

McKibben, Bill. (2009, August 26). *A breakthrough moment: IPCC's Dr. Pachauri Supports 350 target*. Post Carbon Institute. Retrieved from http://www.postcarbon.org/blog-post/41008-a-breakthrough-moment-ipcc-s-dr-pachauri.

Melley, B. (2021a, September 17). *Fighting fire with fire to protect sequoia trees*. Associated Press. Retrieved September 20, 2021, from https://apnews.com/article/fires-environment-and-nature-forests-trees-wildfires-16c2303399942a4015d3a780d2eec4e3?utm_source=Sailthru&utm_medium=email&utm_campaign=Sept17_MorningWire&utm_term=Morning%20Wire%20Subscribers.

Melley, B. (2021b, November 19). *Fires take staggering toll on giant sequoias* (p. A-4). Associated Press in Omaha World-Herald.

Melley, B., & Metz, S. (2021, August 25). Crews struggle to stop fire bearing down on Lake Tahoe, Associated Press in the Seattle Times. Retrieved August 25, 2021, from https://www.seattletimes.com/nation-world/nation/utm_.

Miller, R. W., Rodriguez, D, Randazzo, R., & Wilkins, T. (2021, November 19). *Caldor fire jumps highway amid evacuation order; Scorches its way toward sequoias* (p. A-4). Associated Press in Omaha World-Herald.

Miller, S. (2012, August 26). Blaze approaches lake Tahoe. *USA Today*, 2-A.

Min, S.-K., Zhang, X., Zwiers, F. W., & Hegerl, G. C. (2011). Human contribution to more-intense precipitation extremes. *Nature, 470*, 378–381.

Mitchell, A. (2019, May 20). 'Earthworm Dilemma' has climate scientists racing to keep up. *New York Times*. Retrieved May 25, 2019, from https://www.nytimes.com/2019/05/20/science/earthworms-soil-climate.html.

Murphy, H. (2021, August 6). System of currents is slowing, study finds. *New York Times*, A-5.

Murphy, P. (2012, September 2). *NBC Evening News*.

Myers, S. L. (2021, July 22). A somber toll as record rain swamps China. *New York Times*, A-1, A-6.

Myers, S. L., Bradsher, K., & Buckley, C. (2021, September 1). Extreme weather challenges city life in China. *New York Times*, A-1, A-8.

National Geographic. (2020, September 17). The science connecting WildFires to climate change.

Nguyen, D. (2021, August 2). *Warming rivers in West Killing Fish* (p. A-6). Associated Press in Omaha World-Herald.

Nicas, J. (2021, December 3). Transforming Brazil's Fertile Northeast to Desert in Slow Motion. *New York Times*, A-1, A-10.

Nir, S. M.. (2021, September 3). Cornered in rooms or cars by torrents of water, dozens are found dead. *New York Times*, A-12, A-13.

Normile, D. (2013). Clues to Supertyphoon's ferocity found in the western pacific. *Science, 342*, 1027.

O'Connor, M.R. (2021, November 15). Towering infernos: What is it like to fight a megafire? *The New Yorker*, 30–36.

Olmos, S., Fountain, H., & Romero, S. (2021, July 23) Heat wave, poor policies, and erratic flames create a catastrophe. *New York Times*, A-11.

Pall, P., Aina, T., Stone, D. A., Stott, P. A., Nozawa, T., Hilberts, A. G. J., Lohmann, D., & Allen, M. R. (2011). Anthropogenic greenhouse gas contribution to flood risk in England and Wales in Autumn 2000. *Nature, 470*, 382–385.

Parks, S., & Abatzoglou, J. (2020). Warmer and drier fire seasons contribute to increases in area burned at high severity in western US Forests From 1985 to 2017. *Geophysical Research Letters, 47*(22), e2020GL089858.

Patel, K. (2021, November 19). *Snow-free Glaciers in Winter. NASA Earth Observatory. Sequoias* (p. A-4). Associated Press in Omaha World-Herald.

Pearce, F. (2021, November 25). Nature plants doomsday devices. *The Guardian* (London). Retrieved November 27, 1998, from http://go2.guardian.co.uk/science/912000568-disast.html.

Pegg, J.R. (2008, February 1). U.S. Lawmakers Urged to Lead Global Warming Battle. *Environment News Service*. Retrieved February 16, 2008, from http://www.ens-newswire.com/ens/feb2008/2008-02-01-10.asp.

Pennisi, E. (2009). Calcification rates drop in Australian Reefs. *Science, 323*, 27.

Piangigini, G. (2021, July 27). Residents evacuated as wildfires ravage Sardinia in 'disaster without precedent.' *New York Times*, A-10.

Pianigiani, G. (2021, August 12). Sicily Registers Record-High Temperature as Heat Wave Sweeps [an] Italian Island. *New York Times*. Retrieved August 14, 2021, from https://www.nytimes.com/2021/08/12/world/europe/sicily-record-high-temperature-119-degrees.html.

Pierre-Louis, K., & Plummer, B. (2018, January 4). Global Warming's Toll on Coral Reefs: As if They're 'Ravaged by War'. *New York Times*. Retrieved January 4, 2018, from https://www.nytimes.com/2018/01/04/climate/coral-reefs-bleaching.html.

Pierre-Louis, K., & Popovich, N. (2019, June 10). How dengue, a deadly mosquito-borne. Disease, could spread in a warming world. *New York Times*. Retrieved June 11, 2021, from https://www.nytimes.com/interactive/2019/06/10/climate/dengue-mosquito-spread-map.html.

Plumer, B. (2021, October 13). Fossil-fuel use may peak soon, but perils of climate change still loom. *New York Times*, A-7.

Plumer, B., & Friedman, L. (2021a, November 5). Dozens of countries pledge to stop using coal power. *New York Times*, A-10.

Plumer, B., & Friedman, L. (2021b, November 13). Negotiators strike a climate deal, but world remains far from limiting warming. *New York Times*. Retrieved November 14, 2021, from https://www.nytimes.com/2021/11/13/climate/cop26-glasgow-climate-agreement.html?

Plumer, B., Schagrin, W. C., & Tabuchi, H. (2021, December 17). Examining the role of climate change in a week of wild weather. *New York Times*. Retrieved December 20, 2021, from https://www.nytimes.com/2021/12/17/climate/wind-storms-tornadoes-climate-change.html.

Popovich, N., & Choi-Schagrin, W. (2021, August 12). Deaths from heart wave likely underreported. *New York Times*, A-20.

Pristine Lake Tahoe. (2012, August 31). *USA Today*. Retrieved September 1, 2012, from https://www.usatoday.com/story/news/nation/2021/08/31/caldor-fire-lake-tahoe-evacuations-highway-89/5661531001.

Pyro Cumulous Clouds from Wildfires. (2021, July 20). *NASA Earth Observatory*. Retrieved July 25, 2021, from https://earthobservatory.nasa.gov/images/148678/an-unusually-smoky-fire-season-in-sakha?src=eoa-iotd.

Qin, A., & Chien, A. C. (2021, October 13). China's north sees flooding wreak havoc and kill 28. *New York Times*, A-7.

Rappeport, A. (2021, December 17). Climate change an 'emerging threat' to U S Financial Stability, Regulators Say. *New York Times*. Retrieved December 19, 2021, from https://www.nytimes.com/2021/12/17/us/politics/climate-change-us-financial-threat.html.

Record Rainfall Floods India. (2005, July 28). *New York Times*, A-12.

Research Shows More People Living in Floodplains. (2021, September 28). *NASA Earth Observatory*. Retrieved September 30, 2021, from https://earthobservatory.nasa.gov/images/148866/research-shows-more-people-living-in-floodplains?src=eoa-iotd.

Revkin, A. C. (2009, September 4). Global warming could forestall ice age. *New York Times*. Retrieved September 6, 2009, from http://www.nytimes.com/2009/09/04/science/earth/04arctic.html.

Riebesell, U. K., Schulz, G., Bellerby, R. G. J., Botros, M., Fritsche, P., Meyerhöfer, M., Neill, C., Nondal, G., Oschlies, A., Wohlers, J., & Zöllner, E. (2007). Enhanced biological carbon consumption in a high CO2 ocean. *Nature, 450*, 545.

Rignot, E., Mouginot, J., Morlighem, M., Serrousso, H., & Scjheuchi, B. (2014). Widespread rapid grounding line retreat of pine Island, Thwaites, Smith and Kohler Glaciers, West Antarctica from 1992 to 2011. *Geophysical Research Letters*. https://doi.org/10.1002/2014GL060140

Roberts, L. (2017). Nigeria's invisible crisis. *Science*, 18–23.

Romero, S. (2021a, August 4). Saguaros Like It Hot. But Maybe Not Quite This Hot. *New York Times*, A-13.

Romero, S. (2021b, August 25). Too hot for a cactus. *New York Times*, A-10.

Romero, S. (2021c, September 8). Paddling, skiing, hiking- and coughing. *New York Times*, A-16.

Romm, J. (2007). *Hell and high water: Global warming—The solution and the politics—and What we should do*. William Morrow.

Royte, E. (2021, July). Too hot to live. *National Geographic*, 40–65.

Santora, M. (2021, October 20). Africa's last remaining glaciers are melting away. *New York Times*, A-9.

Scales, H. (2021, November). An icy world in meltdown. *National Geographic*, 101–121.

Schiermeier, Q. (2011). Increased Flood Risk Linked to Global Warming; Likelihood of extreme rainfall may have been doubled by rising greenhouse-gas levels. *Nature, 470*, 316.

Schmall, E. (2021, October 21). Heavy rains kill dozens in India and Nepal. *New York Times*, A-10.

Schwartz, N. D. (2014, September 9). Boom in energy spurs industry in the rust belt. *New York Times*. Retrieved September 10, 2014, from http://www.nytimes.com/2014/09/09/business/an-energy-boom-lifts-the-heartland.html.

Sengupta, S. (2018, January 18). Warming, water crisis, then unrest: How Iran fits an alarming pattern. *New York Times*. Retrieved January 18, 2018, from https://www.nytimes.com/2018/01/18/climate/water-iran.html.

Sengupta, S. (2021, September 17). United Nations Warns of 'Catastrophic Pathway' with current climate pledges. An Accounting of promises made by countries in the years since the Paris accord found that they are not enough to avoid drastic impacts from climate change. *New York Times*. Retrieved September 19, 2021, from https://www.nytimes.com/2021/09/17/climate/climate-change-united-nations.html.

Sengupta, S., Friedman, L., & Plumer, B. (2021, November 12). Glasgow climate talks are down to the wire on money, ambition and fossil fuels. *New York Times*, A-1.

Slash Global Warming Gases Now Urge 1,700 Scientists, Economists. (2008, June 2). *Environment News Service*. Retrieved June 5, 2008, from http://www.ens-newswire.com/ens/jun2008/2008-06-02-02.asp.

Smialek, J. (2021, October 8). Fed's Brainard signals focuses on climate change. *New York Times*, B-2.

Smoke Replaces Ice at Lake Winnipeg. (2021, May 18). *NASA Earth Observatory*. Retrieved June 20, 2021, from https://earthobservatory.nasa.gov/images/148340/smoke-replaces-ice-at-lake-winnipeg?src=eoa-iotd.

Straneo, F., & Heimbach, P. (2013). North Atlantic warming and the retreat of Greenland's outlet glaciers. *Nature, 504*, 36–43.

Sumner, T. (2014). No stopping the collapse of West Antarctic Ice Sheet. *Science, 344*, 683.

Tabuchi, H. (2021a, September 2). Lack of power hinders assessment of toxic pollution. *New York Times*, A-14.

Tabuchi, H. (2021b, September 14). Cleaner jet fuels may help climate, but experts worry. *New York Times*, B-7.

Tarter, N. (2021, December 13). The age of megafires, the mail. *The New Yorker*, 3.

Taub, B. (2021, April 10). We have no choice. *The New Yorker*, 36–49.

Toynbee, A. (1973, September 16). The genesis of pollution. *New York Times*, n.p.

Trenberth, K. E., Dai, A., Rassmussen, R. M., & Parsons, D. B. (2003). The changing character of precipitation. *Bulletin of the American Meteorological Society, 84*, 1205–1217.

Troiasnovski, A. (2021, July 19). As Frozen Land Burns, Siberia Trembles: In Russia's Far Northeast, People Take Arctic Temperatures in Stride, but 100-degree days are Another Matter Entirely. *New York Times*, A-6.

Tunkersley, J., Rogers, K., & Friedman, L. Nations cut deals to slash methane and save forests. *New York Times*, A-1, A-8.

U.S. Power Plant Carbon Emissions Zoom in 2007. (2008, March 18). Environment News Service. Retrieved March 22, 2008, from http://www.ens-newswire.com/ens/mar2008/2008-03-18-04.asp.

Unusually Smoky Fire Season in Sakha. (2021, August 8). *NASA Earth Observatory*. Retrieved August 15, 2021, from https://earthobservatory.nasa.gov/images/148678/an-unusually-smoky-fire-season-in-sakha?src=eoa-iotd.

Vigdor, N., & Lukpat, A. (2021, October 25). 2 unusual storms converge and pummel california coast. *New York Times*, A-18.

Voiland, A. (2012, October 12). A record-breaking year for fire in Sakha. *NASA Earth Observatory*. Retrieved October 14, 2012, from https://earthobservatory.nasa.gov/images/148943/a-record-breaking-year-for-fire-in-sakha?src=eoa-iotd.

Voiland, A. (2021, October 5). What's behind California's surge of large fires? *NASA Earth Observatory*, n.p.

Wakabayashi, D., & Hsu, T. (2021, October 8). Google says it will ban false claims on warming. *New York Times*, B-2.

Warming Past, Present. (2009, September 4). Washington Post in Omaha World-Herald, 5-A.

Warrick, J. (2014, September 9). CO2 levels in atmosphere rising at dramatically faster rate, U.N. Report Warns. Washington Post. Retrieved September 10, 2014, from http://www.washingtonpost.com/national/health-science/co2-levels-in-atmosphere-rising-at-dramatically-faster-rate-un-report-warns/2014/09/08/3e2277d2-378d-11e4-bdfb-de4104544a37_story.html.

Water Shortages in India. (2019, June 19). NASA Earth Observatory. Retrieved June 21, 2019, from https://earthobservatory.nasa.gov/images/145242/water-shortages-in-india?src=eoa-iotd.

Weisman, J., & Friedman, L. (2021, December 21). Behind a 'no, Manchin's role backing coal. *New York Times*, A-1, A-14.

Weiss, K. R. (2018). Drying lakes: Warming climates, drought, an overuse are draining some of the world's biggest lakes. Threatening habitats and cultures. *National Geographic*, 108–131.

Westbrook, G. K., Thatcher, K. E., Rohling, E. J., Piotrowski, A. M., Pälike, H., Osborne, A. H., Nisbet, E. G., Minshull, T. A., Lanoisellé, M., James, R. H., Hühnerbach, V., Green, D., Fisher, R. E., Crocker, A. J., Chabert, A., Bolton, C., Beszczynska-Möller, A., Berndt, C., & Aquilina, A. (2009). L15608. Escape of methane gas from the seabed along the West Spitsbergen Continental Margin. *Geophysical Research Letters, 36*. https://doi.org/10.1029/2009GL039191

Williams, P., et al. (2019). Observed impacts of anthropogenic climate change on wildfire in California. *Earth's Future, 7*(8), 892–910.

Williams, P., et al. (2020). Large contribution from anthropogenic warming to an emerging North American megadrought. *Science, 368*(6488), 314–318.

Wilson, S., Palley, S., & Brulliard, K. (2021, September 1). Caldor fire continues to rage outside South Lake Tahoe. Washington Post. Retrieved September 5, 2021, from https://www.washingtonpost.com/nation/2021/09/01/caldor-fire-south-lake-tahoe/.

Winfield, N., McHugh, D., & Ritter, K. (2021, November 1). *Leaders make mild pledges on climate.* Associated Press in Omaha World-Herald, A-1, A-2.

Yaffa, J. (2022, January 17). Letter from Siberia: The Great Thaw. *The New Yorker*. Retrieved January 26, 2022, from https://www.newyorker.com/magazine/2022/01/17/the-great-siberian-thaw?utm.

Zachos, J. C. (2008, March 18). Target atmospheric CO2: Where should humanity aim? *Draft*. Retrieved March 19, 2008, from http://www.columbia.edu/~jeh1/2008/TargetCO2_20080317.pdf.

Zalasiewicz, J., & Williams, M. (2016). *Ocean worlds: The story of seas on earth and other planets.* Oxford University Press.

Further Reading

Bhatia, A., & Katz, J. (2021, September 16). Hot summer nights get hotter, and more dangerous. *New York Times*, A-12.

Borenstein, S. (2015, March 3). *Syria's civil war linked partly to drought, global warming.* Associated Press. Retrieved March 10, 2015, from http://hosted.ap.org/dynamic/stories/U/US_SCI_WARMING_DROUGHTS?SITE=AP&S.

Breitburg, D., Levin, L. A., Oschlies, A., Grégoire, M., Chavez, F. P., Conley, D. J., Garçon, V., Gilbert, D., Gutiérrez, D., Isensee, K., Jacinto, G. S., Limburg, K. E., Ivonne Montes, S. W. A., Naqvi, G. C., Pitcher, N. N., Rabalais, M. R., Roman, K. A., Rose, B. A., Seibel, M. T., Yasuhara, M., & Zhang, J. (2018). Declining oxygen in the global ocean and coastal waters. *Science, 359*, eaam7240. https://doi.org/10.1126/science.aam7240

Calder, K., & Wickett, M. E. (2003). Oceanography: Anthropogenic carbon and ocean pH. *Nature, 425*, 365.

Caldeira, K. (2016, February 24). Reversal of ocean acidification enhances net coral reef calcification. *Nature, 530*. Retrieved March 2, 2016, from http://www.nature.com/nature/journal/vaop/ncurrent/full/nature17155.html.

Canadella, J. G., Le Quéréc, C., Raupacha, M. R., Fielde, C. B., Buitenhuisc, E. T., Ciaisf, P., Conway, T. J., Gillettc, N. P., Houghton, R. A., & Marland, G. (2017). Contributions to accelerating atmospheric CO2 growth from economic activity, carbon intensity, and efficiency of natural sinks. *Proceedings of the National Academy of Sciences*. https://doi.org/10.1073/pnas.0702737104.

Cave, D., Bubola, E., & Sang-Hun, C. (2021, May 22). Long slide looms for world population, with sweeping ramifications. *New York Times*. Retrieved June 23, 2021, from https://www.nytimes.com/2021/05/22/world/global-population-shrinking.html.

Hughes, T. P., Anderson, K. D., Connolly, S. R., Heron, S. F., Kerry, J. T., Lough, J. M., Baird, A. H., Baum, J. K., Berumen, M. L., Bridge, T. C., Claar, D. C., Mark Eakin, C., Gilmour, J. P., Graham, N. A. J., Harrison, H., Hobbs, J.-P. A., Hoey, A. S., Hoogenboom, M., Lowe, R. J., McCulloch, M. T., Pandolfi, J. M., Pratchett, M., Schoepf, V., Torda, G., & Wilson, S. K. (2018). Spatial and temporal patterns of mass bleaching of corals in the anthropocene. *Science, 359*, 80–83.

Johansen, B. (2014, February 10). Ice free, human free, eventually. *News-Times*. Danbury, CT. Retrieved from https://www.newstimes.com/opinion/article/Bruce-E-Johansen-Ice-free-human-free-5222315.php.

Kolbert, E. (2014). *The sixth extinction: An unnatural history*. Henry Holt. 302–306.

Kolbert, E. (2015, December 7). Unsafe climates. *The New Yorker* [Talk of the Town], 23–24.

Orr, J. C., et al. (2005). Anthropogenic ocean acidification over the twenty-first century and its impact on calcifying organisms. *Nature, 437*, 681–686.

3

Climate Crisis: Code Red for Humanity and Our Home Planet

A Scorecard for an Apocalypse

Contents

© The Author(s), under exclusive license to Springer Nature Switzerland AG 2023

B. E. Johansen, *Global Warming and the Climate Crisis*,

https://doi.org/10.1007/978-3-031-12354-2_3

3

🎓 Learning Objectives

1. Students should learn the difference between short-term trends and long-term changes that are maintained over time (with a usual range of fluctuations), that is, when studying climate change, the difference between weather and climate.

2. "A September, 2021 United Nations report on climate was a revealing read in light of several older ones. The tone of the new one was considerably more urgent. We are now seeing the outcomes of what was, 20 to 30 years ago, scientific theory." This is a quotation from the text below. This United Nations report should be available online. Perhaps a class should evaluate the report against the assumptions and commentary above.

3. Here is another example: "Ryan Pendell, writing in the Omaha World-Herald, (December, 17, 2021, 7) pointed out that on a world-wide scale, the hottest years on record have been 2021, 2020, 2019, 2018, 2017, 2016, 2015, 2014, and 2013. That's nine of nine." Is 9 years of 9 in a row enough to establish a trend (a change in climate) as opposed to a string of haphazard episodes? Does the fact that these temperatures are measured on a worldwide scale raise a case for a change in climate vis-a-vis a short-term statistical accident?

4. A large proportion of the incident reports in this chapter pose cases that could fit analysis of longer trends. For one example, students should study increases in the number and size of wildfires, increases in temperatures over large areas and long intervals of time, increases of rain and snowfall, and so forth, to indicate whether these reports are episodic stories or more meaningful changes in the "plot."

Climate scientists are taught not to mistake one, two, or even three or more episodes of extreme weather for a bona fide change in climate. Enough of a change, affecting a change in averages over time, is required. This chapter chronicles weather episodes—drought, extreme precipitation, and others, mainly during the second half of 2021. This description, by design, is a chronicle of episodes, inviting the student (or other readers) to test the difference between "weather" and "climate." Weather is the story; climate is the plot. Do the events recounted here indicate long-lasting climatic change? Because the story keeps unfolding, only a record in which averages are measured over a significant time may tell—or not—if the trend maintains movement in a different direction. At this point, meteorology becomes a sort of intellectual detective work the ways in which this information can be used to support legal arguments which may, in years to come, be used to formulate deliberations which can be used to define environmental law and penalties for violations of it. By defining real legal doctrines with precedents and penalties, environmental law may escape the funny money status that it has held wherein land, sea, and atmosphere have been defined as free goods that may be freely exploited without meaningful penalties, except those defined by legal property rights held by human beings, corporations, or nation-states. When the Earth and its component parts have legal status—that is rights, ipso facto, in and of themselves—we are on our way to having a body of environmental jurisprudence.

3.1 Scientific and Legal Background

» "We are still knocking on the door of climate catastrophe" (Borenstein & Jordans, 2021, A-12).

UN Secretary General Antonio Guterres, at World Climate Summit, Glasgow, Scotland, November 13, 2021.

» "We probably didn't know what we were doing thousands of years ago as we hunted some large mammals to extinction. But we know what we are doing now. We know how to tread more lightly upon the Earth. We could choose to use less energy, eat less meat, clear fewer forests. And we choose not to. As a result, for many forms of life, humanity is the apocalypse" (Green, 2021, 17).

» "I live in a wounded world, and I know [that] I am the wound" (Terry Tempest Williams in Green, 2021, 273).

The point is often made in the study of climate change that one event does not make a trend. That is the difference between weather and climate. The confluence of many events over wide areas during a relatively short time does, indeed, change long-term trends. That is how weather becomes climate and how it is now rapidly changing.

So how much story do we need before it becomes plot? Ryan Pendell, writing in the *Omaha World-Herald* (2021, 7), pointed out that on a worldwide scale, the hottest years on record have been 2021, 2020, 2019, 2018, 2017, 2016, 2015, 2014, and 2013. That is nine of nine.

Abuse of the Atmosphere: More Powerful Storms

Here we offer a calendar of events, most of which occurred during the second half of 2021 and early 2022, many of which carried the imprimatur of a quickly changing climate across the world. A September 2021 United Nations report on climate was a revealing read in light of several older ones. The tone of the new one was considerably more urgent. We are now seeing the outcomes of what was, 20–30 years ago, scientific theory. Omaha in one summer (2020) had two "100-year" storms, one with hurricane-strength winds, thunder and lightning, and the second, 10 days later (August 7) mainly heavy, flooding rain. Around the world, such reports were becoming more frequent.

3.1.1 Climate Change Diplomacy

The 2021 International Panel on Climate Change's (IPCC) 3000-page report is relevant not only because of the specific ways in which it states the relationship between humankind, greenhouse gases, global warming, and the damage that this

3

rough climate regime may cause, but also the ways in which this information can be used to support legal arguments which may, in years to come, be used to formulate deliberations which can be used to define environmental law and penalties for violations of it. By defining real legal doctrines with precedents and penalties, environmental law may escape the funny money status that it has held wherein land, sea, and atmosphere have been defined as free goods that may be freely exploited without meaningful penalties, except those defined by legal property rights held by human beings, corporations, or nation-states. When the Earth and its component parts have legal status—that is, rights, ipso facto, in and of themselves—we are on our way to having a body of environmental jurisprudence. This chapter follows the weather with an emphasis on phenomena that indicate climate change. Over the United States and Canada, this period (and several previous years) has been characterized by drought in the US West and an unusual amount of storminess in the East. A student of meteorology and climate can tell that this is part of a single pattern (as are some other unusual events, such as a heavy rainfall at the peak of Greenland's ice cap).

3.1.2 Distortion of Atmospheric Circulation (Jet Stream)

"Climate Change is alternating and weakening the jet stream, narrow bands of wind that circle the Earth, flowing west to east. Those changes allow key weather-producing systems of high and low pressure to stall in place" (Borenstein, 2021a, 2021b). Thus, for the last four decades (about 1980–2020, continuing today), high pressure has dominated the Western United States, often producing drought and heat; low pressure has dominated the East, producing rain, snow, and below-average temperatures.

These patterns dominate, but every so often, they will flip, briefly, but not long enough to change the overall pattern. Other parts of the Earth also have patterns that stall in place, producing similar results. The US Pacific Northwest, Western Canada, the Arctic, Antarctic Peninsula, and Siberia are among the fastest warming places on Earth. In the late spring and early summer of 2021, Siberia experienced a truly astounding heat wave at about the same time that the US Pacific Northwest reached its highest temperatures on record. A few other areas were cooler than average. Global warming is not an equal-opportunity disaster.

3.1.3 Necessary Changes in Legal Systems

The four most important lessons from the latest IPCC's climate report lay the groundwork for legal doctrines which already begun to evolve, helter-skelter. It states that if emissions of greenhouse gases are not sharply reduced in a short time, warming and damage from it will continue and accelerate during years to come. These arguments have been stated since at least 1980 by individuals or groups of scientists, but only until recent times have documents bearing the imprints of about

195 states, for all intents and purposes the entire population of the Earth, signed them. The United States was a gigantic exception under the presidency of Donald J. Trump.

The next step (much more important than signing a document) is actual reformation of the world's energy infrastructure to favor Earth-favorable energy sources and economic penalties for continued use of fossil fuels. The development of legal doctrines that favor environmentally useful (and necessary!) changes in world's legal structures to enforce progress toward a sustainable Earth will become necessary as quickly as possible. This is part of what the latest IPCC's newest report is telling us. The use of fossil fuels must be defined as pollution and, as such, must become a crime. Assertion of such as a form of freedom of expression will become as illegal as the murder of one person by another or racing an illegal oil-powered automobile the wrong way down an interstate highway at 200 miles an hour.

Murder and excessive speeding both carry penalties for good reasons. Outlawing the use of fossil fuels also has a good reason because the destruction of a sustaining Earth should no longer be allowed out of ignorance vis-a-vis what science tells us. Taken as a whole, the IPCC's report indicates to us that the outlawing of fossil fuels must begin soon and emphatically. This bus is reaching its final stop. No joke. No fantasy. No "hoax" from doctrines vis-a-vis the likes of Trump and his cohort. The change must be worldwide, but wealthy nations must take the lead to allow development of greater equity. As nonfossil energy sources become the major type of energy, economic differences will even out eventually.

3.2 Effects of Global Warming Around the World

The secretary general of the United Nations, Antonio Guterres, has called the most recent IPCC's report a "'code red'" for humanity. Essentially all of the Earth's increases in temperatures after the mid-nineteenth century (that is, since widespread use of coal and oil began) has been driven by humanity's increase in the use of fossil fuels, the report said. After less than 200 years, fossil fuels have put our Earth on death watch, meaning that their continued use will condemn Earth and its inhabitants to a lethal, miserable future that will continue to become more lethal and miserable as long as dependence on fossil fuels continues. These words may be condemned by some as radical and dangerous to the status quo. Yes, they are, but they are also very necessary, worth saying loudly and acting upon quickly. No funny business. No "clean coal." That road leads to a world that no sane person would want to leave to his or her children.

Ulla Chrobak of *Popular Science*, Aug 9, 2021, phrased the terms of the crisis like this: "The Intergovernmental Panel on Climate Change (IPCC) doesn't hedge in its latest assessment of the state of our planet. It is unequivocal that human influence has warmed the atmosphere, ocean and land," its authors write. "Widespread and rapid changes in the atmosphere, ocean, cryosphere, and biosphere have occurred."

3

"Climate Change 2021: the Physical Science Basis," part of the IPCC's sixth assessment cycle, provides a thorough update to the state of climate science since the IPCC's fifth assessment in 2014. The assessment builds on the latest data to provide a snapshot of how much humans have shifted the climate since industrialization and what is in store in the future. The new report has 234 authors, references to more than 14,000 scientific papers, and was subjected to review from numerous climate experts and government officials. "The most important takeaway is that climate change is now certain, and it's here," said Diana Bernstein, a climate scientist at the University of Southern Mississippi. "If we don't address it immediately … it's going to make everybody's lives miserable" (Chrobak, 2021).

Since the Industrial Revolution [during the nineteenth century], human-generated emissions of greenhouse gases, primarily carbon dioxide, have caused surface temperature of the globe to warm by 1.1 °C, considering only temperatures over land (which is less than a third of Earth's surface); the average global warming is even more—about 1.6 °C. Each of the last four decades has been warmer than any decade preceding it dating back to 1850.

Carbon dioxide has risen to about 420 parts per million in the atmosphere, a level that has not been as high since about 2 million years ago. While it is true that the planet has always had warmer and cooler fluctuations in climate, those shifts occurred very slowly over many thousands of years, in glacial and interglacial cycles. The pace of warming today, on the other hand, is without parallel in the history of our species. What does that mean for us? As Laura Gallardo, co-author of the IPCC's report and a climate scientist at the University of Chile, has said: "The planet is going to survive … as long as the sun is more or less alive," she says. "However, this short time framework in which things are changing, makes it extremely difficult for us to adapt and change in time" (Chrobak, 2021).

Even if we manage to immediately cease burning fossil fuels immediately (which is unlikely), the long-lived nature of carbon dioxide in the atmosphere means that Earth will continue for 50 years (on land) to about 150 years (in the oceans). Global temperatures will creep upward until at least mid-century, the assessment found. That makes keeping warming within international targets such as 1.5 or 2 °C *very* challenging. "Global warming of 1.5°C and 2°C will be exceeded during the 21st century unless deep reductions in CO2 and other greenhouse gas emissions occur quickly in the coming decades," wrote the IPCC's authors.

Humanity's best hope is to cut emissions to net zero (with any human-caused emissions offset by some form of carbon removal) and ideally go negative. Still, even under the IPCC's "very low emissions" modeled trajectory, it is estimated that temperature increases to about 1.6 °C of warming by the mid-century before dipping back to about 1.4 °C by 2100. With just a temporary overshoot of 1.5 °C, we can avoid crossing many dangerous climate tipping points.

In developing their projections, IPCC's scientists narrowed their range of "climate sensitivity," which is how much the planet is expected to warm if the carbon dioxide concentration in the atmosphere doubles from preindustrial levels to 560 parts per million. The likely range of warming from a doubling of CO_2 is now 2.5–4 °C, with a "best estimate" of 3 °C, narrowing from 1.5 to 4.5 °C—a range that has not been updated since the 1970s.

3.2.1 Major Findings from the 2021 IPCC's Report

The extra CO_2 that we have added so far is responsible for much more than a bump in global temperatures. Here are a few major findings from the IPCC's report:

1. Almost all the world's glaciers are retreating, and ice loss is causing polar regions to warm very quickly due to the loss of heat-reflecting frozen cover. Heat waves are hotter and more frequent, and the hottest days in a decade are 1.2 °C hotter than they were between 1850 and 1900. Ocean heat waves have doubled in frequency since the 1950s, and CO_2 emissions are also driving acidification and a decline in oxygen in seas and oceans. Global mean sea level has risen by 20 cm since 1901 and will continue to rise up to about a meter by 2100. Precipitation has intensified, with heavy downpours becoming more frequent. The proportion of major hurricanes has increased and the latitude of their peak intensity has shifted northward. Even with heavier rains, we still have more drought—once-in-a-decade droughts happen twice as often now. With increased temperatures, water evaporates faster and plants also use more of it.

2. In 2021's report, one can see how climate change will affect any given region. The latest IPCC's assessment had an increased focus on regional outlooks for climate change, including the incidence of extreme events. "Eight years ago we weren't completely sure whether some of the extreme events we're seeing were due to human-caused climate change," said co-author Jessica Tierney, a geoscientist of the University of Arizona. "Now we're pretty sure that that is the case. So some examples might be the extreme heat waves that we're seeing around North America right now. The new report says that those heat waves were unlikely to occur without human climate change."

3. If we manage to stop polluting the atmosphere with greenhouse gases, temperatures will eventually stabilize—although we will continue warming for at least a few decades. There will be some more permanent changes to the climate. For hundreds or even thousands of years to come, changes to the ocean, ice sheets, and sea level will not be restored. The ocean will continue to absorb heat, leading to continued ice melting and sea-level rise. Even if humans totally reverse course and move to negative emissions, the seas will remain elevated for perhaps thousands of years, rendering many coastal zones unsafe or impossible to occupy. "Climate is not a linear thing," said Gallardo.

It is understandable to feel deflated by this news. With even 1.5 °C of warming, which is all but inevitable now, droughts, floods, tropical storms, and heat waves will intensify and impact millions of people. But things can also get a lot worse—the frequency and magnitude of these extreme events grow even with a 0.5 °C rise, the assessment says. "Limiting warming to the 1.5 degree C. target of the Paris Agreement would require immediate, rapid, and large-scale reduction in greenhouse gas emissions," Mathew Barlow, a co-author and climate scientist at the University of Massachusetts (Lowell campus), said. "However, regardless of any specific temperature target, every degree matters: Reducing emissions will reduce impacts."

3

We will very likely reach or exceed a 1.5 °C temperature increase in the next 20 years. The only way to avoid this level will be a rapid shift to carbon-free energy; widespread electrification by wind, solar, or other nonfossil fuel sources; and fundamental redesign of both cities and farms. For the United States, a big step could be coming this fall, said Bernstein, passing a clean energy standard and clean energy tax incentives through an upcoming infrastructure bill via a process known as budget reconciliation.

For climate change, the takeaway is to go big or lose the only home we have. The planet will persist without us, but if humanity wants to stay, it needs to fight. This ICCC's report was issued during the summer of 2021, a time which, as if on cue, nature served us a menu of weather disasters which sparked conversation and debate worldwide about global warming and necessary human responses. The events that sparked this "wake-up call" are listed in approximate chronological order below.

3.2.2 Historic Heat in Alaska (2019)

An upper-level ridge of high pressure that slid over Alaska in June 2019 unleashed a heat wave of astonishing intensity. With temperatures soaring into the 80s and even 90s °F in some parts of Alaska, several all-time and daily temperature records fell. Anchorage, Kenai, and King Salmon broke all-time records on July 4, 2019. In Anchorage, the record was not just broken; it was obliterated. The city reached 90 °F (32 °C) on Independence Day; the previous record had been 85 °F (29 °C) on June 14, 1969. Daily temperature records have been kept for Anchorage since 1952. This heat has also been unusual for how long it has lingered. Anchorage faced 6 consecutive days where temperatures exceeded 80°, the longest stretch on record. The city broke daily high-temperature records eight times between June 23 and July 8. The normal daily high temperature for Anchorage in July is 62 °F (17 °C) (◘ Fig. 3.1).

Lightning-triggered wildfires had been burning around Fairbanks in Alaska since June 21, 2019 (also see ◘ Fig. 3.2). A second cluster began burning south of the Koyukuk Wilderness ▶ on July 5. Fires spread more quickly in hot weather because the amount of heat needed to warm fuels to the ignition point is lower. Fires generally burn with the most intensity in the afternoon, when temperatures are typically warmest.

Some of these areas, especially Germany and neighboring countries, including the United Kingdom had record rains and floods in 2021, following the 2019–2020 winter in Europe, which was the ▶ warmest on record, with little snow. The spring was also drier and warmer than usual, with a historic heat wave in the middle of May. For the third year in a row, Europe was facing potential water woes.

According to the ▶ Copernicus Climate Change Service (C3S), meteorological drought conditions started in Eastern Europe in early spring 2020 and migrated across the continent with drier-than-normal weather in April and May. Tributaries and main stems of some of the continent's rivers—such as the Elbe, Warta, and Danube—fell below its usual seasonal flow. In late May and June, surface soil

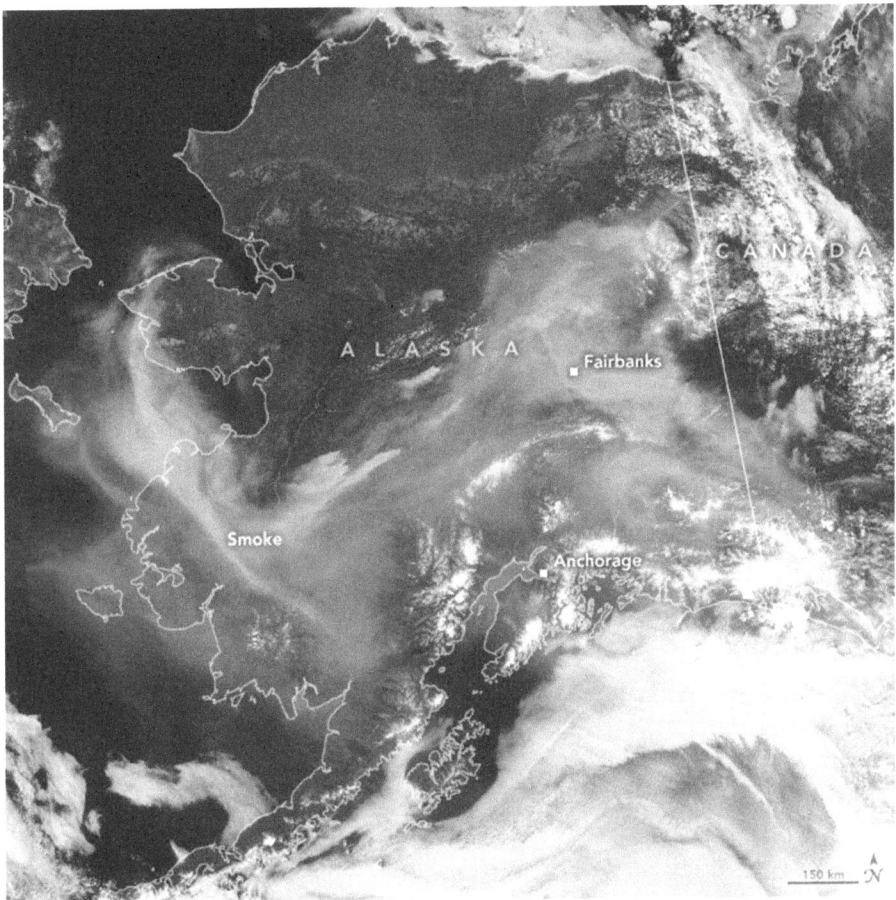

◘ Fig. 3.1 Smoke from Alaska wildfires, July 8, 2019. Source: NASA Earth Observatory

moisture and waterways in some areas rebounded a bit after heavy rain showers. All of this occurred as 2020 continued to be one of the ► hottest years on record globally.

In their ► seasonal review and forecast, C3S meteorologists predicted below-average precipitation for Southern and Eastern Europe in June, July, and August. The ► GEOGLAM Crop Monitor, a ► Group on Earth Observations initiative to monitor agricultural conditions and food security around the world, put much of Central and Eastern Europe and Southwestern Russia under a "watch" for potential drought impacts on wheat production.

In recent years, Central Europe has experienced a series of droughts caused by exceptionally stable weather patterns and high temperatures that can both ► be linked to climate change, said Wolfgang Wagner, a remote sensing scientist at the Technische Universität Wien. The fact that some regions have experienced drought conditions in several consecutive years has already caused significant damage to forests (due to bark beetle infestation) and declines in groundwater levels.

3

◻ **Fig. 3.2**   Alaska. July 8, 2019. Source: NASA Earth Observatory

After 6 years of rainfall deficits, the ▶ Czech Republic reported that nearly 80% of its wells were recording mild to extreme drought. The country's soil moisture in May was at least 30% below normal. Some climatologists called it the country's worst drought in 500 years. In neighboring Germany, underground water storage has also ▶ been depleted in recent years. In 2021, the pattern changed from drought to deluge, as much of this area received very wet conditions with major flooding.

In Ukraine, the water level in the Desna River reached its lowest point in 140 years of observations—a full 5 m below normal for springtime. The Dnieper River watershed, the source of half of the country's water needs, received just 70% of its usual rainfall from September through May. In early June, before the rains of 2021, reservoirs around Kyiv stood at their lowest levels in nearly a century. ▶ Extreme rain in late June 2021 significantly improved surface conditions in the near term. However, much of it ran off (rather than percolating down into the groundwater) because parched soil ▶ does not readily soak up water.

Polish climatologists in 2020 reported some of its worst droughts in a 100 years, with agricultural drought in 11 of 16 provinces. More than 40 rivers and streams fell substantially below normal levels by May 2020, low flows that came just months after the Vistula dropped to its lowest level on record in 2019. The low water levels posed trouble for the electric power industry, which sometimes cannot draw enough water for cooling.

"From the global food security and agricultural commodities perspective, Europe is important because it is one of the largest wheat-producing regions in the world, and also a major maize-producing region. Both wheat and maize are major food security crops," said Brian Barker, a leader of the GEOGLAM group and scientist at the University of Maryland. "The persistent rainfall deficits, combined with the above-average temperatures since winter, have negatively affected large areas across Europe, reducing forecasted crop yields compared to the five-year average in a number of countries."

3.2.3 A Looming Worldwide Food Crisis

The Intergovernmental Panel on Climate Change (IPCC) issued a report in August 2019 describing a looming food crisis that among other things tied lack of food provoked by climate change to increasing migration across borders. The report said that food shortages are more likely to affect poor countries than rich ones, increasing the number of climate refugees.

This is already happening, in Syria, for example, and, as the report points out, in Central America (Guatemala, Honduras, and El Salvador); a drought exceeding historical proportions played a major role in creating food shortages which have had a direct role in increasing the number of people seeking asylum in the United States at the Mexican border that President Trump had been exploiting to increase support for his cherished border wall. Between 2010 and 2015, the number of refugees seeking entry to the United States from these three countries has risen 500% (Flavelle, 2019, A-8).

Roughly 500 million people worldwide live on land that has recently turned to desert, and soil is being depleted between 10 and 100 times faster than it has been forming (Flavelle, 2019, A-1). These migrations are redefining politics around the world as nationalist regimes that oppose immigration rise to power. In the meantime, spreading agriculture has been causing draining of wetlands, exposing stores of carbon dioxide which, once released, cause CO_2 levels to rise, causing temperatures to rise even more quickly.

Rising levels of carbon dioxide and other greenhouse gases increase temperatures in the atmosphere and oceans, melting ice and reinforcing a process that increases the probability of extreme weather, including floods, droughts, and violent storms such as hurricanes, overtime shrinking the amount of land available for agriculture. Rising temperatures also decrease the nutritional value of many foods. In combination, these factors, along with increasing populations, will increase famine. Cynthia Rosenzweig, a senior research scientist at the NASA Goddard Institute for Space Studies as well as a lead author of the IPCC's study, said that "A particular danger is that food crisis could develop on several continents at the same time" (Flavelle, 2019, A-1).

3.2.4 **Water Temperature Crisis in the Great Lakes**

When the usually chilly Great Lakes feel like bath water, you know it has been a warm summer. When the water is that warm at the beginning of July [2021], you know that some records are in jeopardy. Like the rest of the planet, the Great Lakes were experiencing record heat in the summer of that year. July 2021 was the hottest month on record anywhere on Earth, at any time.

Just a few months after an unusually warm winter left the Great Lakes ► mostly free of winter ice, the surface temperatures of the lakes are now near or above records for this time of the year. Late June and early July brought several weeks of persistent high atmospheric pressure, clear skies, hot air temperatures, and light winds, allowing surface waters to warm significantly in the summer sunlight.

According to ► NOAA's reports, surface temperatures were 6–10 °F (3–5 °C) above normal for early July. The ► average lake water temperature across all of Lake Erie—the shallowest lake—was 74.29 °F on July 15; it was 59.83 °F across Lake Superior, the largest lake. On the weekend of July 10–12, water temperatures for beachgoers in the shallower coastal areas of Lake Michigan, Huron, Erie, and Ontario ranged from 75 °F to 85 °F. The only area below normal was the eastern portion of Lake Superior.

Jia Wang, a climatologist at the NOAA Great Lakes Environmental Research Laboratory, noted that the dearth of winter ice cover and early melting has likely caused portions of the Great Lakes to stratify (and form layers) sooner than usual. Such ► layering of warmer water near the surface can reinforce temperature patterns and keep cooler water from upwelling through the ► thermocline. "Then the surface water gets warmer," Wang said. "This is a seasonal positive feedback between the stratification and the surface water temperature."

NOAA's oceanographer Richard Stumpf reported that the annual bloom of cyanobacteria (often called blue-green algae) in Lake Erie arrived a few weeks earlier than normal, likely due to the warm water. He noted that the dominant species, *Microcystis aeruginosa*, prefers water temperatures above 68 °F (20 °C), with growth rates increasing with the temperatures. However, Stumpf and colleagues are only ► forecasting a moderate bloom for the lake this summer. While water temperatures are high, the amount of nutrient runoff (mainly from fertilizers) has been somewhat lower this year due to less snow and rain in winter and spring.

Beyond the warm water temperatures, lake levels also rose to unusually high levels in some places. In May 2020, Lakes Erie, Michigan, and Huron (and nearby Lake St. Clair) set records for water heights, according to the ► Army Corps of Engineers; in June, Lakes Michigan and Huron repeated the pattern, standing a full 5 inches (12 cm) above the record. The lakes started July above record levels.

Long-term NOAA's data show that the Great Lakes usually reach their ► peak temperatures in August. If water temperatures stay high into autumn, it is possible that evaporation and lake-effect precipitation could be enhanced when colder seasonal winds blow over the lakes.

3.2.5 A Heat Wave Scorches the Middle East (2021)

With ▶ meteorological summer just underway, some parts of the Northern Hemisphere were already feeling the heat in early June 2021. In particular, the early-season heat has been scorching countries across the Middle East. Air temperatures in the region on June 6, 2021, showed temperatures around 50 °C (122 °F). Local ground stations recorded temperatures above the 50 °C mark in at least four Middle Eastern countries, including Iran, Kuwait, Oman, and the United Arab Emirates (UAE). According to news reports, Sweihan in the UAE hit 51.8 °C (125.2 °F) on June 6, 2021, which was the country's highest temperature on record for the month of June. Countries in Central and South Asia also were reported to have seen extraordinarily high temperatures for that time of the year or anytime during any year (◘ Fig. 3.3).

Meteorologists at ▶ The Washington Postreported that the heat wave was a result of a "heat dome." The phenomenon occurs when high pressure in the mid-to-upper atmosphere acts as a cap, trapping warm air as it rises and pushing it back toward the surface to warm even more. A notable occurrence of the phenomenon caused temperatures to soar across the Middle East in July and August 2015. The heat wave this year comes about a month before the hottest temperatures of the season typically arrive. A heat dome also propelled temperatures as high as 121 °F in the usually cool and moist Pacific Northwest at the end of June in 2021.

For another glimpse that what is coming (in some places) is already here at about the time (June 5, 2021), swapping "Midwest" (United States) for "Mideast." For the second year in a row, drought and heat parched much of the United States from the Rocky Mountains to the Pacific Coast. Following one of the planet's ▶ warmest years on record, and with precipitation ▶ well below average in the Western United States, scientists and government agencies were watching for diminished water resources and potentially severe fire seasons. And thus it came to pass. In fact, July 2021 was the hottest month in the record of weather observation at anytime and anywhere. Temperatures reached record highs by large amounts.

According to the June 10 report from the ▶ US Drought Monitor, 88.5% of the land area in the West—defined as California, Nevada, Arizona, New Mexico, Utah, Idaho, Montana, Oregon, and Washington—was experiencing some level of drought, with 55% being classified as "extreme." An estimated 90% of Utah was under extreme drought conditions, with 64% rated "exceptional" (the driest classification). Similar conditions are reported across Arizona (87% extreme), California (85%), and Nevada (76%). More than 58 million people were living with the unusually to extreme drought conditions in the region.

3.2.6 Shrinking Lakes, Warming Temperatures, and Political Upheaval

Changing atmospheric patterns aggravated by global warming are playing a major role in drying lakes worldwide, but over-irrigation and other forms of mismanage-

3

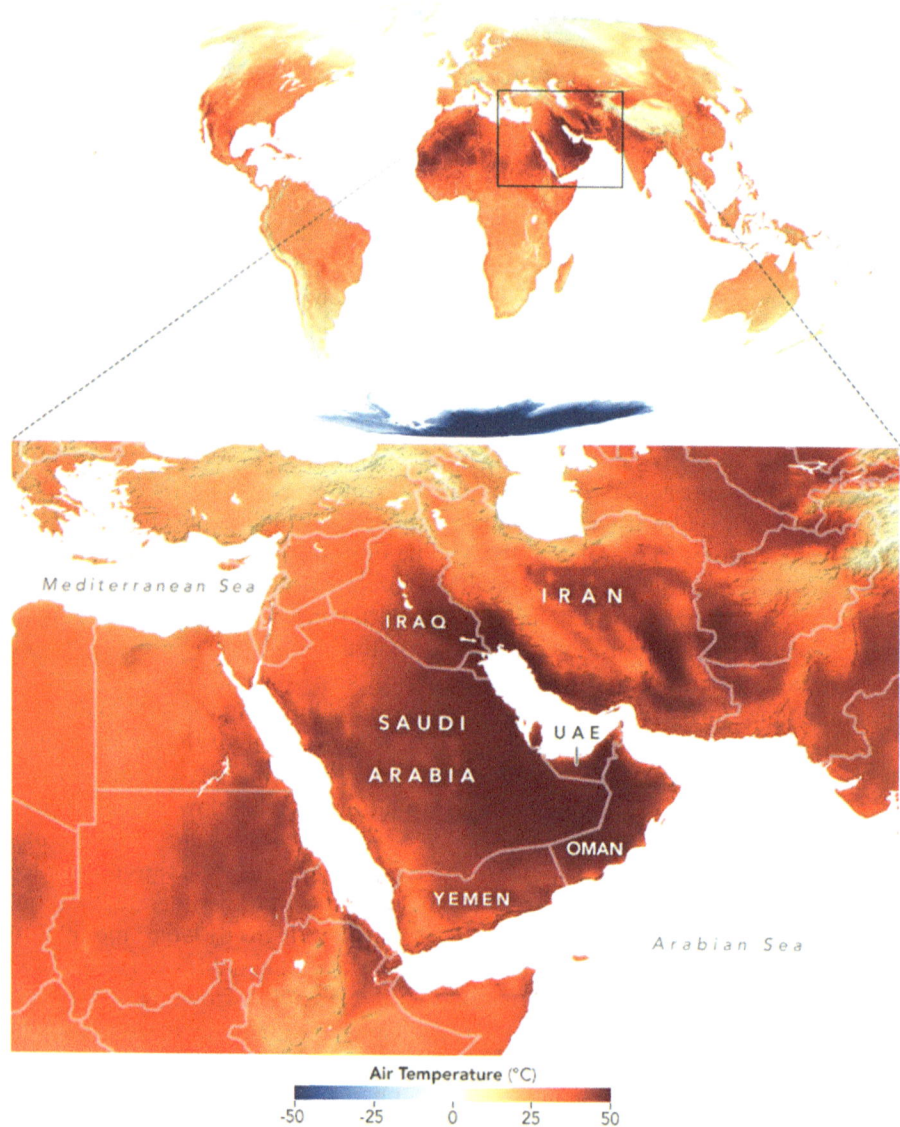

ment play a role as well. For example, in Lake Tanganyika, which is near the equator (and nowhere near a Hadley Zone of the type that spreads drought into midlatitudes), fish that feed four million people are in peril. Problems are similar in and near Lake Poopó, in Bolivia, also near the equator, which has shrunk from 3000 km² to nearly nothing in 30 years.

"Around the globe, climate change is warming many lakes faster than it's warming the oceans and the air," wrote Kenneth R. Weiss in *National Geographic* (2018,

114). "The heat accelerates evaporation, conspiring with human mismanagement to intensify water shortages, pollution, and the loss of habitats for birds and fish" (Weiss, 2018, 114). Lake Titicaca, at 12,500 feet, also in Bolivia, has shrunk as well. In eastern China's Lake Tai, cyanobacterial blooms boosted by warming water threaten water supplies for two million people. In Lake Tanganyika, fish that feed four million people are in peril.

The Caspian Sea and Iran's Lake Urmia, once the second largest lake in the Middle East, has shrunk 80% in 30 years, sped by warming temperatures, evaporation, and diversion of water for irrigation with illegal wells. Urmia's waters once were turquoise; now they are bloodred with algae blooms.

During the last four decades, or since 1979, Iran has built 600 hydroelectric dams, 20 a year on average, for irrigation of farms as well as electric power. Many lakes have been sucked nearly dry, a trend accelerated by intensifying drought. Mismanagement has caused many of the dams to fail, mechanical ghosts in an environmental apocalypse. Its rural economies declined; disgruntled former farmers moved to cities and took part in demonstrations and riots. As elsewhere in the strife-torn Middle East (Syria, to cite a prominent example), climate change contributes significantly to political upheaval.

In Iran, "Once some farmers found they no longer had water for their crops—because aquifers had been overused, or water had been diverted to big agribusinesses tied to the regime, or too many dams had been built and then warmer temperatures shrunk the lakes behind them and nearby wetlands—many of the farmers migrated to the margins of cities in search of employment, food and water," Thomas Friedman wrote in the *New York Times* (2018).

The *Tehran Times*, quoted by Thomas Friedman in the *New York Times* January 8, 2018, reported that the head of drought and crisis management at Iran's Meteorological Organization reported that "Nearly 96 percent of Iran's total area [was] suffering from different levels of prolonged drought." Lake Urmia once had been 87 miles across, but dam building and diversion of water into irrigation schemes that enriched some people, such as Revolutionary Guards, who provided contractors, people close to the Ministry of Energy, and large agribusinesses to siphon off the water, which is a salable commodity in Iran, without regard for national needs or for the future.

A *National Geographic* essay in August 2015 (quoted by Friedman) said that "Now the lake, once one of the largest in the Middle East, looks more like a gigantic crime scene." "It's dried salts now mix with sand, fueling toxic windstorms." Friedman continued: "The Iran story is repeating itself across the Middle East—environmental stresses mixing with resentment over corruption and mis-governance, sparking uprisings. And it is only going to get worse."

Somini Sengupta summarized the situation in the *New York Times* (2018): "In short, a water crisis—whether caused by nature, human mismanagement, or both—can be an early warning signal of trouble ahead. A panel of retired United States military officials warned in December, 2017 that water stress, which they defined as a shortage of fresh water, would emerge as "a growing factor in the world's hot spots and conflict areas....With escalating global population and the impact of a changing climate, we see the challenges of water stress rising with time," said a report by CNA, a research organization based in Arlington, Virginia.

Iran, already hot and dry, is projected to become worse to the point that within a few decades, several millions of Iranians may find their homeland uninhabitable due to a combination of human-caused circumstances that is afflicting many of a hotter world's expanding deserts.

Fire and Ice: The Climate Change Connection

Record wildfires in and near Los Angeles occurred the same day (December 8, 2017) that a sloppy snowstorm blanketed Houston, Texas. A jet stream that arched over Alaska and then plunged southward over the middle of the United States and northeastward along the East Coast intensified raging fires in Southern California (with temperatures in the 80s °F), at the same time that white-out snows stretched from Houston and Atlanta to Buffalo and Boston. For a day or two, there was more snow on the ground in Houston than in Anchorage, Alaska. Scientific evidence has been accumulating that this grotesque jet stream is shaped by rapid melting of Arctic ice (*see* ▶ *Chap. 2 for more discussion of jet stream distortions and its effect on climate and weather*).

3.2.7 Melting Ice Changes Atmospheric Circulation

Ivana Cvijanovic et al. (2017) wrote in *Nature Communications* that "From 2012 to 2016, California experienced one of the worst droughts since the start of observational records. As in previous dry periods, precipitation-inducing winter storms were steered away from California by a persistent atmospheric ridging system in the North Pacific....Sea-ice changes lead to reorganization of tropical convection that in turn triggers an anticyclonic response over the North Pacific, resulting in significant drying over California....We conclude that sea-ice loss of the magnitude expected in the next decades could substantially impact California's precipitation, thus highlighting another mechanism by which human-caused climate change could exacerbate future California droughts."

Jennifer Francis, an atmospheric scientist then based at Rutgers University and the University of Wisconsin-Madison, joined atmospheric scientist Stephen Vavrus, writing in *Geophysical Research Letters*, proposing a theory that associates rising temperatures in the Arctic with changes in weather conditions (and the persistence of particular patterns). Francis and Varvus wrote (2012) that "These effects are particularly evident in autumn and winter consistent with sea-ice loss, but are also apparent in summer, possibly related to earlier snow melt on high-latitude land. Slower progression of upper-level waves would cause associated weather patterns in mid-latitudes to be more persistent, which may lead to an increased probability of extreme weather events that result from prolonged conditions, such as drought, flooding, cold spells, and heat waves."

"The Arctic is warming at two to three times the rate of the rest of the globe," said Francis, "As it warms, there's less contrast between the temperature of Arctic air and the atmosphere farther south. As a result, the jet stream weakens. A strong jet stream tends to flow fairly directly, west to east. A weakened jet meanders at a slower pace, looping north and south. The consequences: A weakened jet stream is more likely to form atmospheric blocks, which tend to create "stuck" weather patterns. The meandering allows Arctic air to plunge southward or warm air to surge northward. Combined, these two factors stack the odds in favor of prolonged hot or cold spells and contribute to stalled storm systems."

This kind of jet stream whiplash tends to build high pressure into California in winter that is more typical of summer. It also accentuates high pressure inland, which sends dry Santa Ana winds down valleys from inland deserts, which become warmer and drier as they descend, creating perfect conditions for wildfires. Severe wildfire seasons such as the one that scorched California during the fall of 2017 probably will occur more frequently as warms and Arctic Sea ice melts. "This is looking like the type of year that might occur more often in the future," said A. Park Williams, a climate scientist at the Lamont-Doherty Earth Observatory in Palisades, N.Y., who was quoted by Henry Fountain in the *New York Times* (2017).

Most years, fire season in Southern California ends in October with arrival of a rainy pattern. By December 2017, that change had not occurred, and jet stream whiplash was a major cause. By 2015, persistent distortion of the jet stream also was provoking some unusual temperature readings in Alaska, such as 86 °F in Fairbanks and 91 °F in the hamlet of Eagle, Alaska, during the third week of May, warmer, during the afternoons, than much of the US East and South—warmer than Dallas or Houston had been all year to that date. This is not always the case. In 2021, record heat in the summer was followed by a cool, very wet, fall, during which floods and mudslides plagued areas that had been badly scorched the preceding summer.

Foliage grows rapidly during wet periods and then dries to explosive tinder when heat and drought return. After 5 years of drought, the winter of 2016–2017 brought soaking rains and near-record snows in the mountains, followed by a return of drought the next year. "For fires, sequencing is really important," said Alex Hall, a climate scientist at the University of California, Los Angeles. "The sequence we've seen over the past five or six years is certainly very similar to the changes that we project as climate change continues to unfold" (Fountain, 2017).

🟠 Drought Compounds Itself

Drought often compounds itself. According to the same *New York Times'* report, "Climate change may affect fires in the state [California] in other ways. While there is conflicting evidence as to whether Santa Ana and Diablo winds are becoming more frequent, Dr. Hall said that they should become drier as the planet warms, because warmer air over the high desert of Utah and Nevada has lower relative humidity and will become drier still as it descends into California. Drier air leads to more desiccation and greater fire risk." Damage from fires also has been increasing because urban areas have expanded, placing larger areas at risk.

3

3.2.8 **July 14, 2021: Summer Heat on Steroids**

Drought extremes involved much of the southwestern United States during the summer of 2021. The heat waves and droughts of the 2021 summer strained even the Saguaro cacti, which are among the largest, oldest, and toughest cacti that usually are masters of desert survival.

Lightning across the Santa Catalina mountains near Tucson, Arizona, ignited fires that scorched the cacti, which grow to 60 feet and live for 300 years, or until the worst fires in several centuries cut several of their lives short.

Meanwhile, smoke from the fires in the Southwest turned Omaha's air hazy and gritty from fires northward into Canada. We could see it, feel it, and taste it. This is the first time I have heard weather forecasts leading with "smoke," not haze, or sunny, or snow, or rain, thunderstorms, or tornadoes, just "smoke" as far as the eye could see.

3.2.9 **August 1, 2021**

Solar and wind power are now less expensive per kilowatt hour of electricity than coal, oil, or natural gas. The *New York Times* now carries all the climate change news that is fit to print. The newspaper now has a climate desk, which puts it on a par with the White House.

3.2.10 **August 5, 2021: Turkey's and Greece's Fires, Drought, and Deluge**

Fires and smoke forced evacuations in Turkey, as the same thing happened in Greece. More than 110,000 hectares (424 square miles) had burned in Greece by August 10, 2021, more than five times the yearly average from 2008 to 2020. Firefighters were called out to keep flames away from the site of the first revival of the Olympic Games, as this years' games took place in Tokyo. Extreme heat scorched Southern Europe, fanning wildfires that are making a caldron of dried brush.

Turkey's president, Recep Tayyip Erdogan, was strongly criticized for mismanagement of the fires as Turkey's people also suffered from an economic depression and the coronavirus. The fires also were aggravated by widespread drought and record-breaking heat. Several thousand people evacuated from their homes and valuable livestock died as coastal resorts' businesses were devastated.

Turkey as a whole had suffered at least 170 fires by August 5, with 59 or more deaths by August 16. Many people were left to fend for themselves in the absence of firefighters. Opponents of Erdogan said that he had no master plan for fighting future fires or other climate-related issues (Gall, 2021, A-6). Opposition leader

Kemal Kilicdaroglu said that "When people whose souls are so hurt call for help, instead of understanding them, [the ruling party] labels the people as terrorists and collaborators in a tactic that only incompetent governments will resort to" (Gall, 2021, A-6).

Occasionally, a deluge upstaged the fires for a day. "The flood we experienced is the heaviest one that I have ever seen," said Suleyman Soylu, Turkey's interior minister, at a press conference in 2016. Eight days later, flash floods killed at least 59 people in Turkey.

3.3 Drought and Deluge Worldwide

3.3.1 August, 2021: Drought and Deluge in England, Germany, China, etc.

Severe flooding in Germany, England, China, etc. is a once-in-500-year (or more) event; small streams turned to raging rivers with severe flooding in China (once in 1000 years, so it was being said); eight inches of rain fell in one hour, in Zhengzhou, a city of five million people in Central China, the heaviest on record for that period of time. Riders were imprisoned in subway cars as they were swamped by torrential rains in Zhengzhou, as also subways also were inundated by heavy rains in London.

Record drought continued in western North America, followed by a one-day devastating monsoon that flooded Phoenix and nearby areas in Arizona. After a record drought, Tucson, Arizona, experienced its heaviest rains in decades, after which the drought returned.

Drought and heat in Iran led to large groups of protesters shouting: "We are thirsty!" Security forces restrained demonstrators by force, provoking irritated anger from residents, who called for an accounting by senior leaders such as Ayatollah Ali Khamenei, the supreme leader of Iran. The protest had simmered for several weeks but exploded on social media.

As deluges hit some areas, drought continued in others, such as Iran. Mohammad, 29 years of age, an Arab street vendor, used only his (very common) first name when he was interviewed by the *New York Times:* "We kept shouting: 'We want water, just water. We don't have water. They answered us with violence and bullets'" said Fassihi. 2021. People wore skimpy clothing or none at all, as they gathered in public outdoor places in New York City to cope with the summer's heat; people also were reported to be sleeping on Baghdad's gritty city roads (at 125 °F) to escape broiling interiors of homes without air-conditioning. This was part of a heat wave over much of the world during the summer of 2021. Temperatures reached 124 °F in Baghdad, perhaps higher in uninhabited areas of Saudi Arabia. It included a heat wave over most of the United States Southwest also during the summer of 2021.

3.3.2 August 8, 2021: Monsoon Rains in Phoenix

In the midst of this worldwide record heat, monsoon rains in Phoenix, Arizona, briefly turned streets into rivers, while drought continued in the rest of the US West. About 86% of the US West suffered the worst drought in known memory while the rest briefly drowned, in a classic example (drought/deluge) of a globally warmed future that has been anticipated by scientists for many years. The heat was accompanied by large wildfires that spread over Western North America across the United States. Skies were a dirty dun brown in Omaha on what otherwise would have been a sunny day.

A new word had entered the media (and the Forest Service) with names and classifications for huge fires, such as "Dixie," which became "megafires." A "megafire" burns more than 100,000 acres. A "gigafire" burns more than one million acres.

Another phrase that came up nearly every summer was "the worst fire season in history." With heat waves and intensifying drought, fires have burned hotter, often with intensity and speed accelerated by dry foliage, into communities of houses that did not exist until a few years previously. Residents of houses in tinder-dry forests start 87% of fires by accident, increasing costs of extinguishing them. Because of the fires' increasing damage, insurers are finding many homes in fire-prone areas unprofitable and are refusing to renew policies. Fires of 150 square miles used to be uncommon in the US West. By 2021, some of the fires were five times that size.

3.3.3 August, 2021: Firefighters Look Small Against Walls of Scorching Flames

Drought and fires persisted in Oregon ("fireballs jumping from treetop to tree-top"), Washington, California, Canada, and Siberia. Such acknowledgments seem belated to those who came to speaking terms with the science of "global warming" (a.k.a. "climate change") decades ago. In California, six of the state's biggest fires occurred in 2020 and 2021.

The Dixie Fire became the second largest in California history as it contributed, along with blazes in other Western states, to Salt Lake City and Denver's worst air pollution in the world (dirtier than New Delhi and Beijing on some days). Air traffic stalled in Denver—ground stop for smoke. The pollution stains the sky across the United States—from nearly all of the West to Omaha and the rest of the mid-continental plains to New York City; Washington, DC; and Boston. Additional fires in Canada added to the gray monotone in the sky or on the ground or both. In California, firefighters became outmatched—tiny figures against towering; roaring; shifting monsters of flashing electric yellow, orange, and red; undulating; and flashing masses, as some of the fires grew to three times the size of New York City, uncontainable, with no end in sight.

The rising smoke became more toxic as it ascended, reacting with sunlight and chemicals in the air. The smoke became more toxic as it ascended into the higher reaches of the atmosphere, forming pollutants in the PM 2.5 range that can be easily swallowed and absorbed by the body, especially the lungs. Such pollutants used to be rare outside of a few urban areas. With the spread of annual fires, they have become commonplace. These smallest pollution particles are the most efficient form of pollution at inciting diseases of the lungs. Lung diseases were being diagnosed in people who had never smoked cigarettes. Many people found their COVID-19 masks also useful for wildfires.

By September 11, the Dixie Fire was nearing 100,000 acres in size and was 60% contained. Firefighters were receiving some aid from light rain showers. Likewise, the Caldor Fire was approaching 230,000 acres and was about 60% contained.

At the same time, wildfires scorched Greece, nearing suburbs of Athens. More than 100 fires were active at one time on the Greek mainland and nearby islands; Olympia, where the first modern Olympics was held, continued to be threatened as the 2021 Olympic Games ended in Tokyo. Fires in Greece were likened to a biblical apocalypse by people being evacuated from the islands by ferry as they watched nearby forests being engulfed by wind-whipped, swirling flames lighting up the night sky onshore (◘ Fig. 3.4).

The disastrous flooding in Europe at the same time was another reminder that the climate crisis does not respect borders: carbon (and ► wildfire smoke) floats above them and roaring rivers crash through them. (And not just in Europe—there was also serious flooding in ► India and ► Arizona and in ► China, and we will doubtless see it somewhere else, simply because hot air amps up the hydrological

◘ **Fig. 3.4**   Caldor Fire races across southern Lake Tahoe. Source: NASA, Public Domain

cycle: more evaporation and more ▶ torrential downpours.) It is natural to think that we should try to solve the crisis in a similarly borderless way, except that the entities that organize our political lives, nation-states, are defined by borders, and it is unlikely that this system will wither away during the decade or so that scientists have given us to halve our emissions.

3

3.4 The Summer of 2021 Burned into Memory

3.4.1 July 11, 2021: Toasted in Death Valley

A temperature of 134 °F in Death Valley is probably the highest reading on Earth in several decades (two slightly higher readings are of doubtful authenticity). Death Valley also set a record July 11 the same day for the highest 24-h cycle, 118.2 °F (Borenstein, 2021a, 2021b, A-15). "The heat wave story cannot be viewed as an isolated extreme event, but rather part of a longer story of climate change with more related, widespread, and varying impacts," said Climate Scientist Jennifer Francis of the Woodwell Climate Research Center on Cape Cod (Borenstein, 2021a, 2021b, A-15).

3.4.2 August 2021: Fire Season Now Year-Round

Seth Borenstein, a veteran science writer with the Associated Press, wrote "From 1991 to 2020, summers in the Rockies and westward have on average become 2.7 degrees F. warmer. The West is warming faster than the rest of the United States and the globe" (Borenstein, 2021a, 2021b, A-15). The heat in the Pacific Northwest as more and more have been prompted by heat domes," hot air sinking from stagnant high pressure, such as one remarkable such zone over British Columbia and the US Pacific Northwest that was responsible for highs of 121 °F, in inland BC, and 116 °F, in Portland and Eugene, Oregon, among many others during late June 2021.

Pennsylvania State University Climate Scientist Michael Mann was quoted by Associated Press Senior Science Writer Seth Borenstein as saying "We've shown [that] climate change has made summer jet-stream patterns more common." What would have been a moderate drought in southwestern North America intensified into a multi-month heat emergency [and] into mega-drought territory" (Borenstein, 2021a, 2021b, A-15). From 2011 to 2020, on average, 7.5 million acres burned annually in wildfires, more than double the average of 3.6 million acres a year from 1991 to 2000. That accounts both for a larger than previous number of fires and larger average size. Fire seasons are beginning earlier and lasting longer. In some areas, "fire season" is now a year-round phenomenon. Air pollution from fires also has become a long-standing problem coast-to-coast in the United States and throughout much of the world. Fires sometimes make their own weather. Smoke also contains hundreds of chemical compounds, many of which aggravate the lungs.

On August 7, in the evening, the Omaha area experienced a violent thunderstorm—downpours (two and a half inches in one hour), with small hail, wind gusts to 70 miles an hour, and a lightning show that must be seen to be believed, a strobe effect so frequent and so strong that my wife warned me to avert my eyes out of danger of epileptic seizures.

Sunday came up under the same dun gray sky that had become a hallmark of the season—smoke aloft from a stratospheric shield, a featureless sky with no blue at all, the sun is a smear of dim yellow and brown—wildfires' smoke to our north and west pouring into the higher elevations, part of a featureless trail of toxins that reached from California to Manitoba to the US East Coast. This had become summer's sky year after year, so common that few people remarked at it.

"These fires are going to be burning all summer," said University of Washington Wildlife Smoke Expert Dan Jaffe. "In terms of bad air quality, everywhere in the country is going to be worse than average this year" (Brown, 2021, A-15). Smoke from Western fires enveloped Manhattan Island, often turning the sun red, during July and first half of August 2021.

This is our gritty new world. Forests and towns burned to ashes as "wildlife smoke experts confirm in our Sunday papers that the grit we are tasting is, indeed, part of a continent-wide smoke stream, now a million tons lofted into the atmosphere each year" (Brown, 2021, A-150). The smoke may aggravate lung function, immune system problems, COVID-19, and influenza (◘ Fig. 3.5).

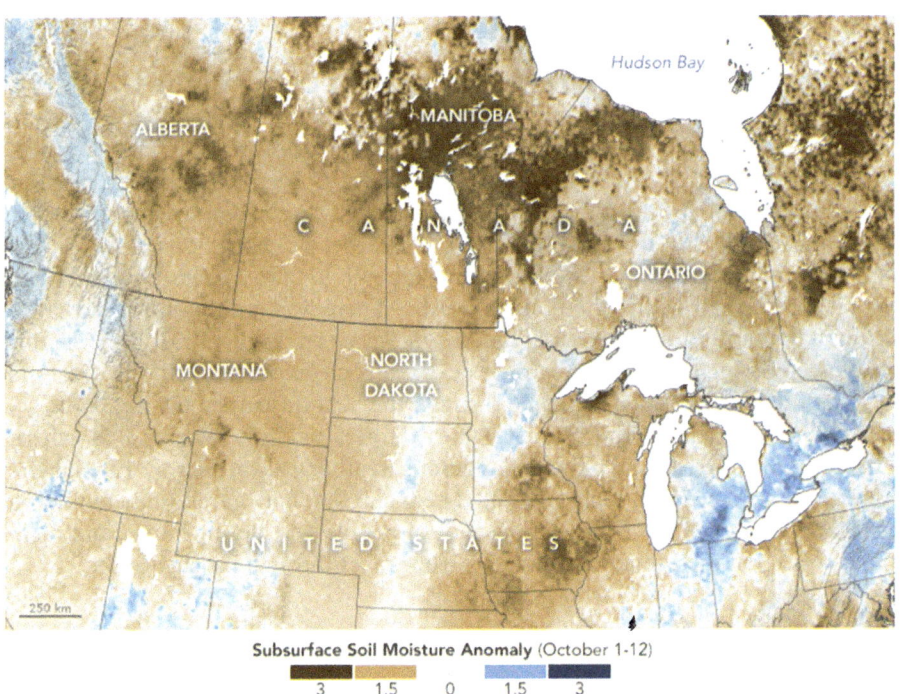

Subsurface Soil Moisture Anomaly (October 1-12)

◘ **Fig. 3.5** Drought index, early fall, 2021, much of southern Canada and the northwestern United States. Source: NASA, Earth Observatory

3

3.4.3 August 2021: Sweating at the Tokyo Olympics

The Olympics in Tokyo began July 23 with commentators remarking that the sweaty heat in Tokyo (90s °F and humidity up to 80%) would strain athletes. It was said to be the hottest Olympics in history. On August 9, Triathlon athletes collapsed of heat illness in humid heat, as officials wondered for how long summer games would be held as global temperatures rise. Paris, a designated host of the 2024 games, has been increasingly besieged by heat waves in recent years; Los Angeles, which is scheduled to host the 2028 Olympics, will be doing so at the peak of the usual wildfire season.

As the 2021 Olympic Games opened in late July, Tokyo was in the midst of a humid heat wave that pushed temperatures above ▶ 34°C (93°F) several days in a row (Heating Up in Tokyo, 2021).

For meteorologists and climatologists, the arrival of sweltering temperatures in Tokyo in August came as no surprise. Tokyo has always had hot and humid temperatures, but after ▶ decades of global warming, increases in the frequency of heat waves, and ▶ extensive urbanization, summers in Tokyo are as uncomfortable as ever—▶ and sometimes dangerous—for the city of 14 million.

The challenges are not unique to Tokyo. The other two finalist cities for these Olympic games—Madrid and Istanbul—were also sweating through heat waves during the last week of July 2021. "This is the summer that… climate change emerged from the abstract to the 'now'" said one observer in 2021.

3.4.4 July and August 2021: Under a Heat Dome in the US West

Hundreds of lightning bolts have ignited about 50 new fires ▶ in parched Oregon, where the United States' biggest blaze is likely to keep burning for months. Residents of the Northwest recalled the surreal heat wave in that area at the end of June: highs of 116 °F in Portland, 108 °F in Seattle, etc. Lents, one of Portland's poorest communities, experienced a record high of 124 °F. The heat wave cooked Walla Walla's famous sweet onions to mush. "There was nothing we could save," farmer Fernando Enriquez Sr. told the *Seattle Times* on August 15, 2021. Even this year's seeds, which would have grown next year's crop, were wiped out. Growers were hit unevenly, but ▶ none came through the heat unscathed (◼ Fig. 3.6).

Gusty winds drove the United States' largest wildfire toward a Northern California county seat, and more ▶ dangerous weather was on the way. Firefighters were keeping an eye on a lone, emaciated bear cub that may have lost its mother in the blaze. Meanwhile, in France, thousands of people were fleeing homes and vacation spots ▶ on the French Riviera as a blaze races through nearby forests.

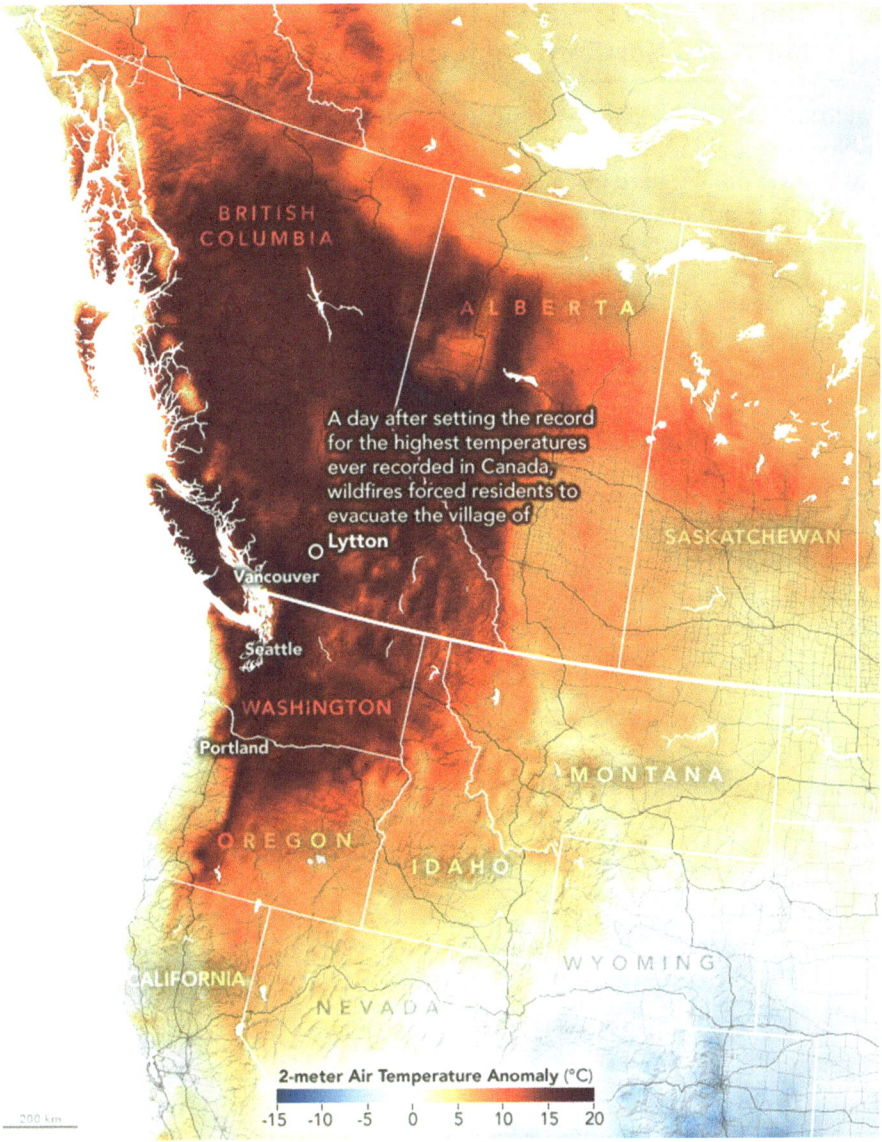

❏ **Fig. 3.6** Probably the most intense "heat dome" in recorded history broke high-temperature records throughout Washington, Oregon, British Columbia, and Alberta during the early summer of 2021. Many of the temperatures, as high as 121 °F, were the hottest by large margins ever recorded at their respective locations. Source: NASA, Earth Observatory

3.4.5 August 6, 2021: "We Lost Greenville": Dixie Fire Wipes Out Much of a Northern California Community

California's ▶ largest active wildfire wiped out most of the Greenville community in Northern California, sweeping through at a dangerous speed and endangering anyone who did not heed evacuation orders, officials said.

"Fierce winds pushed the ▶ Dixie Fire well east of its previous footprint, and flames quickly overtook businesses and homes in Greenville, a Plumas County community of about 1000 people roughly 150 miles north of the state capital, Sacramento. Flames raced into the community from a wooded mountainside," Greenville resident Teresa Clark ▶ told Sacramento TV station KXTV, in a Cable News Network (CNN) accounted By Jason Hanna and ▶ Joe Sutton.

"Within two hours, our town was gone," Clark, who left before the fire swept through, told KXTV on Wednesday. "We were sitting outside town about a mile away, and you could hear propane tanks just exploding" (Hanna & Sutton, 2021).

First responders on Wednesday evening were still trying to get everyone out. Exactly how many was not clear. "Right now, there are still a lot of people unfortunately in Greenville who did not evacuate. And so, we are having to deal with that … and get all those folks out," said Jake Cagle, the operations sections chief for California's fire incident management team. The ▶ climate crisis is making more destructive wildfires increasingly common. The Dixie Fire is one of 12 large active fires in California that have seared more than 475,000 acres so far, according to the ▶ National Interagency Fire Center.

By August 6, 2021, for the year, California fires had burned more than 8000 square miles, three times the amount by that date in 2020, itself a very bad fire year.

Much of Greenville was in charred ruins early in August—burned and fallen to the ground as thick smoke largely made for a dark, gray atmosphere—daytime video showed.

In the community's commercial center at Main Street and State Route 89, nearly every business appeared to have collapsed or been gutted, with flames still flickering in the debris up and down the sides of both streets, video recorded by ▶ Brandon Clement showed.

3.4.6 During Early August 2021: Wildfires Reignite in Turkey

In the midst of a severe heat wave and following months of very dry weather, Turkey again faced some of its worst wildfires in years. Over the past 7 days, more than 130 wildfires were reported across 30 Turkish provinces. Most of the fires ignited along the Mediterranean and Aegean Sea coasts, several in resort areas around Antalya, Mugla, and Marmari. The ▶ European Forest Fire Information Service reported that more than 136,000 ha (525 square miles) have burned in Turkey already this year, about three times the average for an entire year.

As of August 3, at least nine wildfires were still burning across Turkey, fed by strong winds, air temperatures above 40 °C (104 °F), and low humidity. Croatia, Iran, Spain, Russia, Ukraine, and Azerbaijan provided equipment and personnel to help Turkish firefighters bring the blazes under control.

Much of Southern Europe had been baking for weeks under extreme heat not seen since the 1980s. National temperature records were set in both Greece and Turkey in the summer. Air temperatures reached 45 °C (113 °F) in Greece and surrounding areas, and the heat was forecast to continue for several days. Fires also were burning in Greece and Lebanon (◘ Fig. 3.7).

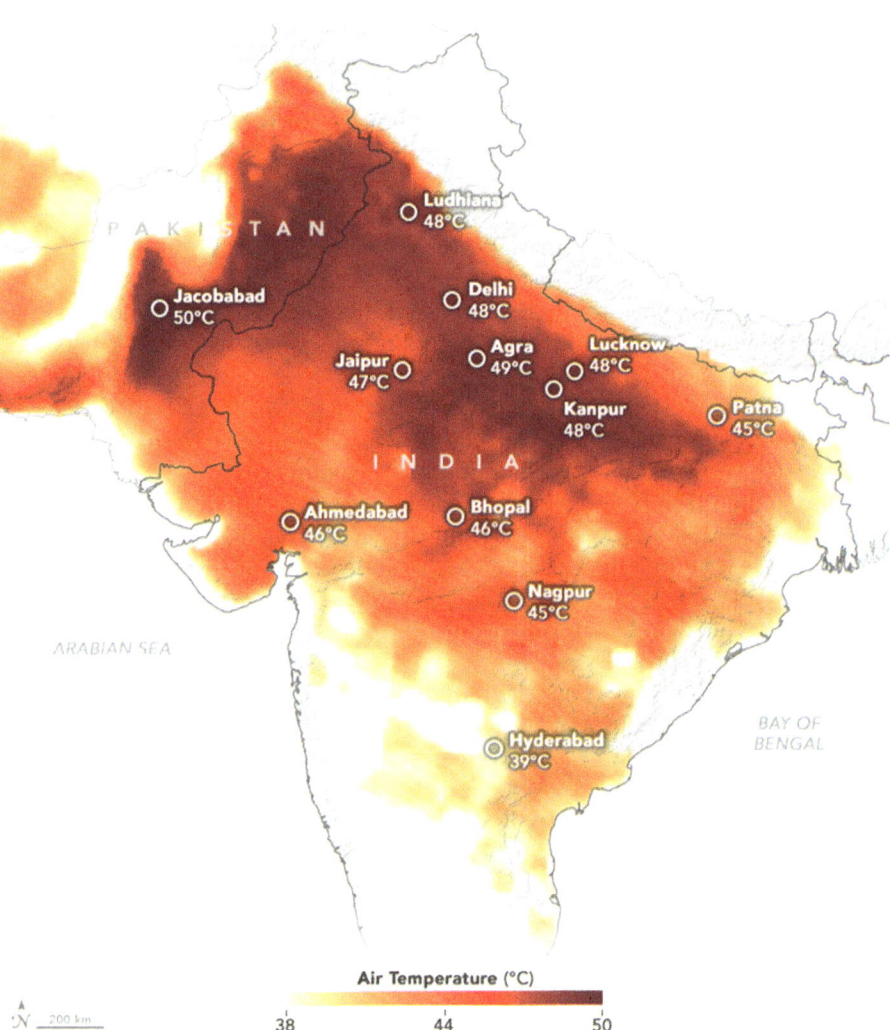

◘ **Fig. 3.7**   As with many other locations worldwide during the summer of 2021, India, which usually experiences very hot pre-monsoon temperatures, had dangerous levels of heat. Source: NASA

3.4.7 Early August 2021: Pyrocumulous Clouds Form from Wildfires

Scientists have been tracking thunderstorms created by wildfires. However, the ferocity of the storms that grew out of huge, hot fires in Canada during the summer of 2021 surprised them. In 2000, atmospheric scientists from the US Naval Research Laboratory (NRL) ▶ first reported that smoke plumes from intense wildfires could spawn towering thunderstorms that channeled smoke into the stratosphere, as high or higher than the cruising altitude of jet aircraft. These ▶ pyrocumulonimbus, or pyroCb, events surprised and shocked scientists at the time. Prior to that discovery, only explosive volcanic eruptions and extreme thunderstorms and tornadoes were thought to be capable of building cloud masses so high.

Although the workings of these smoke-infused storm clouds have come into clearer focus, their increasingly extreme behavior in recent years has surprised and worried some scientists who track them. The latest encounters with these fire-breathing smoke clouds came in North America in June and July 2021 during an unusually hot fire season that arrived early in Canadian and US forests.

Pyrocumulous clouds are one very visible way in which large wildfires can change the weather, creating huge, hot downdrafts that can be dangerous to firefighters. The first recorded "fire tornado," or "firenado," was spawned from a pyrocumulous cloud near Canberra, Australia, in 2003. Since then, such storms have been recorded in Western Russia, Europe, Africa, and North and South America. Others have even been sighted above the Arctic Circle. One such formation grew to be 620 miles across. In 2019, during widespread wildfires over much of Australia, 18 pyrocumulous formations were recorded in a single week (O'Connor, 2021, 31).

Michael Fromm, David Peterson, and a team of colleagues from NASA and several other institutions have used the ▶ Advanced Baseline Imager (ABI) on the NOAA-NASA ▶ GOES weather satellites, as well as sensors on other satellites, to identify 61 pyroCbs in North America in 7 months as of July 29, 2021, about the halfway point of the fire season.

Their observations included a remarkable ▶ outbreak of ten pyroCbs along the Saskatchewan-Manitoba border ▶ on July 16. It was more of the wildfire smoke storms than scientists have ever observed in North America on a single day since they started tracking all of them with satellites in 2013. In the ABI image shown above, all of the marked clouds ended up generating PyroCbs, though some were still in the ▶ pyrocumulus (pyroCu) stage when the image was taken. The image below shows an example of a small pyroCb rising above the McKay Creek fire on June 30, 2021 (◘ Fig. 3.8).

The July outbreak came 2 weeks after another unusual event—what Fromm called a "monster pyroCb" that exploded on June 30 above the ▶ Sparks Lake fire in Western Canada. A ▶ storm cell grew over a forest fire in British Columbia and spread across more than 160,000 km² (62,000 square miles), an area slightly larger than the state of Georgia.

Pyrocumulonimbus cloud

Active fire
(IR signature)

N   2.5 km

◨ **Fig. 3.8** ▶ Pyrocumulus (pyroCu) clouds boil out of fires, indicating their intense heat. Source: NASA

As it spread out, there fire sent a chimney of smoke up to 16 km^2 (10 miles), according to data collected by the ▶ Multi-angle Imaging SpectroRadiometer (MISR) on NASA's Terra satellite. Meanwhile, a GOES satellite observed the storm unleashing ▶ extraordinary bursts of lightning. After watching satellite clips of the storm blowing up and smoke spreading widely as updrafts hit the stratosphere, one meteorologist characterized the cloud's behavior as ▶ "absolutely mind blowing" (Pyro Cumulous Clouds From Wildfires, 2021).

Scientists said it was the largest single pyroCb cloud they have ever observed in North America. The North American Lightning Detection Network recorded nearly ▶ 113,000 cloud-to-ground lightning ▶ strokes during the event, a large amount for a storm in Canada. One meteorologist calculated that this one pyroCb event produced about ▶ 5% of Canada's total annual lightning all at once. Because smoke particles in pyroCbs limit the size of water droplets, the thunderstorms pro-

3

duce minimal rain. So the burst of lightning may have ▶ sparked new fires, accelerated their spread, and reenergized the meteorological engine that created it in the first place.

All of this played out during an unusually severe and odds-defying heat wave that pushed temperatures in British Colombia to ▶ record levels. Those extreme temperatures would have been ▶ "virtually impossible" without global warming, according to scientists from the ▶ World Weather Attribution initiative. "We're seeing pyroCbs occur almost nightly now, and we're only halfway through the fire season," said Peterson. "This could get much worse before it gets better (Pyro Cumulous Clouds From Wildfires, 2021).

With extreme fires and pyroCbs becoming more common, Fromm and Peterson find themselves asking what it all means. "We are ▶ running projects to make it easier to forecast where pyroCbs are going to pop up. The hope is that we can improve the systems that keep firefighters, pilots, and people as safe as possible," said Peterson. "But we're also watching how much smoke reaches the stratosphere and developing methods to quantify what it means for Earth's radiative balance and climate" (Pyro Cumulous Clouds From Wildfires, 2021) (◼ Fig. 3.9).

3.4.8 August 2021: Sizzling in Siberia

As soon as the snow began melting in May 2021, fires ▶ broke out in the ▶ Republic of Sakha (Yakutia) in Eastern Russia. Over the spring and summer, blazes proliferated and intensified. By August, fires had consumed large swaths of the region's larch forests. Plumes of smoke blanketed skies, turning ▶ day into night, closing airports, and prompting talk of an apocalypse in the city of Yakutsk.

The ▶ Moderate Resolution Imaging Spectroradiometer (MODIS) on NASA's ▶ Aqua satellite acquired a natural-color image of large clouds of smoke spreading over Sakha on August 8, 2021. Plumes of this size and opacity had been common for weeks, leading to poor air quality for many of the 280,000 people who live in the nearby city of Yakutsk. A few days before the image was captured, Russia's aerial fire protection service ▶ reported 178 active wildfires in Sakha. More than 2600 people and 380 pieces of equipment were fighting fires at that time (◼ Fig. 3.10).

In an area as large and remote as Sakha, satellites offer one of the best options for monitoring wildfire activity and tracking the severity of seasonal fire outbreaks. According to Mark Parrington, a scientist of the European Centre for Medium-Range Weather Forecasts, the 2021 Sakha wildfires ▶ have set a record for estimated carbon emissions for the period from June 1 to August 1. Parrington monitors fires for the ▶ Copernicus Atmosphere Monitoring Service (CAMS) using a satellite-based data record that began in 2003 (Unusually Smoky Fire Season in Sakha, 2021).

Counted together with major fires in North America, Africa, and Europe, the Sakha fires helped push global wildfire emissions to record levels in July 2021, ▶ according to CAMS data. CAMS estimates near real-time wildfire emissions from its ▶ Global Fire Assimilation System (GFAS), which aggregates observations of fires acquired by NASA's Aqua and Terra satellites.

☑ **Fig. 3.9** On January 4, 2020, wildfire smoke inundated Southeast Australia, blowing off the mainland, northwest to southeast. Based on prior events—including a "super outbreak" in December 2019 and January 2020 when pyroCbs were even more numerous—scientists know that the smoke these events injects into the stratosphere can spread out and behave like a massive shade, leading to temporary regional cooling. With enough smoke in the stratosphere, the cooling effect could be sizable, perhaps even enough to cause global temperatures to decline, much like the eruption of ash from Mount Pinatubo famously ▶ did in the 1990s due to the volcanic particles that lingered in the atmosphere. The Australian fires delivered about one-tenth the mass of cooling aerosols into the stratosphere than Pinatubo did. "If pyroCbs become more frequent, the climate impacts could really start to add up," said Fromm. "It's something we will be watching closely" (Pyro Cumulous Clouds From Wildfires, 2021). Source: NASA

"Based on current official burned-area reports, Sakha is on track to have an extreme year of fire, but it won't surpass previous extreme years if the fires are extinguished by the end of August," said earth scientist Jessica McCarty of Miami University (Ohio). "If large fire events continue into September and October, we could see burned area totals ▶ surpassing 2020" (Unusually Smoky Fire Season in Sakha, 2021).

3

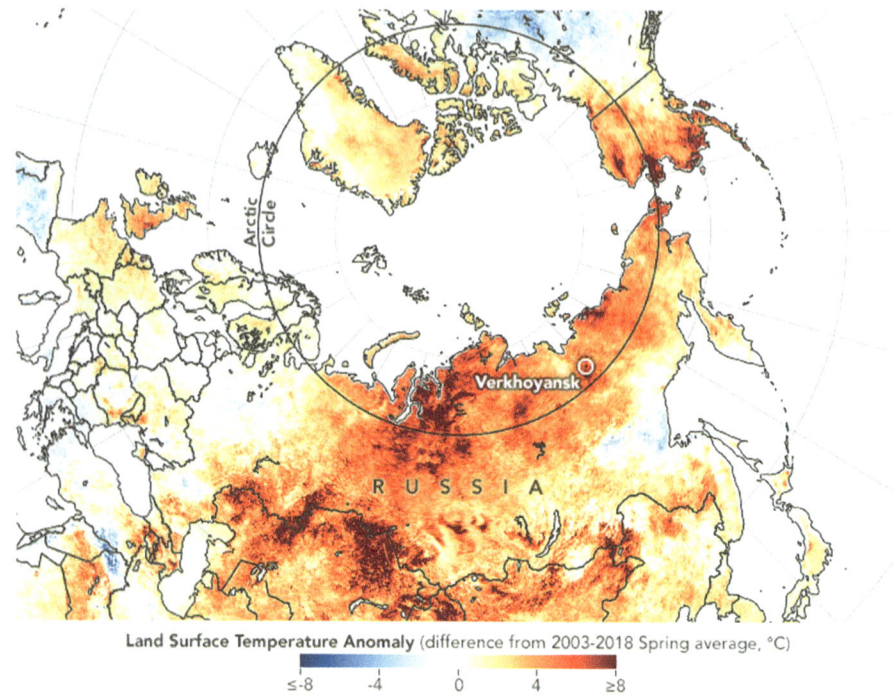

Land Surface Temperature Anomaly (difference from 2003-2018 Spring average, °C)

≤-8   -4   0   4   ≥8

◘ Fig. 3.10   Siberia, a land of legendary cold, has been warming up, too. During 2020, nearly all of Siberia was above average by varying degrees (orange to red on the map). The center of the heat wave was Verkhoyansk, which recorded the first 100 °F reading above the Arctic Circle. Source: NASA Earth Observatory

Vast areas of Yakutia have burned in the past 20 years, according to ▶ one recent analysis. About half of the fires in the region are caused by lightning. ▶ One team of researchers estimated that people start about one-third of Sakha's fires with discarded cigarettes, sparks from vehicles, negligence with campfires, and arson. Other common causes include power line failures, crop fires, and logging.

3.4.9 July and August 2021: Megafires with Names

Once upon a time, not so long ago, even the largest of wildfires did not have names. Now, there are more names than one can keep track of, an indication of their size, number, and ferocity. During July 2021, for example, the Bootleg Fire began with a bolt of lightning on a very dry Oregon mountainside. As has been typical during an intensifying drought, within a few days, the Bootleg Fire had its distinctive moniker (borrowed from a small creek in the area) and was the size of Los Angeles. Brittle Ponderosa Pines and heat made what had been a small fire a roaring monster, consuming 600 square miles and still burning. The fire became so large that it created its own weather and spread its smoke aloft, entirely across North America, contributing (with smoke from other fires) a gray and orange sunset in New York City. Some people used their COVID-19 masks to block wildfire smoke. "What we are seeing here today is the convergence of several smoke plumes," said Nancy French, a wildlife scientist at Michigan Technological University (Olmos et al., 2021, A-11).

Global warming has created a perfect hot, dry ecosystem for violent fire, as government policies have suppressed every fire, no matter how small, leaving half-charred underbrush to be consumed by larger fires. "All of this has created a recipe for catastrophic fire," said James Johnston, a researcher of the Oregon State University College of Forestry. "We've been baking this cake for some time" (Olmos et al., 2021, A-11).

3.4.10 August 11, 2021: "The Fire from Hell"

An explosive fire thunderstorm cloud rose 40,000 feet from California's Dixie wildfire, still the state's second largest in recorded history. The "firenado's" lightning spawned several spinning vortexes. The fire, which had seared almost 500,000 acres by August 11, was making its own weather. At the same time, Siberia, known for extreme cold in winter, was being devastated by wildfires that covered more land area than all of Earth's other concurrent wildfires combined.

Henry Fountain wrote in the *New York Times:* "A towering cloud of hot air, smoke, and moisture that reached airliner height (30,000 to 35,000 feet) and spawned lightning. Wind-driven flames raced across the landscape, often leapfrogging firebreaks," and spinning off rare fire tornadoes (Fountain, 2021, A-21). This description came from the Bootleg Fire, close to the California-Oregon border during mid-July 2021, but it became more common as the flaming fire season of 2021 continued—a new kind of fire that traveled under its own momentum and swallowed landscape and buildings as it moved. Usually, weather controls the size and momentum of weather. In this case, it was the opposite. Much of the time, there exists no way that such a monster can be controlled by relatively small squads of firefighters with water-carrying aircraft and relatively puny firefighters with chemicals, fire hoses, and axes. These fires carry burning embers and drop them in front and behind firefighters, igniting new blazes as they burn. The fire's lighting also can spawn new fires, making it almost impossible to contain with new burning brush springing up in front of and behind firefighters. Some of them call such incendiary beasts apocalyptic names or simply.

3.4.11 Greece: "A Natural Disaster of Unprecedented Dimensions"

On the Greek Island of Evia, south of Athens, flames crackled through pine forests and villages as temperatures reached 115 °F. More than 20 other countries offered firefighting aid. Greece's prime minister declared the fires on Evia and other parts

3

of Greece "a natural disaster of unprecedented dimensions" (Horowitz, 2021, A-1). Greece's Supreme Court ordered investigation of reports on criminal gangs, citing "the excessively large number of fires of unusual intensity and expanse" which had started at the same time (Kitsantonis, 2021, A-6).

While some Greeks saw fires as an act of God, others saw them as "another inevitable episode of Europe's extreme weather brought on by the man-made climate change that scientists have now concluded is irreversible" (Horowitz, 2021, A-1). A decade or more ago, the idea of "tipping points" was popular, used to describe points at which climate disasters would accelerate beyond humankind's ability to contain or snuff them out. By the summers of 2021, informed opinion was increasingly accepting a new idea: that "tipping points" had been crossed, and only the most strenuous, urgent, and unified human intervention might stand a chance of reversing them.

3.4.12 Mid-August 2021: Fires Scorch Evia, Greece

For several weeks, as flames scorched Evia and other parts of Greece, as well as Sicily, blazes again grew in Turkey, killing at least six people. Record heat caused by forests suffused with tinder, augmenting the fires, which consumed more than 200,000 acres of forest land. Residents of small villages across Turkey formed human chains with fire hoses, aiding firefighters, even after officials ordered them to evacuate. Even as fires appeared beaten down, some of them rekindled. The officials then joined the fight, instead of issuing useless orders. Bulldozers were called out to dig firebreaks. Everyone tried to save houses—theirs and neighbors. Some succeeded; others did not. More than 20 other nations send aid to shore up Turkey's firefighting forces. People are arrested for starting fires by neglect or arson.

Fires also burned in unusual places, such as Norway, Finland, and Sweden, at the same time. The Western half of North America continued to burn with unprecedented intensity, as fires also burned in many other places. British Columbia was battling nearly 300 fires at once.

Depending on the flow of the jet stream, which global warming seems to have more-or-less frozen in place, flooding rains fell in such places as the US East Coast, Germany, Belgium, and the Netherlands. Other countries, such as Italy, suffered from a smorgasbord of fires, flooding rains, damaging hail, and temperatures as high as 118 °F.

"We lived in paradise," said Babis Apostolou, 59, tears in his eyes, as he looked over the charred land surrounding his village, Vasilika, on the Northern tip of Evia. "Now it's Hell." "The sun struggled to penetrate the dense [smoke] haze with a sickly orange hue" (Horowitz, 2021, A-6).

West-northwest of Evia, at about the same time, temperatures rose to as high as 48.8 °C (119.8 F), in the ancient city of Syracuse, probably the highest temperatures anyplace in Europe since weather records have been kept, 400–500 years in some areas. The temperature was recorded by a mechanical observation station.

The old record, 48.0 °C, was set during July 1977, in Athens, Greece. "We are used to torrid summers, but I have no memory of such unbearable heat," said Francesco Italia, mayor of Syracuse. "It is so humid that you just can't be outside after a certain hour" (Pianigiani, 2021, A-4). Humidity makes temperatures seem higher and contributes to heat illness and stroke. The humid heat rotates counterclockwise around low-pressure systems that sweep from the Sahara desert to the south and pick up moisture over the Mediterranean Sea.

At about the same time, wildfires also ravaged Sardinia, also close to Italy, where many villages and tourist hotels were evacuated. "It is a disaster without precedent," said Governor Christian Solinas, who invoked a state of emergency. Fires moved more than 25 miles from the interior of the island to the beachfront town of Porto Alabe. A thousand-year-old olive tree that was used as a symbol by the village of Cuglieri was destroyed by the flames (Piangigini, 2021, A-10).

Snuffing the flames was not helped by temperatures that reached 104 °F for 2 consecutive days, with an abundance of dry vegetation that served as perfect wildfire fuel. The dry, hot weather also extended into October some years. Local people compared the situation in Sardinia to that of the Western United States and Canada vis-a-vis searing heat and record heat and drought. "These new phenomena are connected to climate change," Micillo said. "As Italy is, no doubt, becoming hotter" (Pianigiani, 2021, A-10).

3.4.13 August 2021: Air-Conditioning—Where Does the Hot Air Go?

The Climate Impact Lab, a consortium of climate researchers, estimates that by 2099, economic development will increase access to both health care and air-conditioning, saving millions of lives. By the middle of the twenty-second century, according to an estimate by the International Energy, the number of air-conditioning room units will increase from 1.6 billion in 2021 to 5.6 billion.

This is a major problem for the atmospheric overload of greenhouse gases because, given the present state of air-conditioning technology, most systems use a liquid refrigerant (hydrofluorocarbons) that are potent greenhouse gases, which, "if they are improperly disposed of, have a global warming potential thousands of times greater, molecule for molecule, as carbon dioxide" (Royte, 2021, 56).

This is not quite as bad as it sounds because if Earth was home to as many hydrofluorocarbon molecules as carbon dioxide, all of us—plant and animal—would have been fried lifeless long ago. We have not, but hydrofluorocarbons are still far from an atmospheric free pass. Today's air-conditioners do not make heat vanish. They merely dump it outside (Royte, 2021, 56). According to a study cited by Royte, air-conditioning in Phoenix, Arizona, dumps enough heat to raise outdoor temperatures 2 °F, raising the amount of hydrofluorocarbons used even more. In addition, air-conditioners use huge amounts of electricity, about 6% of world-

3

wide consumption. Despite publicity lavished upon relatively small solar and wind power production, worldwide most electricity is still produced by burning dirty, old-fashioned coal. Most assuredly, air-conditioning technology is going to have change to address basic environmental problems and keep the lights on and air-conditioners, washing machines, and our electric cars. As pointed out in this chapter, an electric car is only as environmentally correct as its power source. If your power company burns coal to generate your power, you are plugging the atmosphere with coal; you are doing worse than gasoline.

And one more thing: as the number of human beings and their power requirements grows, temperatures rise. In the Southwestern United States, temperatures in the triple digits now arrive weeks before they did a century ago. Without drastic and lasting reductions in greenhouse gas emissions, the heat-related death toll could rise to 100,000 per year in the United States by the year 2100. This is only four generations away. If 100,000 a year fails to move one's mental goalposts, try 1.5 million a year, which is a possible heat-related death toll for India by the year 2100. (see also ◘ Fig. 3.8, above).

Extreme heat is more than irritating. Short of killing people, it can produce premature, underweight, and stillborn babies; it also makes people more violent. That is where we are headed. The years 2015 through 2020 globally have been hottest in the history of worldwide record keeping. As this book is being written, 2021 looks hotter still. Within 30 years, given present projections, agricultural areas in the US South will be too hot for outside work for several days of an average summer. Large parts of the Earth's tropical and desert areas also may be too hot for tending crops, according to a study in the *Proceedings of the National Academy of Sciences*.

More than a million acres burned in British Columbia between April and August 2021, spurred largely by unusual heat. Fires were especially severe in late June, as a heat wave of historic intensity raised temperatures on June 28, to as high as 121 °F in one small town, Lytton, which the following evening was leveled by fire initiated by a wind-driven "dry" (i.e., rainless) thunderstorm. The temperature was the highest on record in Canada. One fire about 200 miles north of the Washington State border burned between mid-July and mid-August (Isai, 2021, A-6). The weather was so hot that helicopter engines were failing, stalling the pace of firefighting.

3.4.14 August 12, 2021: The Heat Death Toll in British Columbia and the Pacific Northwest

The *New York Times*, working with several databases, estimated the heat-related death toll in the Pacific Northwest (Washington, Oregon, and British Columbia) during its unprecedented heat wave in late June 2021 to 600, three times earlier official estimates. This estimate does not include British Columbia, where 600 or more people also were said to have died.

3.4.15 August 13, 2021: Smoke in Your Eyes

The official weather forecast for Seattle on August 13 read: "Thick smoke. High 96." Our condolences to everyone who moved there believing that that he or she was checking into paradise.

3.4.16 August 15, 2021: Six Inches of Rain for El Paso in 2 Days and Then Back to Drought

Spain reported widespread high temperatures of 117 °F. Portugal reported widespread ravaging fires. El Paso, Texas, emerged from its worst drought in memory, at least for a time, with six inches of rain in 2 days. The downpour caused damaging flooding, also only for a day or two. After that, it was back to drought. The average annual precipitation in El Paso is 8.5 inches.

3.4.17 August 15, 2021: Meanwhile, Along the Gulf Stream

Water in the Atlantic Ocean, including the Gulf Stream, circulates in a complex system that influences weather worldwide. Scientists are investigating whether this circulation is slowing as temperatures rise. Rapid rises could accelerate sea-level rise and produce stronger hurricanes over the US East and Gulf of Mexico coastlines. Tropical monsoon seasons may change and reduce rainfall over parts of Africa.

Niklas Boers, a researcher at the Potsdam Institute for Climate Change Impact Research in Germany, believes that this circulation system may be approaching a tipping point at which it would abruptly transition to a much slower state. However, Boers' work is not universally accepted among other experts, some of whom advise more study.

3.4.18 August 5, 2021: "Agricultural Emergencies" in Manitoba and Alberta

As in the western half of the United States, agriculture on Canada's plains and prairies have been devastated by drought. Smoke from fires in British Columbia hovers over cattle and farm fields. Fields meant for cattle pastures "have been stunted, parched, or mowed down by an infestation of grasshoppers, with farmers running out of alternatives for food. Many rural municipalities in the provinces of Manitoba and Alberta have declared agricultural emergencies.

Farming families are contemplating what was once unthinkable: selling off some or all of livestock that it took many generations to breed. This year's drought could spell the end for those whom raising cattle is an income, a tradition, and a way of life...." "We can't starve these animals," Michael Duguid said of the cows

3

on his farm in the heavily agricultural Interlake region roughly 100 miles north of Winnipeg, Manitoba's capital. "'They have the right to eat and drink water just like us, and the humane thing to do is to let them go'" (Isai & Gundlock, 2021, A-4).

Many farms in the region have already sold off all or most of their animals, some on ranches that have been in families for a century or more. Many farmers read and watch the news and realize that they are part of an international meteorological emergency that probably will persist and intensify over time.

3.4.19 Smoke Replaces Ice at Lake Winnipeg

On ► May 7 and ► May 14, 2021, winter ice still covered most of Lake Winnipeg and the smaller lakes nearby, and walleye were starting to snag on fishing hooks. By May 19, most of the ice was gone, temperatures hovered between 30° and 33 °C (86°–91 °F), and the parched landscape was blanketed with smoke and fire. It seemed like Canada's Manitoba Province had jumped almost instantly from winter to summer.

On May 18, 2021, the Moderate Resolution Imaging Spectroradiometer (MODIS) on NASA's Aqua satellite acquired this natural-color image of smoke billowing from several intense wildfires west and southwest of Lake Winnipeg. Strong westerly winds drove the plumes across Hudson Bay and the Canadian interior as far as Quebec. Other fires (not shown) burned to the west in Saskatchewan and to the east in Ontario. Researchers examining ► NOAA GOES satellite images detected hints of ► pyrocumulus "fire cloud" formation south of Lake Winnipeg.

Much of southern Manitoba and Saskatchewan had an abnormally dry autumn, winter, and spring weather, so much of the region entered a spell of ► severe to extreme drought. John Pomeroy, a water resources researcher at the University of Saskatchewan, ► told CBC News that soil moisture is about 40% of normal. The ► western United States and ► much of Mexico are enduring similar conditions.

Following a record-breaking spring heat wave and several days of gusty winds, the Environment Canada declared fire danger to be "extremely high" in southern and central Manitoba. Provincial and regional governments have banned campfires in parks and closed many hiking and biking trails. Leaders of the Misipawistik Cree and Lake St. Martin First Nations encouraged residents to evacuate some areas due to encroaching fires, and several homes were destroyed. Parts of highways 5, 6, and 307 in Manitoba were shut down due to smoke.

3.4.20 August 2, 2021: Forest Fire Smoke Detected at the North Pole

Forest fire smoke was detected at the North Pole for the first time in recorded history because of substantial wildfire activity at lower latitudes. Some small wildfires had reached as far northward as the west coast of Greenland.

3.4.21 August, 2021: "The Murdoch Legacy"

Tim Hewes, an Anglican minister, working with the Christian Climate Action staged a one-man protest at the front doors of News Corp. corporate headquarters in London. Hewes sewed his lips together in silent protest of News Corp.'s lack of reporting on climate change by Rupert Murdoch's media, which include holdings in the UK, United States, Australia, and others. Hewes wore signs on his chest reading: "The Murdoch Legacy: The Six Mass Extinction" and "Murdoch Did This: Muted Climate Science." The company was silent on this protest as Rev. Hewes accused it of being about climate change news. Photo and caption describing Hews' action appeared in *Harper's Magazine*, November 2021, p. 17.

3.4.22 August 2021: US Congress' Bipartisan Cooperation on Climate?

During the summer of 2021, with evidence of global warming reaching new levels world-wide, the US Congress struck a multibillion-dollar deal to fund solutions to a wide range of updates to a large number of infrastructure projects. In this deal were an unusually large number of climate change projects, considering the hiss-and-spit attitude attributed toward each party by the other. Tens of billions of dollars were directed toward protection from floods, protection from wildfires, finding and purifying new sources of water for areas reacting to floods, and relocating small communities (such as some along Alaska's Northwest coast that are sliding into the seas as storms become more violent). For the first time, both parties seemed to be willing by words as well as deeds that the United States needs to address such actions as infrastructure and that major problems need to be addressed.

For the first time, as well, members of both houses of the US Congress were hearing and heeding constituents' voices related to climate change. "It's difficult to oppose solutions to crises that your constituents are suffering through, said Shalini Vajja, a former Barack Obama aide who advises cities on how to prepare for climate crises. And as threats become more frequent and dangerous, "the constituency for climate resilience is now everybody" (Flavelle, 2021a, 2021b, 2021c, 2021d).

3.4.23 July 31, 2021: The Hottest Single Month in US Recorded History

NOAA announced that July 2021 was the hottest single month in US recorded history, all is more remarkable because (as Jim Hansen pointed out) it occurred during a La Niña period, which usually favors below-average temperatures. "This new record adds to the disturbing and disruptive path that climate change has set for the globe," said Rick Spinrad, a NOAA's administrator (Jimenez, 2021, A-15).

3

The summer of 2021 was the hottest in recorded human history. It was too hot, even, for Arizona's Saguaro Cacti, which can grow 200 feet tall and live for 200 years. During the summer of 2021, lightning started a fire that totally burned or singed several thousand saguaros. The lightning did what people are not. The cactus is such an icon in the Southwest that sawing down or burning one of them may be punishable by several years in prison. However, buffalo grass or other competitors are affected by scarce water. Fires also evaporate water and with droughts have become longer and more intense, as fires become more frequent, and with warming temperatures, saguaros are in peril (Romero, 2021a, A-13).

> **Six Inches of Rain in 2 Days and Then Back to Drought**
>
> Occasional rains during "monsoon" seasons (generally June through August) have not been keeping up with deepening drought. Heavy rains are more of a tease than a solution in an area in which 85–90% of the land is in "severe" or "exceptional" drought. During early August 2021, for example, El Paso, Texas, received 6 inches of rain in 2 days, about 70% of the city's annual average. Most of it ran off, creating ravines that were no help with the drought.
>
> Extreme weather's marriage to climatic warming is evidenced not only to heat and drought but also to heavy precipitation, rain, or snow. Sometimes temperatures and precipitation alternate back and forth, but more often they follow well-defined movements in a well-defined and often "stuck" jet stream. Drought and heat in the Western United States and Canada have often been coupled with heavy, flooding rain, and snow in US East Costal area. Heavy rain in Germany, London, etc. can be coupled with the drought and fires in Italy, Greece, and Turkey. Take out a map, overlay precipitation patterns, and locate the jet stream. A satellite photo of North America illustrated the jet stream as it dove from Alaska over the Eastern Pacific Ocean, drawing heat off the continent, and then turned northeast over Wyoming and Colorado, drawing in cold air.
>
> Some of 2021's fires burned for more than a month and consumed more than half a million acres (heretofore a very bad fire season for all the fires in many states). Some people regarded these juxtapositions of flames and mud and drought and deluge as Biblical ironies, when they were nature's signature in a warming world.

3.4.24 August 2021: Some (Squid) Like It Hot

While salmon and other cold-water fish were dying en masse in the Pacific Northwest, "market squid" were thriving. Along the western US Coast from Central California to Northwestern Washington State, the squid had increased about 500% in 22 years, according to a journal of the *American Fisheries Society* (Turner, 2022, A-8) during January 2022. Increases in squid numbers were associated with maritime "heat waves," also known as "blobs," which have been appearing more often off the coast.

3.4.25 July Through September 2021: Salmon Die as Habitat Warms

The dramatic and devastating warming of the US West Coast during 2021 (which has been discussed above) included massive numbers of salmon, a cold-water fish. Fishing people noticed first that young salmon were dying by the thousands as the fish attempted return to their spawning grounds on the Klamath River, as well as other rivers and streams, at the end of their annual migration back to where they were born. Thousands of fish died in one California river and an entire run died in another. Once the temperature reaches a certain level, the fish suffocate. Soon, salmon runs were dying the length of the Pacific Coast, Washington, and Idaho to Central California.

Salmon are born upstream after spawners return to the same spots where eggs had been laid season after season. Today, some salmon mature in the wild, but others are grown in hatchery fish farms. Native American peoples whose economic and ceremonial lives depend on return of the salmon found their entire way of living destroyed as water levels fell and their waters warmed, allowing parasites to grow on them. Salmon are a staple food for indigenous peoples all the way up the West Coast coast of North America from Central California to Alaska, where rivers were so warm that some of the salmon were swimming through the Bering Strait and into the Arctic Ocean, surprising Inuit peoples there who were looking for seals and other species of fish which were dying as Arctic ice was melting and their habitats became too warm.

The loss of one salmon run (group migration) damages its entire habitat for years to come; in California, the commercial salmon fishing industry has been worth about $1.4 billion a year to people for whom fishing supports a livelihood (Nguyen, 2021, A-6). Sports fishing has been shrinking as well. Sales of salmon plummeted in grocery stores as prices reached $35 a pound after the unprecedented heat of 2021 killed large numbers and eliminated much of the next generation (Nguyen, 2021, A-6).

In 2021, federal fishery officials watched runs shrink in rivers all along the Pacific Coastline, some by 80% or more. Part of the problem was aggravated by human-constructed lakes, such as one near Mt. Shasta in Northern California, where natural streams have been cut off. In a few years, naturally spawned salmon will disappear, as warming water and drought accelerate the entirely human-made habitat that kills them.

"Here in Los Angeles, we have had periods of extreme droughts, and periods of extraordinary flooding….and ecological degradation, and a pandemic, and, of course, heat waves and wildfires, said D.J. Waldie, a cultural historian and author in Southern California. "But they didn't all come on the same summer day" (Hubler, 2012, A-2).

3

3.4.26 August 17, 2021: Lakes Mead and Powell—Major Water Sources for the US Southwest, Fall to an All-Time Lows

Lake Mead, on the border between Arizona and Nevada, long has been one of the US West's largest water sources, supplying 40 million people in the US Southwest. It has been losing water during a two-decade drought and by mid-summer of 2021 was down to 34% of its capacity. The lake, near Las Vegas, Nevada, at that time was at its lowest level since it began to fill after the completion of Hoover Dam during the 1930s (◨ Fig. 3.11).

As its water level has declined, the lake has left a "bathtub ring" of mineral deposits along its shores, indicating how much water it has lost. "Even in the occasional recent year with good snowpack in the Rocky Mountains, the amount of runoff into the [Colorado] River has declined. Researchers say that [global] warming is largely to blame [for 40 to 60 per cent of the decline], as soil has become so dry that it soaks up much of the melting snow like a sponge" (Fountain, 2021, A-13). Allocations of water to cities, towns, farmers, and ranchers have been reduced, by an average of 18%, with more cuts in prospect as Lake Mead's decline continues.

Water also is stored in Lake Powell, which, like Mead, is part of the Colorado River system, which flows generally southward, through Mexico, into the Gulf of California. In recent years, however, the river has dried upon before it hits the gulf, another sign of the drought's severity and the amount of water being drawn from it upstream. The water is divided between Mexico and several US states (Arizona, Colorado, Nevada, New Mexico, Utah, and Wyoming). The river begins in the Rocky Mountains, which are also low because drought has reduced snow runoff, before it takes an irregular 1500-mile path into the Gulf of California.

In late July 2021, the Great Salt Lake in Utah reached its lowest water level on record, and it has continued dropping since then. On July 23, the US Geological Survey (USGS) gauge at the Saltaire boat harbor at the southern end of the lake recorded the average daily level at 4191.3 feet (1277.5 m) above sea level, the lowest mark since measurements began in 1875. The previous low was set in 1963.

Prolonged drought coupled with water withdrawals has caused a dramatic drop in Lake Powell's water level. The northern part of Lake Powell is actually a deep, narrow, meandering reservoir that is steadily becoming smaller as it extends from Arizona upstream into Southern Utah.

3.4.27 The Climatic Cost of Excess Rains in China

During the last half-century, China has plowed under farmland and hamlets to construct the signatures of urban life and huge ranks of apartments and factories that have eliminated rural poverty for hundreds of millions of the country's 1.4 billion people. At the same time, weather extremes caused in large part by climate change have contributed to new problems. Most of these problems have been ignored for years, making and facing them more daunting as extremely heavy

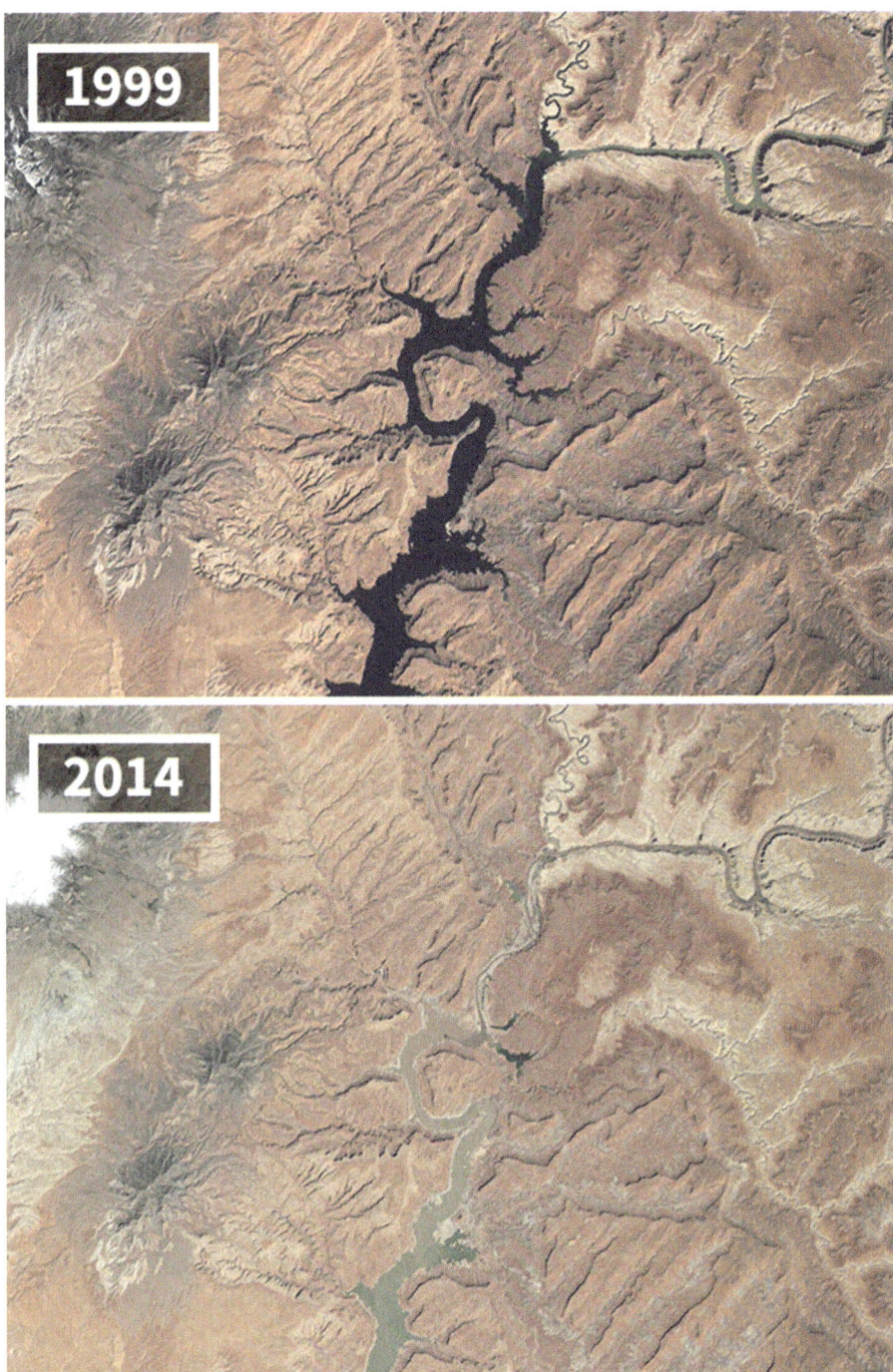

◘ **Fig. 3.11**   Lake Powell in 1999 (above) and 2014 (below). Note decrease of water levels. Source: NASA

rainfall has contributed to flooding that taxes sewer systems and water runoff. Severe storms brought all of this to a head during July of 2021, killing at least 70 people. The flooding has not only been generated by changes in weather and climate but also to "a general manifestation of urban problems," according to Kong Feng, a professor of public policy at Tsinghua University in Beijing (Myers et al., 2021, A-8).

China's rains during the summer of 2021 were called a "once-in-a-millennium event" but have been occurring much more often than that. Students of meteorology could question such assertions because most weather records extend only 100–400 years into the past, nowhere near the 1000 years that comprise a millennium (one thousand). This is true throughout the world, not solely in China.

Briefly stated, construction of so many very dense urban areas has reduced the amount of space that can accommodate spillover and absorption of excess water. By mid-2021, China had 93 cities with populations of a million or more (Myers et al., 2021, A-8), as leaders have ignored environmental issues in favor of economic growth. By 2020, however, the results of this ignorance was resulting in constant (mainly urban) crises. China has, for example, increased green spaces within cities to absorb excess water. Central authorities also have pledged to reach carbon neutrality by 2060 by cutting emissions of greenhouse gases, most notably carbon dioxide. Carbon dioxide already emitted affects climate for at least a century, so such cuts will need to be maintained and intensified for several generations to have meaningful effects.

3.4.28 India, 2021: Heat Waves and Monsoon Deluges

Three signature weather events on the Indian subcontinent are intense heat waves, heavy monsoon rains, and deadly cyclonic storms similar to hurricanes and typhoons. Recent events and professional projections indicate that both will become more extreme as the area warms with the rest of the Earth.

India experienced two strong cyclones during the first half of 2021; many cyclones also come ashore from the Bay of Bengal and rose into the Himalayas, as very heavy rains condense as moisture rose along mountainsides along the south faces of the Himalayas, the highest mountains on Earth. Indian heat waves are some of the most intense anywhere, with temperatures sometimes reaching 110–120 °F. A nation of almost 1.3 billion, India is widely impoverished, and most people cannot afford air-conditioning; millions also have no electricity. One intense heat wave in 2021 killed several thousand people.

Monsoon rains are a usual seasonal late spring and early summer event, but their absence or overabundance can cause death for many thousands of people. The absence of enough rain can cause farmers to go bankrupt, causing many to kill themselves when their farms are foreclosed. Rain that is too heavy overwhelms all systems, urban and rural. In one instance, the City of Mumbai received 37 inches of rain (roughly 3 feet) in 2 days. Much of Bangladesh (enclosed within

India on the northern edge of the Bay of Bengal) is only a few feet above sea level, subject to floods from the massive delta of the Ganges River system. As sea levels rise and melting snow from the Himalayas rushes down the Ganges' tributaries, much of Bangladesh may cease to exist.

3.4.29 2021: Great Britain—The Ocean Is Very Large and Very Wet

In Great Britain, rains are usually relatively light, but sometimes they pour, given proximity to the vast Atlantic Ocean and North Sea. When a repeating pattern sets up in the atmosphere, the heavy rain can become deadly. In the midst of 2021, Britain had two intense storms in 2 weeks, producing flooding rains that stranded people on subways (as dirty water rose around them) and kept ambulances from reaching flooded hospitals. The heavy rain followed a very unusual heat wave that pushed temperatures to the upper 90s °F in London (very unusual there) and had officials advising Londoners to stay inside with generous access to cool water (air-conditioning is rare). Londoners also professed a high degree of agreement that the whiplash weather could be traced to a warmer atmosphere that raises the moisture level in rainstorms (Kwai, 2021, A-129).

3.4.30 System Collapse in the Kalahari?

The Kalahari Desert is already red-flagged as a climate hot spot. Modeling by climate scientists at the University of Cape Town (South Africa) suggests that within a decade, when the global temperature could exceed the rise of 1.5 °C (2.7 °F) that the United Nations Paris Agreement seeks to avoid, average temperatures in Botswana—just north of South Africa—will already have warmed by nearly 4 °F. Beyond 5.4° in average warming globally, which would indicate 7.6 °F, "the science anticipates system collapse for the Kalahari....Decades of farming have thrown the region into disarray, and now it seems that the freight-train effects of planetary heating are bearing down, too" (Jourbert, 2021, 117–118).

According to the South African Weather Service, 2015, 2016, and 2019 were the hottest years on record since at least 1950. In January 2016, at Augrabies Falls, about 150 miles southwest of Tswalu was more than 119 °F. "Among the highest temperatures ever recorded so high above sea level in the Southern Hemisphere [and] the second highest reliably measured temperature in Southern Africa. Without climate change, the 2015-16 heat wave would have been about a 1-in-10,000-year event, [according to] recent analysis" (Jourbert, 2021, 134).

During August 22–30, 2021, more fires and deeper deluges created more life-threatening problems across the United States. The Caldor Fire threw a smoke plume across parts of two states (California and Nevada). Swift-flowing floods also killed more than 20 people in Tennessee. This pattern (drought and fires in the

West, deluges and floods in the East) confirmed the stagnant level of the jet stream which has become associated with global warming. Within 2 weeks, the same jet stream pattern, swinging north over Greenland, helped to produce rain, for the first time in observable history, at the highest point of the island's topography.

3.4.31 August 2012: Siberia Burning, Again

Wildfires have become a life-threatening threat in Magaras, in Siberia, where the mayor of this village stands on permafrost that is rapidly becoming less permanent as it melts, a state of affairs which is being accelerated by warmth and wildfires. On unusually warm mornings when the air fills with irritating smoke, all able-bodied men are now routinely called to the town hall to fight a fire that has reached a highway that is a vital link to the outside.

For the third straight year, as of 2021, people in northeastern Siberia have found themselves inundated by fires of a size and ferocity that no one would have imagined a decade ago. In an area with a record of the coldest winters on Earth, except for Antarctica, temperatures near 100 °F, hot orange glows sat sunset because of smoke-filled skies, and a gritty haze constitute a hellish nightmare.

Siberia's fires have followed once unimaginable heat, both earmarks of extreme global warming which is happening more rapidly than any other place on Earth. During 2020, uncontrollable fires rampaged over 60,000 square miles in this area, leaving a Florida-sized mess of melted permafrost and singed pine trees. That was four times the amount of land burned in the United States during the same year (Troiasnovski, 2021).

These fires (and many others around the world) also accelerate global warming by releasing huge amounts of carbon dioxide into the atmosphere. The fires also remove from the land trees which themselves usually breathe in carbon and expire carbon dioxide. Thus, day by day and fire by fire, Russia's vast boreal forests are being turned from a greenhouse gas sink (which draws CO_2 from the atmosphere) to a CO_2 "source" and to a carbon dioxide source, a switch in Earth science roles that also has often befallen the Amazon Valley in recent years. Forests burning around the world are enduring the same conversion, adding CO_2 to the air as they burn.

Rain is not unusual over much of the Earth in the summer, unless you are at in a deep desert—say in the middle of the Sahara. Rain is not even 100% unknown in our globally warmed world on Greenland's immediate coast, close to the ice cap. Small wildfires have even been spotted along Greenland's coast, where the ice cap has receded. Something that is phenomenally new, however, is rain at the highest point of its ice cap, two miles above sea level, where it fell on Saturday, August 13, 2021, at a summit research station, where liquid precipitation fell with temperatures slightly above the freezing point. The station is occupied year-round under auspices of the National Science Foundation.

The Summit research station sits roughly two miles above sea level and 500 miles north of the Arctic Circle. The rain joins the list of global warming firsts, such as the first day in three digits °F above the Arctic Circle the previous summer.

"It's incredible, because it does write a new chapter in the book of Greenland," said Marco Tedesco, a researcher at Lamont-Doherty Earth Observatory of Columbia University. "This is really new" (Fountain, 2021). As illustrated by a remote sensing, scientists were surprised to observe a large wildfire in ▶ August 2017 in western Greenland near Sisimiut. Two years later, another sizable fire burned in the same region. Satellites first detected evidence of the fire, as you can see in ◲ Fig. 3.12.

The later fire burned in Qeqqata Kommunia, about 18 km (11 miles) northwest of Sarfannguit and just east of Sisimiut. It occurred ▶ near a hut on the Arctic Circle Trail and was likely started accidentally by a hiker, noted University of Miami Scientist Jessica McCarty. The fire appeared to be burning in an area with mossy wetlands (called ▶ fen) and heath shrublands.

▶ Warm, dry weather helped set the stage for the fire. ▶ Meteorological data shows that the region has been unusually hot and dry in recent months. And it was particularly warm the day that the fire burned (◲ Figs. 3.13 and 3.14).

The pattern that brought rain to Greenland's summit shares some attributes with unusual jet steam patterns that have contributed to bouts of drought and deluge rainfall at lower latitudes, as explained by Henry Fountain of the *New York Times* (2021, A-8). "The jet stream, rather than flowing in its normal pattern from west to east, dipped southward over northeastern Canada. That brought low-pressure over warmer waters, where it picked up heat and moisture. The jet stream then looped …northward, bringing that air to southwestern

◲ **Fig. 3.12** Wildfires in western Greenland, August 2017. Source: NASA

3

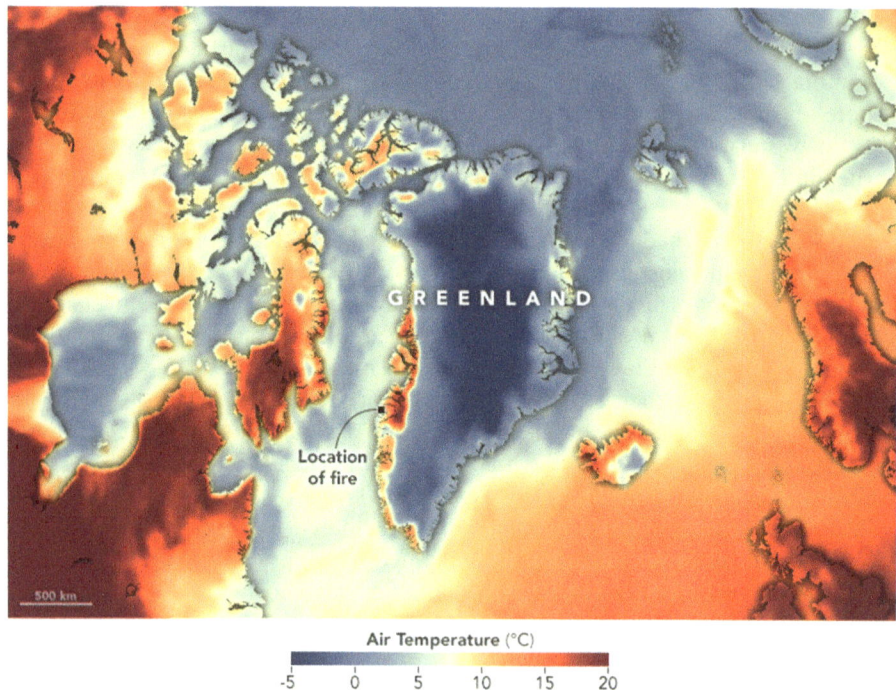

Air Temperature (°C)

-5   0   5   10   15   20

🔲 **Fig. 3.13**   A NASA illustration placed the second Greenland fire (July 10, 2019) in geographic context. The map shows air temperatures at 2 m (about 6.5 feet) above the ground. This fire was short-lived in comparison to the 2017 blaze, which persisted for 2 weeks. In this case, firefighters ▶ extinguished the blaze by ▶ July 11, the day after it had begun. Source: NASA

🔲 **Fig. 3.14**   Another unusual place for a wildfire: South Korea, summer, 2021. Source: NASA

Greenland; it swept over the ice sheet. The warm air and the moisture-laden clouds caused temperatures to rise at Summit and the precipitation to fall as rain rather than snow."

3.4.32 August 21–23, 2021: Deluges from Tropical Storm Henri

As a typical tropical storm (maximum winds 60 miles per hour), Henri did not make many headlines when it hit New York City and New England. As a rainfall record-breaker, it did. In parts of New York City, all-time daily records of 8–10 inches fell. More than 60,000 people were asked to leave a Central Park concert before the likes of Bruce Springsteen and Paul Simon could sing. A nationwide television audience also missed the performances. Remnants of Henri merged with Fred, another dying, but still very wet tropical system, after Fred had dumped torrential rains on Tennessee and North Carolina, causing flooding that some observers likened to a tidal wave, killing more than 35 people. The storm set a daily state record for rainfall in Tennessee, 17 inches. The storms were pouring rain on the ground saturated by previous heavy rains, part of a stagnant jet stream pattern (discussed in this book's ▶ Chap. 2 on the science of climate change). The same pattern was part of punishing, record heat and drought in Western North America, as well as the first recorded rain to fall at the highest point of Greenland's ice cap (Carlowicz, 2021).

3.4.33 Giant Fires Burn Toward Lake Tahoe

According to the Earth Observatory, as August 2021 came to an end, 42,647 fires had scorched 4,879,574 acres across the United States in 2021—close to the 10-year average for the first 8 months of a year, according to ▶ statistics from the National Interagency Fire Center. By August 31, some of the 83 fires that were still actively burning on August 31 were especially fierce; two of them had crossed the crest of the Sierra Nevada for the first time in recent record-keeping.

The largest blazes burned in Northern California, most notably the Caldor and Dixie Fires, which together scorched nearly one million acres (1500 square miles) by the end of August (Arc of Fires in the U.S, 2021). Smoke from the Caldor Fire near Lake Tahoe was visible in an image, captured by NASA on the afternoon of August 30, 2021. By the end of August, the Caldor Fire was only 14% contained. The Dixie Fire, also in Northern California, had burned about a million acres, including several hundred homes and other structures. By the time the Dixie Fire was extinguished, later in September, it had cost about $610 million to fight, leaving California fire executives to wonder how many such fire seasons their budgets could withstand.

The back-to-back seasons of 2019–2020 and 2020–2021 had been the worst, and most expensive to contain, requiring, among other things, millions of dollars for flame retardant, a command center that "looked like a small town that had sprung up overnight" (McDonald, 2021, A-10), with parking spaces for 500 fire engines and 194 water trucks, about 200 bulldozers, a resupply warehouse "roughly the size of a Home Depot, roughly 6600 firefighters and others, more or less, many own whom were working around the clock. The size of this enterprise reflected the growth of wildfires during the last decade. In 2010, 72,000 fires in the United States burned 3.4 million acres; in 2020, 59,000 acres burned 10.1 million acres, according to the National Interagency Fire Center (McDonald, 2021, A-10). The drive to control the Dixie Fire was the most expensive such event in the history of the United States (and very possibly, if comparable statistics existed, the world).

The Caldor Fire spread rapidly amid dry and gusty conditions, prompting officials to expand mandatory evacuation orders to the Tahoe Basin, including South Lake Tahoe, a lakeside city of nearly 22,000 people. According to news reports, Caldor is only the second known fire to have crested the ridge of the Sierra Nevada, burning from one side to the other.

Seasonally, summer in Tahoe has been getting hotter, as it also has become longer. Winter is shorter, and milder, with more rain and less snow. The 2021 fire and evacuation of South Lake Tahoe cost the area $93 million in lost revenue in 2 weeks, according to the Center for Economic Development at the University of Nevada—Reno (Levin, 2001, B-7). (No one would have guessed that on December 26, 2021, the same area would be coping with as much as 9 feet of fresh snow.)

Meanwhile, the ▶ Dixie Fire continued to spread. Since igniting on July 13, by August 28, it had burned more than 800,000 acres (or about 100 square miles) and had surpassed the August Complex to become the second largest fire on record in California. The Dixie Fire was nearly 50% contained as of August 31; a large number of wildfires were burning in Washington, Oregon, and Northern California. While the Caldor and Dixie Fires are the largest, other wildfires are also posing hazards to people, property, and ▶ air quality across the US West.

Two days later, the eye of Hurricane Ida passed close to New Orleans (◘ Fig. 3.15).

3.4.34 Hurricane Ida Mauls the Southern and Eastern United States

As California continued to burn, Hurricane Ida (16 years to the day after Hurricane Katrina had made landfall around noon on August 29, 2005) came ashore at Port Fourchon, Louisiana. It was the fifth strongest hurricane (based on wind speed) to make landfall in the continental United States. Ida, an amazingly potent landfalling hurricane, would maul much of the United States' eastern third during the next week (◘ Fig. 3.16).

A few days after Ida came ashore, President Joe Biden visited sites of Ida's wreckage in New Orleans and in and around New York City September 6–7 and told everyone that climate change had become "everybody's crisis." He then toured neighborhoods flooded by the remnants of Hurricane Ida, warning that it is time for America to get serious about the "code red" danger or nature would do it for us.

☑ **Fig. 3.15** Hurricane Ida comes ashore in Louisiana and Mississippi, August 31, 2021, as a strong Category 4, one of the strongest to come ashore in US history. Note the bursts of lights: Dallas and Houston to the west and Atlanta to the east. Source: NASA

In the last 24 hours before its landfall, Ida's central pressure dropped from 985 millibars to 929, and sustained ► winds intensified rapidly from 85 to 150 miles per hour, according to the National Hurricane Center. The storm has undergone with winds increasing by at least 35 miles per hour within 24 hours. The intensification was partly fueled by the ► hot summer surface waters of the Gulf of Mexico, which were about 30–31 °C (86–88 °F). It is easy to see how warming waters affect a hurricane, especially when it is in the bathwater warm gulf of Mexico, where the extra warm Gulf Stream originates. Warm water is a fuel for a hurricane. It is not much of a reach to realize what happens to an intense storm of this sort when water warms because of climate change.

"For me, the most compelling aspect of Ida was its rapid intensification up to landfall," said Scott Braun, a scientist who specializes in hurricanes at NASA's Goddard Space Flight Center. "The storm was very similar to ► Hurricane Opal and ► Hurricane Katrina in that they underwent rapid intensification over a region, or eddy, of deep warm water known as the ► Gulf Loop Current. In addition to providing warm water for fuel, such eddies impede the mixing of colder water to the surface. Such cooling would typically lead to storm weakening, or at least an end to strengthening. Both Opal and Katrina weakened before landfall, mitigating their damage, even though they were obviously still intense. In Ida, near-coast weakening did not really occur" (Hurricane Ida Batters Louisiana, 2021). Hurricane Ida pushed a wall of water—a massive storm surge—onto the coast of Louisiana and Mississippi. Weather stations and media reports noted surges rang-

3

◘ **Fig. 3.16** Hurricane Ida spreads out over US East Coast, still a dangerous storm. At times the remnants of Ida poured several inches of rain per hour on areas that had been hit by ▶ heavy rains from Tropical Storm Henri only a few days before flash floods from Ida. Even before that, the area over which the remnants of Ida passed had been saturated by several other deluges—again, the stuck jet stream at work. Source: NASA

ing from 3 to 9 feet (1–3 m) in places such as Grand Isle, Shell Beach, Lafitte, Barataria, Port Fourchon, and Bay Waveland. Port Fourchon is a ▶ major commercial and industrial hub for the United States, particularly for oil and gas shipments.

According to a report by NASA's Earth Observatory, the strength of hurricane winds often attracts the most attention, but meteorologists warn that water (including storm surges) are the ▶ most dangerous part of a tropical cyclone. Ida, in 2021, was a good example.

Although Ida's winds weakened ▶ after making landfall in Louisiana, the storm system upended daily life for millions of people and proved deadly for some along much of the East Coast of the United States.

Rain gauge data indicated that many of the areas worst-hit by Ida ▶ saw 6–10 inches (15–25 cm) of rainfall within a few hours, a deluge that quickly turned streets into rushing rivers and buildings into islands. Due to the extreme conditions, the National Weather Service (NWS) even issued ▶ flash flood emergency warnings for New York City and parts of New Jersey. These alerts are reserved for exceedingly rare situations when a severe threat to human life and catastrophic damage from a flash flood is happening or will happen soon, according to the National Weather Service (Hurricane Ida Batters Louisiana, 2021).

The flooding from Ida caused more than 50 deaths, in the US Northeast, after Ida was "downgraded" to a tropical depression. Several rivers in Pennsylvania and New Jersey broke all-time records for water levels. In southeastern Pennsylvania, ▶ East Brandywine Creek rose as high as 19 feet, smashing the previous high-water mark of nearly 14 feet, which had been set during Hurricane Floyd in 1999. In Wilmington, Delaware, Brandywine Creek swamped urban areas and led to large numbers of ▶ water rescues. In New Jersey, the ▶ Millstone River at Weston and the ▶ Raritan River at Manville both narrowly broke flood records as well. The rainfall total of ▶ 8.41 inches at Newark, New Jersey, on September 1 was the most for any calendar day in Newark's records (Flash Floods from Ida Swamp the Northeast, 2021) (◘ Fig. 3.17).

At about the same time, remnants of Hurricane Henri, now a tropical storm, soaked large parts of the US Eastern Seaboard with massive rainfall (Courtesy NASA). By the last week of August, the Caldor Fire had gutted hundreds of homes as it advanced toward the resort complex of Lake Tahoe, while several thousand firefighters worked to box in the flames. Tourists who had hoped to boat or swim found themselves looking at thick yellow haze instead of alpine scenery. The Caldor Fire was less than 20 miles (32 km) east of the lake that straddles the California-Nevada state line. The fire was eating its way through rugged timber-lands as it approached the Lake Tahoe basin, California's state fire chief Thom Porter warned. "Ash rained down as tourists ducked into cafes, fishing gear shops on Lake Tahoe Boulevard for a respite from the unhealthy air" (Melley & Metz, 2021).

By August 31, the Caldor Fire had burned to within a few miles of South Lake Tahoe or, as locals said, one ridge away. Porter said that the fire was "knocking on the door" of the Tahoe Basin, "as the last of tourists to evacuate were pummeled with ash in the midst of a choking haze" (Miller, 2012, 2-A). The air was so polluted that "it was… toxic and thick that it spiked past the highest levels on air quality charts" (Fuller & Shawn, 2021, 1-A).

During the first week of September, winds slowed, and 3000–4000 firefighters plus a phalanx of ditch-digging bulldozers were able to battle the fire to near stand-still just short of South Lake Tahoe's hotels, cabins, condos, and casinos. Even so, the toll was eye-popping: more than 700 homes and several businesses burned. The

3

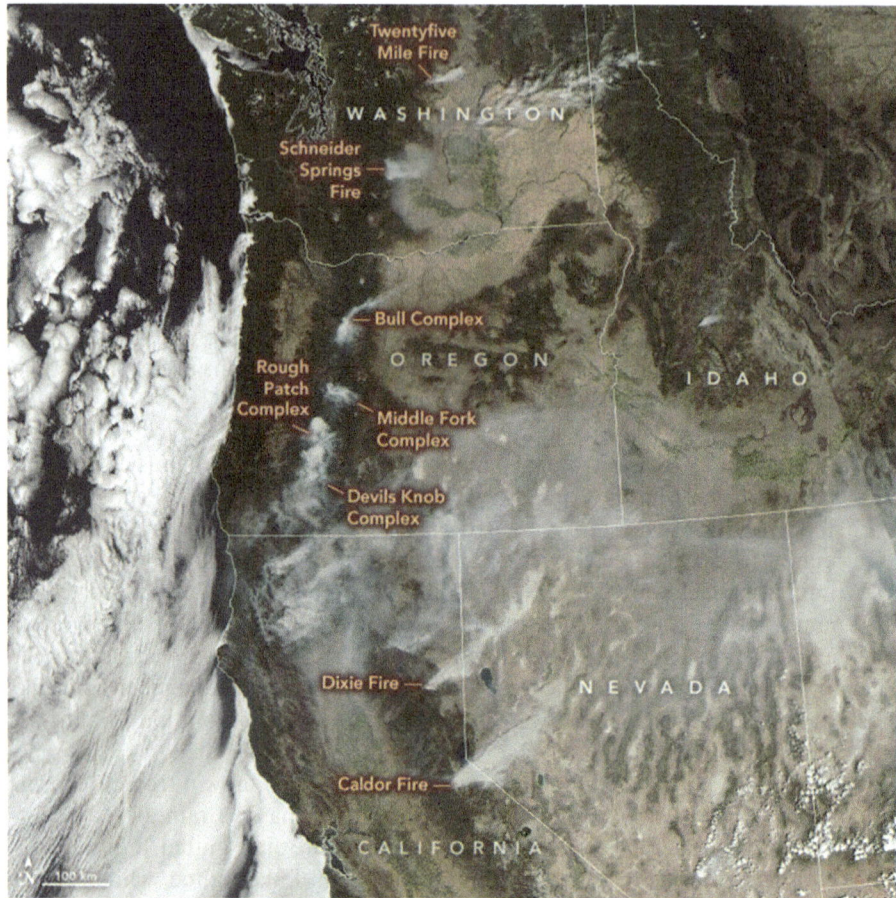

⬛ **Fig. 3.17**   As the remnants of Ida and Henri drowned the US East Coast, firefighters continued to battle the major Western fires that had been burning for several weeks. Source: NASA

ashes were everywhere: "Ash from the Caldor Fire snowed on Lake Tahoe," said one observer (Healy, 2021, A-12). Most of South Lake Tahoe's hotels had been closed and were being used as bunkhouses for exhausted firefighters, who also were supplied by hotel staff who had remained. Wafts of smoke stood nearly still in hotel hallways, hovering, like ragged, dirty ghosts (Healy, 2021, A-14).

Scott Wilson et al. of the *Washington Post* (2021) described the Caldor Fire's advance on South Lake Tahoe:

"It looked like twilight at midday here, then a billow of smoke made the sun vanish entirely on a sharp bend around Inspiration Point. The redwoods and firs, old and graceful and lovely, now resembled nothing so much as fuel for a fire still menacing one of California's great natural wonders and the people who make a living from it."

The Caldor Fire continued to blaze along the granite ridgeline above this abandoned tourist town on Wednesday, smothering the burger joints, Jet Ski rental shops, margarita havens, and million-dollar homes in a thick acrid haze. By mid-day, the fire had burned through more than 204,000 acres of wilderness.

Humidity remained very low on September 1, as the wind varied both in strength and direction, confounding firefighters who had used bulldozers to push a last-ditch firebreak between the Caldor Fire and the avenues, homes, and businesses of South Tahoe. Evacuations were ordered on the eastern side of the lake on September 1, affecting all guests except those involved in fighting the fire. Businesses were closed, including casinos in Stateline, Nevada.

The fire had started August 14, and within 2 weeks, it was only 12% contained, as wind-driven flames consumed drought-stricken fir forests. Owners of forest-shrouded McMansions, purchased as home prices shot up, saw their investments literally go up in smoke. Some residents decided to ride out the inferno on the lake in their sizeable yachts, as more than 3000 firefighters were battling the Caldor blaze, which was throwing embers sometimes a mile at a time toward the Lake Tahoe area, now one of the worst fires, in terms of size and damage, in US history. Some of the firefighters used snowblowers to blast water at the flames and increase humidity. Hotels were used to house evacuees and fire crews who needed food and rest.

"This is what climate change looks like," said Daniel Swain, a climate scientist at the University of California at Los Angeles (Fuller & Shawn, 2021, A-13). People were unable to gather outside because of the smoke and ash. They also had no refuge indoors because of the COVID-19 delta variant. The Dixie Fire, 5 weeks old by August 27, had grown to a 500-mile perimeter to the northwest of Reno. On August 29, authorities ordered evacuations of homes surrounding Lake Tahoe as the fire continued its unrelenting approach.

The red-orange sun was partly obscured, as most Lake Tahoe residents decided to leave town. Schools had closed in Reno, 60 miles from Lake Tahoe. By September 1, the fire was throwing off burning embers in 40-mile-an-hour winds that blew over firefighters' heads, falling hundreds of yards ahead of firefighters. The fire rapidly jumped a last ridge of solid granite that firefighters had hoped would act as a firebreak. That hope has dashed as the fire shot embers across the ridge and began to advance, like a Champion Sprinter, down the other side, still advancing on South Lake Tahoe. About 50,000 people slowly evacuated along narrow roads. Nearly all businesses were locked up, except for one casino, where gamblers were hoping for one last big payout, as giant exhaust fans sucked the fire's smoke out of the gambling parlor.

At the same time, the fire advanced, "lighting up trees and homes, lodges and restaurants in the Tahoe Basin. On Tuesday, the fire spread from treetop to treetop, lofting embers that were able to ignite new spot fires more than a mile away"(Wilson et al., 2021). The Caldor Fire had burned at least 320 square miles, including at least 700 homes, some of which were reduced to useless stumps and smoking shreds. The fire, one of the most aggressive in an area had ever experienced, was still only 11% contained. Even then, 17,000 structures remained threatened, including small towns and wineries.

"This fire has just simply outpaced us. We emptied the cupboards of resources," said Jeff Marsolais, a supervisor of the Eldorado National Forest and an adminis-

3

trator on the fire (Melley & Metz, 2021). Marsolais said, adding that while the blaze had slowed its explosive growth in recent days, "That can change."

The Associated Press reported that the new orders for people to immediately evacuate included part of the tourist city of South Lake Tahoe and about 15 miles (24 km) up the western shore of the lake, a day after communities several miles south of the lake had abruptly been ordered to evacuate as the Caldor Fire raged nearby.

On August 31, the Caldor Fire continued to blaze its way toward South Lake Tahoe, as several thousand residents fled. As it reached almost 350 square miles, the fire was jumping highways, as thousands of firefighters seemed utterly overwhelmed. Stiff, dry winds propelled the fire as temperatures rose to near 100 °F, and the relative humidity fell to about 10%. The fire was still only 15% contained (Miller et al., 2012).

As the fires continued their march, California Governor Gavin Newsom said "We must remain a leader on climate change. California's several-year run of historic fires is the direct result of its shifting climate, one of extremes—from drought to floods, temperatures near freezing in the back country one week and at record highs the next" (Wilson et al., 2021). All of this was taking place with the summer, with the usual fire season only half over.

Meanwhile, the Dixie Fire, still the second largest in state history at 1148 square miles (2973 km^2), was 43% contained. At least 682 homes were among more than 1270 buildings that had been destroyed (Melley & Metz, 2021).

As of August 25, 92 major fires were being fought in 12 Western states, according to the National Interagency Fire Center in Boise, Idaho. The Northern California region was at the center of this disaster zone, as several fires burned several hundred homes and other structures. Several of these fires were still mainly out of control.

The Schuylkill River in Pennsylvania also received major flooding. With a rain ▶ gauge in Norristown recording a record height of nearly 27 feet (19 feet is considered major flooding), rushing waters shut down highways and swamped homes, businesses, and cars in some parts of Philadelphia. In New York City, 3.15 inches of rain fell on Central Park in one hour, smashing the hourly record set just days before during the remnants of Tropical Storm Henri. Floodwater poured into the city's subway system, basement apartments, and the cars of stranded motorists (Hurricane Ida Batters Louisiana, 2021).

In a report from the Intergovernmental Panel on Climate Change, scientists drew an ▶ inequivocable cable link between human activity and global warming. They also pointed to an observed increase in the intensity and frequency of heavy precipitation events in ▶ eastern North America as a consequence of warming. Scientists projected that an ▶ increase in both mean and extreme precipitation is very likely in the future (Hurricane Ida Batters Louisiana, 2021).

The remains of Ida also generated several damaging tornadoes, including one that ripped apart houses and other buildings in Annapolis, Maryland. New York City received 6 inches of rain which paralyzed large parts of its subway system. To scientists, the damage to the country's largest urban area was an unusually vivid example of a stagnant jet stream anchored over the United States.

"Unprecedented is an understatement," said Governor Phil Murphy of New Jersey, whoin, 2012 witnessed damage from tornadoes and extraordinary downpours. Other witnessed called the flood biblical. In South Plainfield, NJ, a 31-year-old Dhanush Reddy was swept into a 36-inch-wide mouth of a sewer pipe during the torrent. His bruised and bloodied body was found September 2 in Piscataway, almost 5 miles away (Nir, 2021, A-12).

In Queens, New York, three Latino immigrants who shared an underground room found water streaming through its one small window and door. Richard Garcia, 50, and Oliver de la Cruz, 33, barely escaped before the torrent put enough pressure on the window and door to steadily tighten them. Once the two had escaped, they realized that the third, Roberto Bravo, a retired construction worker, had been sealed in by the rising waters, screaming for help as the water level reduced the space between the ceiling and his nose. Soon, the screams ceased (Nir, 2021, A-12, A-13).

On the Louisiana coast, and inland along the Mississippi River, inventories of toxic chemical spills were delayed by lack of power. The area is one of the United States' leading sources of toxic manufacturing substances. Two gas pipelines were known to be leaking isobutane and propylene that are flammable and extremely dangerous to human health. A plastic plant was emitting ethylene dichloride, which is also toxic to human health (Tabuchi, 2021a, A-14). These are among more than 138 toxic sites in the path of Hurricane Ida. The area is a primary producer of fertilizer that uses anhydrous ammonia and other chemicals. The Environmental Protection Agency also was forced to delay readings from 23 superfund sites by hurricane damage (Tabuchi, 2021a, A-14).

The landfall of Ida was accompanied by a storm surge of 5–8 feet, which rolled over some of the same coastal islands as Katrina. Because seas have been slowly rising and the land slowly sinking, communities such as Grand Isle, closer to the coast than New Orleans, have been nearly obliterated by repeated wind, torrential rain, and storm surges. New Orleans' damage has been growing over time as well, as land sinks and the sea rises. Ida was being called a poster child for climate change disasters.

Nearly a million people lost power as utility crews were overwhelmed. Some brick buildings crumbled. Seafront houses were swept into the Gulf of Mexico. First light on August 30 revealed miles of homes ripped apart with damage typical of a major tornado, except over a much larger area, mainly with damage by ocean surges.

Ida also posed a major test for New Orleans' post-Katrina $20 billion "storm protection system." Most of the systems held up, but power was knocked out city-wide, clean water was unavailable in large areas, humid heat was stifling, but anyone who thought of cooling off in the floodwaters had to think of alligators that were swimming around in them. One woman watched an alligator rip off a man's arm. She looked for help but returned to find the man and the alligator gone. Another problem was sewage; sewage pumps stopped working, prompting concerns about overflow that could have turned parts of New Orleans into a gigantic string of open-air toilets.

3

As Louisiana's governor, John Bel Edwards, ▶ put it, if you had to draw up the worst possible path for a hurricane to take in Louisiana, it would be something very, very close to what everyone had experienced during Ida.

While the levees may have held against Hurricane Ida, other facets of climate change may render their protection problematic. For one thing, warmer water makes for stronger hurricanes, as discussed in more detail in ▶ Chap. 2. Next, New Orleans is built on marshland, which makes much of the city sink about 5 inches every decade. Add sea-level rise, and remember that much of New Orleans is already below sea level.

Background Information

During 1982, I interviewed for a professor's job at Loyola of New Orleans, where my hosts took me to lunch on the "waterfront." After ordering my meal, I looked up from my menu and realized that I was staring at a wall—and, behind it, the prow of a large ship. It took a few seconds for me to restore my sense of reality. I was looking *up* at a piece of nautical hardware that let me know where I was—several feet *below* the waterline.

This was about 40 years ago. I am very happy that, according to scuttlebutt, the dean did not like my selection of clothing or my haircut that day. At the time, I was annoyed. I did not get the job. Because I was unemployed and in need of an academic job, I was not happy. In retrospect, I was happy that the dean did not like my style, if that is what it was. A job in a sinking city that repeatedly gets sucker-punched by hurricanes was no place to get tenure. I ended up in at the University of Nebraska at Omaha for 38 years and had the luck to dodge an occasional flooding rain or tornado. Professionally and professorially, I have been a profile in good luck.

Even the Army Corps of Engineers does not like the looks of New Orleans' major levee. In mid-2021, the Corps asked for $1.7 billion to augment the levee, which could be unable to withstand future floods by 2023. How long can the cycle of raising the levees to keep them from sinking continue? The point is that New Orleans is unsuitable for occupancy in the long run, no matter how many times our most assuredly anthropogenic Army Corps of Engineers "improves" it.

3.4.35 Heat and Drought Devastate California's Almond Crop

Almond farmers in California's San Joaquin Valley have been pulling up some of their trees and practicing "deficit irrigation" to keep the remainder alive as heat and drought make one of the state's most profitable crops (worth more than $6 billion a year) more vulnerable to changes in climate that have afflicted much of the US West for decades (Chea, 2021, D-7). Farmers of melons, cherries, and asparagus, as well as other crops, are dying of drought and heat as well, as farmers lose their investments. California in 2020 produced 80% of the world's almonds, a crop with large demands for water (Chea, 2021, D-7). About 70% of the almonds have been exported worldwide. Before the drought and heat intensified, almond production rose from 370 million pounds in 1995 to 3.1 billion in 2020. Important California reservoirs also have been drying up. Shasta Lake, three of the state's largest lakes, has fallen to 30% of capacity and Lake Oroville was at 25%, both in mid-2021.

3.4.36 Chile: "Water Towers" Running Dry

Along most of South America's West Coast, which runs along the sharp spine of the Andes, the mountains are called "water towers." As drought has continued and snow levels have risen, the reservoirs are drying up. Every single major city along that coast, from Bogota, Colombia, to Santiago, Chile, depends on the Andes for life-giving water. Under usual conditions, as airflow that rises along the Andes' eastern face and dries as it descends to the West Coast is very dry (except during La Niño conditions, when the airflow is reversed), it adds a major drought, and tens of millions of people go thirsty.

In Chile, the biggest city is Santiago, and the water source is the Maipo River. By the end of a droughty summer, two-thirds of its water comes from shrinking glaciers. Since 2010, the area has been enduring a mega-drought which has been intensifying. By 2019, the area was regularly receiving less than a quarter of its usual annual rainfall.

The enduring nature of the present drought is more and more troubling; by 2021, it had been the longest and most intense in recorded history.

3.4.37 Utah: The Grit Salt Lake; Not What Brigham Young Had in Mind

Once upon a time, two of today's human lifespans ago, Brigham Young, who founded the Church of Jesus Christ of Latter-day Saints (at a time when "Mormon" was considered a slur), stood on the high ground east of a strip of land that today comprises Salt Lake City; the Great Salt Lake, Provo; and highways, strip malls, a large number of tract houses, a $1.5 billion skiing industry (Romero, 2021c, A-16), and an airport that is too large for its surroundings. The area also is home to several churches paying homage to various factions of the church that Young told his Promised-Land-seeking religious migrants. On a map turned upside down, the area does, indeed, look like Jesus's home, except that the mountains are much more spectacular—when and whether one can see them through the haze.

These days, seeing the mountains, have become problematic, especially during 6–8 months that comprise the fire season, when gigantic wildfires 500–1000 miles northwest from Washington, Oregon, California, etc. turn the sky a mud-brown and, some days, put Salt Lake City in a league with Beijing, Shanghai, and New Delhi with some of the filthiest air on the planet.

The Great Salt Lake is becoming saltier as water evaporates, and the air is becoming grittier, especially when strong winds howl across the flatlands and brings gritty air across the urban strip that houses 80% of Utah's population, snugly couched along the foothills of the Wasatch Range. Some days the wind that howls through Brigham Young's Promised Land carries an eye-burning mixture of several toxic chemicals. During the winter, this bowl of land gathers a mixture of chemicals, arsenic, auto, and truck exhaust, as well as coal power plant exhaust and stagnant fog into an atmospheric inversion.

3

Residents of the area are prone to just about any lung disease they can name, in no small part because the air pollution is laced with very small particles that easily lodge in the lungs and play a major role in causing lung diseases. The skiing magazine *Powder* (only partially in jest) warned visitors to pack gas masks with their skis. All of this is aggravated by the fact that a major drought during the last 20 years as well as freshwater diversion for agriculture has shrunken the Great Salt Lake to its lowest level in more than 100 years and opened more gritty salt flats to more and more toxic whirlwinds (Romero, 2021c, A-16).

3.4.38 September 11, 2021: A Hurricane Hits Newfoundland

A very rare hurricane hit Newfoundland, in easternmost Canada, with 80 mile-an-hour winds, amidst power outages and fallen trees. Newfoundland is known for its wintertime blizzards and has been hit in the past by former hurricanes that have lost their ability to generate much wind. The most memorable and nasty winds in Newfoundland are usually generated by winter blizzards. A full-blown Category 1 hurricane is a real rarity, an evidence of a warming Gulf Stream, which flows northeastward up the US East Coast and then well south of Canada. Newfoundland is located at about the same latitude as southernmost Siberia. Larry was not done with its tricks. A couple of days later, its remains rolled over southeastern Greenland as an early-season snowstorm with more than 10 inches in some locations.

3.4.39 September 14, 2021: Front-Page News in Seattle: RAIN!

The *Seattle Times* led its online version with what usually would not be unusual in Seattle: *RAIN!* Read all about it! It was the first rain for this thirsty city in 3 months, aside from a sprinkle. The prospect of a solid storm had people in the Northwest running gleefully for their long-unused umbrellas. Vendors of T-shirts that read "Seattle Rain Festival: January 1–December 31" danced in glee.

Anyone who was worried about Seattle's worst drought ever (plus all those fires and the heat dome from hell) was not prepared at all for November, when the Emerald City (so-called for its verdant foliage, even in winter or summer) achieved its rainiest fall on record. Much of it was cold and raw, with forecasters calling for a rough (snowy and cold) winter. In the dictionary of climate change, look under "drought and deluge"—that is, too much of something and then too much (or little) of something else. And so it was: on December 25, 2021, after a chilly, wet autumn, Seattle, and its lowland surroundings reveled in bragging rights to *real* white Christmas. No half-inch of slushy slop this. The city and its many suburbs were paralyzed.

3.4.40 January 3, 1955: When Pigs Fly

The United States (i.e., major airlines with head offices there) announced plans to cut jet fuel emissions by 20% by 2050 by using biofuels from plants, various kinds of wastes, and other organic matter to replace between 10 and 20% of the jet fuel that the industry uses per year, about three billion gallons. The airlines have set sustainable fuel targets before, and later backed off of them, the latest in 2017. Hold the presses, however. This all may be a public relations pose. Biofuels are a less efficient burn than jet fuel. They can emit more pollution than jet fuel, and the cost of manufacturing them from such low-energy feedstock as corn stubble consumes more energy than refining oil. Add the emissions from fertilizers used to grow organic feedstock, and you have got a poor substitute for bad old jet fuel (Tabuchi, 2021b, B-7).

3.4.41 Fires Encircle California's Sequoia National Park

Flames advanced toward the world's largest grove of giant, ancient trees in the Sequoia National Park, as firefighters established firebreaks and cleared underbrush before the fire could use them as fuel. The grove includes about 2000 massive sequoias, including the General Sherman Tree, the world's oldest and largest tree by volume (◻ Fig. 3.18).

On September 17, the 275-foot Sherman Tree was wrapped at its base in a very large fire-resistant blanket. By September 23, the fire was about a mile away from the Sherman Tree, with the entire complex (several fires burning last once) nearing the suburbs of Fresno. The largest trees, including the Sherman Tree, were saved, but roughly 100 were burned, mainly in the Sierra Nevada, beyond retrieval.

When the fire season was over, 2261–3637 sequoias had been burned to a degree that they could not survive. The previous year, fire killed 7500–10,400 of the trees that were more than 4 feet in diameter. During those 2 years, between 13 and 19% of the world's mature sequoias burned (Melley, 2021, A-4).

Brian Melley wrote for the Associated Press (2021, A-4) that "Blazes so intense to burn hot enough and high enough to kill so many giant Sequoias—trees once considered nearly fireproof puts an exclamation on climate change's impact."

3.4.42 September 23, 2012: British Prime Minister Boris Johnson Becomes a Vanquisher of Carbon Dioxide

In a speech before the United Nations, British Prime Minister Boris Johnson, reversing his prior position, said it was time for the world to "grow up" regarding the urgent actions needed to deal with climate change. Johnson made these state-

3

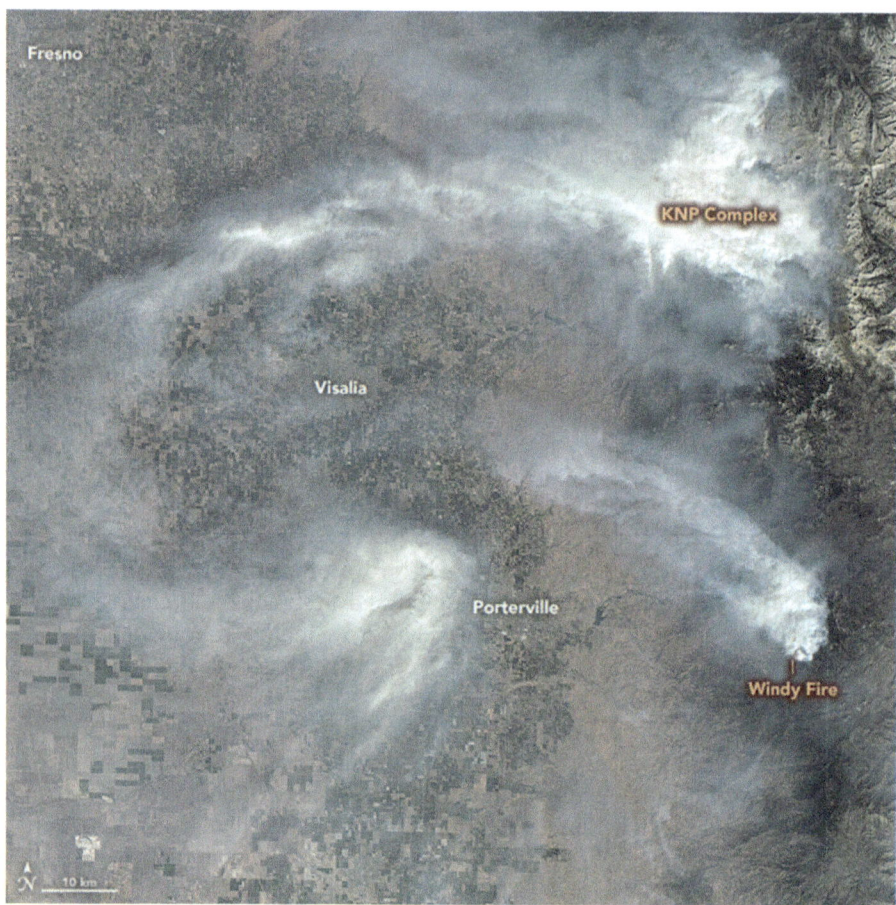

☐ **Fig. 3.18**   During September of 2021, fires encircled California's Sequoia National Park. Source: NASA

ments in the context of a major conference on the subject that he hosted in Glasgow, Scotland, in October, which he called a "turning point" for humanity (Adam, 2021).

Speaking before the United Nations General Assembly, Johnson, who is a Conservative Party member, said that if radical action is not taken, "We will see desertification, drought, crop failure, and mass movements of humanity on a scale not seen before, not because of some unforeseen natural event or disaster but because of us, because of what we are doing now."

Earlier, Johnson had ridiculed "eco-doomsters" and supported climate change deniers. He now described the world as "this precious blue sphere with its eggshell crust and wisp of an atmosphere—[which] is not some indestructible toy, some bouncy plastic romper room against which we can hurl ourselves to our heart's content" (Adam, 2021). Many in the media called his change of position a remarkable conversion.

In his address, Johnson said that if radical action is not taken, "we will see desertification, drought, crop failure, and mass movements of humanity on a scale not seen before, not because of some unforeseen natural event or disaster but because of us, because of what we are doing now." Johnson concluded his time on the podium with a reference to the ancient Greek writer Sophocles who said that he was often quoted—by Johnson—as "saying that there are many terrifying things in the world but none is more terrifying than mankind, and it's certainly true that … our species is uniquely capable of our own destruction."

3.4.43 September 24, 2021: Two Million Acres Burned in the United States in 9 Months

With wildfires continuing, the United States reached two million acres burned in 2021. This was three times the number of acres burned by the same date in 2020, itself a record year at that time. "The fire season across the West has been starting earlier and ending later year by year," said Cal Fire, continuing "Climate change is considered a key driver of this trend."

Given the near absence of rain, the fire season shows no sign of ending. Instead, the drought-afflicted area is moving northward and eastward into the Great Plains.

A mega-fire firefighter commented: "Suddenly, an enormous whoosh rose from the Canyon, and a copse of aspen exploded….[it] rained ash and fire from above…. That day we witnessed a kind of fire that we had never seen before. Decades of wildfire suppression combined with a warming planet had created time bombs all over the West. And we were implying an outdated strategy, spending hundreds of millions of dollars to send college-age crews into tinderbox forests with shovels and axes. A six-foot fire break against a [fire-initiated thunderstorm)….How dangerously ill-equipped we were" (Tarter, 2021, 3).

3.4.44 September 24, 2021: Coastal Flood Insurance Premiums May Rise

The US government announced that as of October 1, 2021, premiums on federal flood insurance, which had been heavily subsidized, would rise sharply to reflect actual conditions, including sea-level rise and increasing storminess in "our new, wet reality." The largest rises factor in the increasing dangers to waterfront homes. "Communities in Florida and elsewhere around the country will see those subsidies begin to disappear in a nationwide experiment in trying to adapt to climate change: Forcing Americans to pay something closer to the real cost of their flood risk, which is rising as the planet warms" (Flavelle, 2021a, 2021b, 2021c, 2021d).

"Subsidized insurance has been critical for supporting coastal real estate markets," said Benjamin Keys, a professor at the University of Pennsylvania's

3

Wharton School. Removing that subsidy, he said, is likely to affect where Americans build houses and how much people will pay for them. "It's going to require a major rethink about coastal living" (Flavelle, 2021a, 2021b, 2021c, 2021d). Coastal Florida, where land is sinking, seas are rising, and hurricanes have been more numerous and intense, faces flood insurance premiums that may rise ten times. As could be expected, owners of upscale properties on coastlines have primed their lobbyists in opposition to this cold bath of climate change reality.

The federal government has been the primary source of flood insurance in the United States since 1968, when the National Flood Insurance Program was created. Private insurance companies have quit issuing such policies because of high risks. The government program is run by the Federal Emergency Management Agency (FEMA), which subsidizes wealthy homeowners on ocean-front homes at the expense of many lower-income people of color who live inland.

In a study published in *Nature* during 2021 (and picked up by NASA's Earth Observatory), scientists found that the proportion of the world's population exposed to floods grew by 20–24%. Although researchers expected an increase in the number of people living in flood-prone areas, the new estimates were ten times greater than what previous models had predicted.

Reviewing 20 years of NASA satellite data, researchers mapped 913 large flood events across 169 countries and compared them with global population data to better understand the impact on people. The team found that from 2000 to 2015 between 58 and 86 million more people could be found living in floodplains around the world. More than 255 million people were affected at least once by major floods in that period. The increase in exposure includes new human developments and migration but also the reclassification of some lands in the wake of large flood events and sea level rise.

"We need to understand why people are moving into floodplains and what ways we can support flood mitigation," said lead author Beth Tellman, a geography researcher at the University of Arizona. "I think satellite and Earth observations can be transformative in how we think about building resilience in a world marked by climate change" (Research Shows More People Living in Floodplains, 2021).

"The people moving into floodplains are probably the most vulnerable, marginalized populations, and they may not have much of an option to move anywhere else," she said. "Some use the term "blue-lining," which refers to the way neighborhoods with racist housing policies toward communities of color ('▶ redlining') often have much higher flood risk. Redlining has led to historic underinvestment in flood mitigation infrastructure and in increased risk" (Research Shows More People Living in Floodplains, 2021).

According to the study, the proportion of people exposed to floods increased in 70 countries across all continents except Antarctica. Increased flood exposure was concentrated in middle- and low-income countries, with many of the countries located in Asia and sub-Saharan Africa. At least 213 million people were shown to be exposed to flooding in South and Southeast Asia alone.

3.4.45 September–October 2021: "Green Slime" That Thrives on Warming Water Invades Toledo's Water Supply and Creates Other Problems in the Pacific Northwest

A "silent" climate change crisis has been unfolding along the Pacific Northwest coast as ocean temperatures rise. The deaths of marine creatures are alarming scientists and coastal Native American tribes whose livelihoods rely on the ocean. The culprit: hypoxic conditions have been detected every summer in recent memory as dissolved oxygen in the waterfalls to dangerously low levels. The 2021 hypoxia season got off to ► its earliest start in 20 years, and it was still suffocating marine life well into September.

Meanwhile, the Great Lakes in late summer 2021 again fell prey to what had become a regular event—invasion by a phenomenon known locally as the "green slime," with its worst concentration along the western shores of Lake Eire. Cities such as Toledo had their drinking water fouled by the "slime" in 2011 and 2014 and every year after that, as water temperatures rise during the late summer. Tourism was reduced and local people were not able to use the lake for fishing, boating, or swimming. The warmer the lake water, the more "slime" usually forms between late summer and early fall.

According to NASA, which has photographed the slime from space, "This harmful algal bloom (HAB), cyanobacteria, has the ability to produce toxins that cause illness or death with humans or pets who come in contact with the contaminated water.…The algae blooms are caused by runoff pollution. Typically the pollution develops from rainfall that washes fertilizer and manure from a farm field into the streams that eventually flow into Lake Erie" (King, 2021).

3.4.46 Corals Continue to Decline Worldwide as Waters Become Warmer

Since global warming has been widely recognized as a worldwide problem (about 1980), the state of corals has been watched as an indicator of the oceans' health. Corals, which contain a plethora of lives and act as home for multitudes of fish and other animal life, may lose their multicolored living tapestry and turn a sterile, dead white within a narrow temperature range—for example, from 87 °F to 90 °F. Some corals also have a capacity to regenerate after they have been killed by overheating. When alive, corals are among the most densely populated ecosystems on Earth. Corals are so sensitive to temperature changes that they may be used as thermometers. Part of a large reef (such as Australia's Great Barrier Reef) may be dying in its north while thriving in its south.

Coral reefs worldwide lost about 14% of their mass between 2010 and 2020, according to a detailed report issued in October, 2021, by the International Coral Reef Initiative, for which more than 300 scientists from 73 countries gathered data. With a sense of urgency, the report said that corals will continue to die until carbon dioxide levels arrested and begin to fall and that this will not happen until about

3

50–150 years after humanity makes major cuts in combustion of fossil fuels ("substantial" here may be taken as "reduce to zero") (Einhorn, 2021, A-11).

In addition to rising temperatures, reefs in many areas are being killed by pollution that runs off land, as well as destructive fishing practices, as well as chemicals from manufacturing on land that reaches the oceans. The first worldwide coral "bleaching" event occurred in 1998, during a strong El Niño event that raised ocean temperatures. After that, most corals rebounded. However, after 2009, "It's a constant decline [at] a global level," said Serge Planes, a research scientist of the Center for Island Research and Observatory of the Environment in Moorea and, in French Polynesia, an editor of the report.

Coral reefs support fish that feed more than a billion people, including many millions in island nations such as the Philippines and Indonesia. As such, they are a critical source of protein. When an aquatic environment nurtures coral reefs, they are among the most productive, diverse ecosystems on Earth. Coral reefs are sometimes called the rainforests of the sea. Coral reefs cover much less than 1% of the world's ocean floors, while at the same time hosting more than a third of the marine species presently described by science, with many species remaining undocumented. Some of these organisms may provide new sources of anticancer compounds and other medicines. Coral reefs also protect shorelines from erosion by acting as breakwaters that, if healthy, can repair themselves.

Record high temperatures on both land and in the world's oceans from 2014 to 2017 intensified coral damage, threatening a third or more of reefs with eradication. Kim Cobb, a marine scientist at the Georgia Institute of Technology, was stunned by damage off Kiritimati Island, near the center of the Pacific Ocean. "The entire reef is covered with a red-brown fuzz," Cobb said. "It is otherworldly. It is algae that has grown over dead coral. It was devastating" (Innis, 2016). Around Kiritimati, within a broad swath at the center of the El Niño, water temperatures rose from the usual 78 °F–88 °F. "We are currently experiencing the longest global coral bleaching event ever observed," said C. Mark Eakin, the Coral Reef Watch coordinator at the National Oceanic and Atmospheric Administration in Maryland. "We are going to lose a lot of the world's reefs during this event" (Innis, 2016). Reefs that had required several centuries to grow were being destroyed in a matter of weeks.

3.4.47 Ocean Acidity Accelerates

A team of scientists replicated oceanic acidic levels at the beginning of the industrial revolution and traced them to the present day, learning that rising acidity has been stunting corals' growth since humankind began reducing oceanic pH. "Acidification-induced reductions in calcification are projected to shift coral reefs from a state of net accretion to one of net dissolution this century," they wrote. "When ocean chemistry is restored closer to pre-industrial conditions, net community calcification increases. In providing results from the first seawater chemistry manipulation experiment of a natural coral reef community, we provide evidence that net community calcification is depressed compared with values expected for pre-industrial conditions, indicating that ocean acidification may already be impairing coral reef growth, the team concluded."

"Our work provides the first strong evidence from experiments on a natural ecosystem that ocean acidification is already slowing coral reef growth," said study lead author Rebecca Albright of the Carnegie Institution for Science. Previous studies had been completed in laboratories. "Ocean acidification is already taking its toll on coral reef communities. This is no longer a fear for the future; it is the reality of today," Albright added. This research was released as an intense El Niño event raised ocean temperatures to record highs around the world, contributing, from 2014 to 2016, to the longest episode of coral bleaching (death) on record.

By the end of the twenty-first century, according to Ken Caldeira of Stanford University, surface acidity around Antarctica will be roughly double preindustrial levels (a 0.2 decrease in pH), threatening life-forms' ability to maintain their shells. Such a level would put about two-thirds of cold-water corals in corrosive waters (Kintisch & Stokstad, 2008, 1029). Acidification will affect corals acutely, dissolving their shells at a time when warming temperatures are already threatening their survival. "While bleaching…is an acute stress that's killing them off…acidification is a chronic stress that's preventing them from recovering," said Joanie Kleypas, a coral reef scientist at the National Center for Atmospheric Research in Boulder, Colorado (Kolbert, 2006, 72).

Because coral reefs are among the richest biological areas of the oceans, acidification poses a major, long-term threat to aquatic life. Thomas Lovejoy, who coined the term "biological diversity" in 1980, compared the effects of ocean acidification to "running the course of evolution in reverse." The two most important biological factors for organisms in the ocean are temperature and acidity, Lovejoy said. The effects of changes in both provoked by human-induced carbon dioxide emissions reach to the base of the oceanic food chain, with profound long-term implications favoring "lower" forms of life such as jellyfish and other invertebrates. In the very long run, some scientists fear human intervention in the oceanic system may be favoring the return to slime as a predominant life-form there, according to German Marine Biologist Ulf Riebesell (Kolbert, 2006, 75).

The limit of acidity tolerance for many corals is now expected at about 550 parts per million carbon dioxide, a level that will be reached, at present rates of accretion, about the middle of the twenty-first century. As CO_2 rises above that level, the oceans will encounter a "reef gap," "when these magnificent, diverse structures disappear from the world" (Zalasiewicz & Williams, 2016, 189). The last reef gap occurred 55 million years ago, during the Paleocene-Eocene Thermal Maximum. The corals did eventually recover, after several million years. Life will continue during such a gap, as coral reefs "will be replaced by 'slime-rock' systems dominated by algal and microbial mats and jellyfish" (Zalasiewicz & Williams, 2016, 191).

3

> **Corporate Attention Toward Climate Change**
> The financial risks of climate change have become well known enough that the US Federal Reserve Bank, the United States' central bank, made it a target of "serious assessment" (Smialek, 2021, B-2). Lael Brainard, a Federal Bank governor, said that the institution will develop climate-related stress tests for banks related to climate change. Brainard also said that banks found to have excessive exposure to climate-related perils will be advised as to how to diminish them. The central bank also will develop methods to "game out" effects of melting ice caps and excessive wildfires may have on banks as a whole. The European Central Bank and Bank of England developed such guidelines before the United States.

3.4.48 October 8, 2021: Google Jacks Ads Claiming Climate Change Is a "Hoax"

Google said it would no longer accept advertisements that contend that climate change is a "hoax" or a "scam." That denies "the long-term trend that the climate is warming, or denies that greenhouse gas emissions or human activity is contributing to climate change" (Wakabayashi & Hsu, 2011, B-2). Climate change denial was added to a list of to-delete content that includes videos about climate change or commentary about tragic events. This move may step up pressure on Facebook, a major rival of Google for advertising dollars, but does not have an explicit policy on barring climate change denial (Smialek, 2021, B-2).

3.4.49 2021: A Record-Breaking Year for Fires in Sakha, Siberia

The larch forests of the Republic of Sakha (Yakutia, in Siberia) grow in a region with some of the world's largest temperature swings, dominated by a deciduous conifer called ▶ Larix gmelinii. This hardy type of larch is capable of withstanding temperatures as low as −70 °C (−94 °F) and surviving in frozen permafrost soils—traits that have given the tree the most northerly range on Earth. During the summer of 2021, fires raged through the larch forests for months, Sakha's most severe fire season in several decades, with more than 8.4 million hectares of forests burned. "That's an amazing amount—nearly four times the average," said Amber Soja, a NASA and National Institute of Aerospace associate research fellow who has conducted field research in the region. It is also record-breaking. More forest area burned in Sakha than in any year since data was first collected in 2000 (Voiland, 2012).

Sakha is almost twice the size of Alaska. "What happens in Sakha, and in boreal forests more broadly, matters tremendously," said Soja. "Boreal forests store more carbon than any other type of forest in the world. Many of the fires here burn for a long time—weeks, even months. Some have burned the same areas in multiple years," Soja explained. "These fires aren't just spreading across the landscape.

They're also burning *down*. They're thawing permafrost, burning through layers of peat in some areas, and releasing stored carbon and methane that has built up over millennia" (Voiland, 2012).

Outbreaks of large fires in Sakha have happened before, including 2004, 2010, 2013, 2019, and 2020. The 2019 and 2020 fire seasons were particularly extreme in Sakha's tundra regions north of the Arctic Circle. As this area baked under extreme drought and heat, it experienced the two earliest and largest fire seasons on the satellite record.

In 2021, there has not been nearly as much burning north of the Arctic Circle; instead more of the fires occurred in forests farther south. "We saw a different part of Sakha burn this year," Soja said. "But the underlying driver—droughts and heat amplified by climate change—was the same" (Voiland, 2012).

3.4.50 Wind and Solar Advancing, but Not Fast Enough

The International Energy Agency (IEA), which tracks the use of energy production and changes in its technology worldwide, issued an annual report on October 12, 2021, that the said use of fossil fuels is expected to peak about 2025, with major advances in technology and use of wind, solar, and other emerging technologies. The bad news, said the IEA, is that these changes are not nearly fast enough to avoid extensive damage from global warming unless governments step up their efforts to end use of fossil fuels, with solar and wind advancing and electric cars' sales increasing. A major problem is developing infrastructure. For example, are electric cars fueled by energy that comes from renewable sources? An electric car drawing energy from a generating plant that burns coal is worse for the environment than one that burns gasoline. What about a network of electric charging stations that is comparable to our placement of gasoline stations?

Wind and solar energy are now the least expensive forms of electricity generation in most markets, and sales of electric cars are increasing rapidly. Despite this progress, the world is still on track for an average temperature rise of at least 2.6 °C (4.7 °F) by the end of the twenty-first century. The United Nations has already deemed such a projected rise "catastrophic," citing more severe storms, wildfires, heat waves, etc. with a rise of 1.11 °C. Since the beginning of an industrial age powered until recently almost entirely by fossil fuels, the IEA projects that to avoid widespread damage around the world, addition of fossil fuels must be newly cut in half by 2030 and *ended entirely by 2050* (Plumer, 2021, A–7).

3.4.51 October 13, 2021: President Biden Embraces Wind Turbines Coast to Coast to Coast

At about the same time that the International Energy Agency called on governments to increase commitments to renewable energy quickly, US President Joe Biden issued a plan to promote installing large-scale wind energy turbines along the entire lengths of the US East, Gulf of Mexico, and Western coastlines. The plan was first announced on October 13, 2021, at a wind energy convention in Boston by Interior Department Secretary Deb Haaland. It involves leasing federal government waters by 2025. The plan is not an edict; fishing people and coastal landowners plan to oppose the idea.

The plan also will be required to pass through a gauntlet of state, federal, and congressional reviews that will raise questions about such things as threats to endangered species and interference with the needs of the military, tourism, and many others established interests.

It is, however, the most ambitious exercise of federal advocacy on behalf of wind energy in US history. Biden said that the turbines will provide good jobs and reduce US use of fossil fuels (most notably coal). "This is a very big, big deal. This is a signal like we have never had before in the United States about where she can go with offshore wind," said Dan Reicher, who served as an assistant secretary of the Energy Department in the Clinton Administration (Davenport, 2021, A-18).

3.4.52 October 12–13, 2021: Torrential Rains Submerge Parts of Urban China for the Second Time in 3 Months

For the second time in 3 months, unusually intense rainfall drenched northeastern China, much like the remnants of tropical storms have battered the eastern United States. At least 28 people died and more than 120,000 were displaced in these latest Chinese downpours. Half of the dead fell into a river from a commuter bus that fell off a bridge. Several coal mines also were shut down during the storms.

Three months earlier, 300 were killed in the central province of Henan during record-setting rains.

3.4.53 October 18, 2021: Climate Change in the Arts

The climate change rubric suffused some unusual venues in the arts after a 2021 summer of vicious wildfires and killing floods. Witness a musical notice in *The New Yorker* (p. 6, October 18, 2021):

Renee Fleming: "Voice of Nature"
CLASSICAL "Voice of Nature: The Anthropocene." A new record from the soprano Renée Fleming, uses Romantic and contemporary songs to chart humanity's evolving relationship with the natural world in the face of climate change….wrapped…in a voice of such loveliness.—*Oussama Lahr.*
Nor are we finished with the October 18, 2021, edition of *The New Yorker*. In the venerable "Talk of the Town" by André Wheeler ("Beach Scene" p. 15). "The Brooklyn Academy of Music….put out a call for people who wanted to be beachgoers in the avant-garde climate crisis opera 'Sun & Sea.'" More than 20 tons of sand were dumped on the floor of a building at BAM, with sunscreen, beach balls, and towels, and a high-calibre hose standing in for an "ocean" for "swimmers" as "classically trained singers keened about the collapse of coral reefs," as a couple walked a dog on the "beach." (p. 15). Now, if carbon dioxide just had a sense of humor…or irony. (Andre (accent over e) Wheeler. "Beach Scene." *The New Yorker*, October 18, 2021.

3.4.54 October 2021: Venice's Lagoon Filling, Again

Residents of Venice, Italy, watched as brackish lagoon water flowed into St. Mark's Basilica and the rest of this 1600-year-old city that is now reminded too often that the seas are slowly rising as the city sinks. Venice could remind some visitors of New Orleans, Louisiana, but without the hurricanes, but with sinking land and rising seas. A new study published by the European Geosciences Union is 50% higher than an older worse-case global sea rise average of 80 cm (2 feet, 7 1/2 inches) forecast by a United Nations science panel. Venice's new forecast for the end of the twenty-first century is 120 cm, which probably will wash away many off the city's famous canals.

St. Mark's chief caretaker, Carlo Alberto Tesserin, told the Associated Press: "Conditions are continuing to worsen since the flooding of November 2019. We therefore have the certainty that in these months, flooding is no longer an occasional phenomenon. It is an everyday occurrence," said Jane Da Mosto, an executive director of We Are Here Venice, added: "We are living with flooding that has become increasingly frequent, so my concern is that people haven't really realized we are in a climate crisis" (Barry, 2021). Venice is trying to cope with the rising waters and sinking land with a $7 billion floodgate system called Moses, which thus far has been overmatched by nature, as floods get deeper. Venetians lug around their rubber boots and chuckle at tourists who do not want to get their feet wet during the ever-more frequent instances of *acqua alta* (high water).

3.4.55 October 20, 2012: Africa's Last Five Glaciers Melting

Africa's last five remaining glaciers are quickly melting and may become memories within a few years, according to a new report issued by the United Nations. One of the four is the iconic Mount Kilimanjaro, in Tanzania, made famous in the novels of Ernest Hemingway. Its thick crown of snow and ice of a century ago has become a scanty tiara of pockmarked, dirty ice. The others are Mt. Kenya, the Ruwenzori range near two peaks, the birders of Uganda, and the Democratic Republic of Congo (Santora, 2021, A-9).

3.4.56 October, 2021: Out-of-Season Rains in India and Nepal

Usually, October in India marks the end of the monsoon season on the Indian subcontinent. However, in October 2021, the counterclockwise circulation of a cyclonic storm that was nearly stationary over the Bay of Bengal slammed moisture against the Himalayas' foothills, producing torrential out-of-season deluges over the area. By October 21, close to 100 people had been confirmed dead, with more expected as waters recede. Before the storm ended, its bands of rain had swept over much of India, southward to Kerala.

The magnitude of the storm had not been expected by India's meteorologists. "Rainfalls in October were reported in the past, too, but not to this intensity," said Ajaya Dixit, an expert on climate change vulnerability in Nepal. "Climate change is real, and it is happening" (Schmall, 2021, A-10).

3

3.4.57 October 21, 2021: The Death Penalty for Forest Arson?

During the fall off 2020, a series of fires blazed through Northwest Syria, said by Bashar al-Assad to have been arson aimed at destabilizing his dictatorial regime. For the alleged arson, 24 men were executed in mid-October 2021. Eleven other men were sentenced to life in prison at hard labor. The men all were accused of terrorism, including the destruction of public and private property. As with many things involving Syria's multiyear civil war and matters of life and death, motives and results were hazy at best. Were the fires, all in the Northwestern part of the country, of natural causes, for which the 35 men, all political enemies of Assad, were being rubbed out? The Middle East has suffered many fires in several countries in recent years because of rising temperatures and drought which have wiped most of Syria's agriculture.

"The idea that 24 people were executed in relation to wildfires just smacks of the farce that Balshar al-Assad has a made of the justice system over the last decade," said Sara Kayyali, a Syria researcher of the Human Rights Watch (Hubbard, 2021, A-9).

3.4.58 October 22, 2021: Climate Change and National Security

The US government has rung alarm bells about the malign results of climate change before, especially its expected effects vis-a-vis military affairs and environment. Never, however, have reports to the president laid out the depth, range, and destructive power of global warming as these did on October 21, 2021, involving plans across nearly all government departments and agencies.

The leading theme of the reports often is national security, a broad umbrella under which many malign influences may gather, from rising problems related to immigration to tensions over resources opening as the Arctic ice cap melts. Senior financial advisors also released discussions of climate change's effects "as an emerging threat....to the U.S. economy" (Flavelle et al., 2021, A-1). Many other subjects were raised and government agencies covered.

Taken together, these many reports encompass a new focus of activity on climate change across the entire US government, a favorite theme of President Joe Biden, which sharply contrasts with that of ex-President Donald Trump, who described climate change as a hoax invented by Democratic radicals, the Chinese, and any number of other so-called enemies. The new emphasis also came a few weeks before a worldwide conference on the subject in Glasgow, Scotland, as well as following a year of climate-caused major disasters—fires, floods, and more—around the world.

Taken together in another way, these are political statements in a political world where carbon dioxide has no politics nor ambitions. It merely holds heat. Denizens of the Earth will have to wait (in a world where we have precious little time) to see whether all of these reports do anything other than produce—what else?—a lot if hot air.

3.4.59 October 24–25, 2012: Drought Turns to Deluge in California

Fire season on the US West Coast ended with a life-imperiling splash as drought turned to deluge. The "pineapple express" moved in from San Francisco to Seattle as it merged with another storm to form a "bomb cyclone," switching peoples' worries from fires to floods rolling off all those hills denuded by some of the worst fires in recorded history. The central pressure of this storm was 942.5 millibars, that of a Category 4 hurricane. By September, days after record heat and drought in the same area, the US West Coast and Canada were swept by heavy snow and flooding.

Most of the formerly forested hills turned quickly to mud, rolling across highways, occasionally inundating the scraps of homes and businesses scorched weeks before. The storm, which looked like a sprawling hurricane on pictures from space satellites, was the most severe to hit the US West Coast in recorded history. The slides of rocks, trees, and mud sounded like roaring freight trains as they tumbler and rumbled down hillsides, pushing some cars off roads. The National Weather Service said that the rainfall, as much as seven inches, posed a danger of "life threatening flash floods. The atmospheric river is aiming a firehose, if you will, into our area" Sean Miller, a meteorologist for the Weather Service, said in Monterey, California, on Sunday [October 24, 2021] (Vigdor & Lukpat, 2021, A-18).

In Blue Canyon, California, 10.4 inches of rain fell in a 24-hour period on October 24–25, breaking a record of 9.33 inches set in 1964, according to the National Weather Service. Over the same period, downtown Sacramento saw 5.44 inches of rain, breaking a record set in 1880. Just one week before, Sacramento had broken another extreme record for the longest dry spell, marking 212 days without measurable rain.

Three to four days later, the same storm dropped several tornadoes on the Midwest and then swung up the northeast coast, feeding on the Gulf Stream, and blew up a second time, another "bomb cyclone," off the south New England coast.

3.4.60 October 2021: Turkey's Flamingos' Lake Dries Up

For centuries, Lake Tuz in Central Turkey has hosted huge colonies of flamingos that migrate and breed there when the weather is warm, feeding on algae in the lake's shallow waters.

This summer [2021], however, a heart-wrenching scene replaced the usual splendid sunset images of the birds captured by wildlife photographer Fahri Tunc—

Carcasses of flamingo hatchlings and adults scattered across the cracked, dried-up lake bed.

The 1665 km² (643 square mile) lake—Turkey's second largest lake and home to several bird species—entirely receded this year. Experts say Lake Tuz (Salt Lake in Turkish) is a victim of climate change-induced drought, which has hit the region hard, and decades of harmful agricultural policies that exhausted the underground water supply.

3.4.61 October 28, 2021: Oil Company Executives Grilled Before Congress

Executives who direct four major oil companies were called before the US House of Representatives' Oversight Committee and critically questioned regarding whether they lied about the effects of combusting oil and natural gas (methane) on climate change. Some of the committee members compared this hearing to those during the 1960s when executives of large tobacco companies contended that their products did not cause lung cancer. By 1994, CEOs testifying before Congress agreed that ingesting burning tobacco did have negative health effects but was not addictive. Similarly, members of Congress in 2021 examined the industry's decade-long "climate disinformation" efforts about the impact of fossil fuels on global warming. Rep. Carolyn Maloney, Democrat of New York, said that the companies had failed to submit documents that may show how the companies have been spending money on advertising and lobbying against the idea that their products have had any role in causing increasing levels of greenhouse gases polluting the atmosphere.

Upon prodding, the executives, Darren Woods of the ExxonMobil, David Lawler of the BP America, Michael Wirth of the Chevron, and Gretchen Watkins of Shell, plus Mike Sommers from the American Petroleum Institute and Suzanne Clark from the US Chamber of Commerce, each said that climate change was real and that all of them were taking steps regarding their companies' directions toward renewable energy. However, at street level, more gasoline and other oil and gas continued to be sold, and carbon dioxide levels continued to rise to higher levels than at any time since the Pliocene, 2–3 million years ago. If CO_2 had a sense of humor, it would have gotten a chortle out of this situation.

3.4.62 October 31, 2021: National Leaders of Carbon Emitters Meet and Produce Little Except Hot Air

In a conference, the Earth's 20 largest nations by size of economic output (G-20, responsible for about three quarters of the world's carbon dioxide emissions) met in Rome, with climate change the major subject of discussion. This conference was followed by the Conference of Parties 26 (COP 26), the latest in a year-long series of diplomatic and scientific meetings exclusively about setting goals for global reductions of greenhouse gases.

Press and participants' evaluation of climate change outcomes at G-20 ranged from "mild" to "tepid." Boris Johnson, a prime minister of Great Britain, called the outcome "drops in a rapidly warming ocean" (Winfield et al., 2021, A-1). The G-20 delegates did agree to stop funding coal-fired power plants in poor countries, where they have been a major source of new power. Otherwise, what came out of this event was not worth the tons of carbon dioxide generated by the jets which brought delegates to the G-20 conference.

Next, it was on to Glasgow, Scotland, where attention was expected to focus on allowable targets for carbon dioxide emissions. In Paris, during 2015, a maximum range of 1.5–2.0 °C was set, without which Earth would descend into climate chaos—that is, sliding down a climatic slope so quickly that abating the rise in temperatures would become impossible. By 2021, many large economies already were overshooting these targets. Some of those countries (Russia and China were notable) already had indicated that they would boycott COP 26.

3.4.63 November 1, 2021: Rifts at COP 26

The COP 26 summit was called to order in meetings that the *New York Times* characterized as being marred by "Rifts" and Finger-Pointing (Tunkersley et al., 2021, A-1, A-7), next to a large photo of men and women, mainly in suits, at rows of tables with microphones that looked as if it must have been at least an acre in size, with speakers ranging from coal advocates from Russia and China, which burn more than half of the world's carbon (and were conspicuous by their absence), to small Pacific islands, whose speakers described the end of their worlds in less than 10 years due to rising seas.

Then there was US President Joe Biden, who had been portrayed as a one-person cheering squad for clean energy at home, striking a middle-of-the-road ambivalence by advancing a case for clean energy in the long run but admitting that we will not be able to disown coal, oil, and gas for some time to come. A matter of practicality is what he said, as many environmental advocates said that practicality was what got us here to begin with. In other words, nearly everyone was advancing his or her own national interests and might come up with no solution at all, which has been typical of COP meetings since 1995—a COP-out that leaves the scientists flummoxed, environmentalists very disappointed, and fossil fuel advocates quietly happy that they had dodged one more dirty bullet.

As autumn continued, some countries (or groups thereof), doubting that they could meet the reductions in greenhouse gases that they had pledged, began to move the goalposts. On February 2, 2022, for example, the European Union said that it would treat some natural gas and nuclear power plants as "sustainable" bridges to a greener future. This statement came with a request for help from the private sector. Natural gas is the second most powerful greenhouse gas of the atmosphere, after carbon dioxide. However, it remains there only a few days or weeks, a small fraction of the time that CO_2 resides there. Nuclear power in the past has been accident-prone, and its cost of construction can be prohibitive. Neither is usually listed as a viable new alternative to fossil fuel. Instead, both are

3

being called "transitional" forms of energy, to goose up some countries' numbers to meet international goals as wind, solar, and other non-fossil fuels are being developed.

3.4.64 November 3, 2021: "Historic" Storm Pummels Alaska; COP26 Takes "Baby Steps"

A summer of scorching heat in the northwestern quadrant of North America ended abruptly with a very sharp seasonal change as a wicked storm pounded Alaska and then moved south and eastward down the Pacific Coast. The storm overwhelmed infrastructure in some mountain areas with as much as 10 feet of snow. At lower elevations, extreme amounts of cold rain flooded many small communities and washed away two-lane blacktopped roads that link many of them.

In Glasgow, representatives of about 100 countries shared traditional greetings and speeches as about 40% of the participants (mainly heads of states) departed, leaving the floor to scientists and negotiators discussing limits on methane emissions and deforestation. Both came with caveats. Methane has always played second fiddle to carbon dioxide in the climate change orchestra. Its level in the atmosphere varies widely, and it washes out within a few weeks, whereas carbon dioxide can remain there for several months or more. Deforestation agreements have been routinely flouted as climate envoys return to home front politics and economic pressures. Examples are Brazil, with its deforestation-friendly federal government, as well as Russia and China, which had all but boycotted the climate change summit. The Global Methane Pledge said that the signatories would cut emissions of this greenhouse gas by 30% by 2030. In addition to China and Russia, Australia and India, both large methane producers, failed to sign the pledge (❏ Figs. 3.19, 3.20, and 3.21).

Most of the representatives of more than 100 countries also pledged to eliminate deforestation by 2030, contradicting recent rises in such countries such as Brazil, where deforestation had surged during the presidency of Jair.

Bolsonaro, who favors economic development overall else of soy, palm oil, cattle ranches, and timber for commercial uses. Bolsonaro also was "conspicuously absent" from the meetings (Andreoni, 2021, A-8). Bolsonaro, in the meantime, was circulating slick videos touting his and Brazil's commitments to green jobs, an end to deforestation, a cut in greenhouse gas emissions by half by 2030. Bolsonaro's description of Brazil as a "green power leader" (Andreoni, 2021, A-8) struck some participants as too Orwellian for them. The meetings had plenty of rhetoric to go around. Boris Johnson, a prime minister of Great Britain, a conservative who had been a late to convert environmentalism, said, of deforestation: Let's end this great chainsaw massacre" (Tunkersley et al., 2021, A-8).

Several scientists, environmentalists, and other activists meanwhile bemoaned the transformation of the series of "conference of parties" meetings from a search for genuine solutions to climate change problems to a gaggle of bankers, industrial leaders, and fossil fuel advocates looking to advance various private agendas,

■ **Figs. 3.19–3.21** As summer turned to autumn in 2021, nature's fire extinguisher, with plenty of imported water and firefighting muscle, helped to snuff the Caldor Fire south of Lake Tahoe. A potent blast of cold air and snow did the same off the northwest coast of North America. Source of all three images: NASA. Public Domain

meanwhile as "green" as possible in the face of glooming disaster. "Private jets brought investors and fossil fuel lobbyists in embarrassing profusion," wrote Christopher Caldwell in the *New York Times* (November 26, 2021). "At Glasgow," wrote Caldwell, "A few self-nominated representatives from very rich indus[tries] laid claim to a special role in in shaping the human future." In the meantime, pledges were made as oil and coal continued to burn, and the clock to disaster kept ticking. Even after copious badmouthing of coal that went on at Glasgow (as well as before and after that), statistics for the year indicated that the world had combusted 9% more coal in 2021 than 2020 (Friedman, 2022, A-18).

3.4.65 November 2, 2021: Krill, Anyone? Temperatures Rise Rapidly on the Antarctic Peninsula

While Prime Minister Johnson and President Bolsonaro were touting their green *bona fides* in Glasgow during 2021, word was going around, courtesy of the November (2021) *National Geographic*, that in February 2020 (summer on the Antarctic Peninsula), a temperature of 64.94 °F had been reported there. Given the location, where temperatures very rarely reach 32 °F even in mid-summer, 64.96 was a heat wave of major proportions, creating major melting of ice. In the meantime, species of penguins that depend on icy seawater and krill were going hungry. Krill is a major base of the food chain in Antarctica, supporting all sorts of sea life, from small fish to whales.

Of late, human beings have discovered krill, which are about 2–3 inches long—not usually under that name as a snack, but in pills that are stuffed with omega-3 fatty acids, which have many beneficial effects vis-a-vis human and animal health. Reaping profits for the krill trade, factory ships have been entering waters around the Antarctica from Japan, China, Chile, South Korea, and other far-off ports, to harvest the krill in areas that once were locked in ice much of the time. The annual haul of krill has risen from about 130 thousands of metric tons in 2012 to 451 in 2020, "Take away krill, and the food chain collapses" (Scales, 2021, 107, 116). A factory ship can suck more than 800 tons a day out of Antarctic waters (Scales, 2021, 116). As ice melts, the number of krill ships increases.

3.4.66 November 4, 2021: The Persistence of Coal, Even Among Those Who Should Know Better

As the first week of COP 26 drew to a close, the convening delegates were reminded of just how tough it will be to dislodge King Coal, the dirtiest of fossil fuels, the cheapest, and the most widely used. All but a lunatic fringe of scientists and nearly all of the political delegates at COP 26 realize that to preserve anything resembling a survivable world, coal will have to be relegated to the Greenhouse Gas Museum.

In our time, banishing coal is proving to be a *bona fide* hard sell—until now, an exercise in wish fulfilment.

Exhibit A of this existential problem surfaced November 4, Thursday, of the first week. The delegates were presented with a pledge to phase out coal. The result had Alok Sharma, the head of the conference, proclaiming that "The end of coal is in sight" (Plumer & Friedman, 2021a, A-10). In that case, Sharma was being a very farsighted individual, considering that only 23 countries in a world of 197 nations had signed a commitment to stop issuing permits for coal-fired power plants. Among the roughly 175 nations that *did not* sign on were China, India, Australia, and the United States which, together, burn about 80% of the world's coal (China and India burn 54%). China and India, with their enormous populations who have little or no electricity, pledged commitment to solar, wind, nuclear, and hydropower but not nearly enough to eliminate coal, their major power source. The United States has copious supplies of coal and uses 20% in its electric power base. In addition, US politics is deeply in thrall to coal political contributions. The most prominent political champion of coal in recent times has been West Virginia Senator Joe Manchin, III, who draws a steady income from oil-based investments. Australia is a major coal exporter that heavily subsidizes many members of the country's political ruling class. Exhibit A here is Scott Morrison, a prime minister. Some of the countries that elected to promise the phaseout of the use coal were surprising, such as Poland, which has rich reserves and mainly a coal-based economy. Of the 23 nations that signed onto the prospective coal ban, 10 burned no coal at all; all 23 signers accounted for only 13% of the world's coal use.

A pledge, of course, is a promise. It is not an accomplished fact. When the authors of the Paris Accord of 2015 recommended limiting future carbon dioxide levels to 1.5 °C over industrial-age levels, they said that to do this, all fossil fuels would have to be phased out by 2035. Such a target was, in 2022, becoming less likely with every passing day, especially as coal consumption recently trended upward after several years of decline. Alok Sharma seemed to have donned a happy face with regard to a very grim situation. May his remark about the end of coal use be enshrined in the Greenhouse Gas Museum.

In the meantime, more than 100,000 protesters streamed by outside the meeting hall in Glasgow; their protests aimed mainly that the negotiations inside were too little and too late to avoid planetary disaster. Greta Thunberg, the young climate activist who attracted a huge following by crossing the Atlantic Ocean from Europe to New York City in a small sailboat, called the meetings a public relations event.

The marchers, from several countries, gathered in a steady cold rain, chanting, signing, playing bagpipes, and carrying signs, as they wound through much of Glasgow's city center. The variety of allegiances taking part in the march indicated that climate change had become a cause behind which other causes, such as income equity, had become allied. Indigenous activists from the Americas took part to demand that world's leaders protect their ancestral lands.

3.4.67 Late 2021: US Congress Passe Billions for Climate Work, but Senate Turns Thumbs Down

An infrastructure bill passed by US House of Representatives contained $47 billion to help cities and towns prepare for floods, fires, extreme storms, and droughts that have intensified due to climate change, the first such allocation in US history. However, the Senate refused to cooperate.

No matter how much money may be allocated to "fix" the climate crisis, many people have been realizing that it may be verging on too late. Elizabeth Kolbert, a staff writer at *The New Yorker* and winner of a Pulitzer Prize for her *Sixth Extinction*, wrote for the magazine as the 2021 climate summit in Glasgow was beginning its second week: "The sad fact is that, when it comes to climate change there's no making up for lost time. Every month that carbon emissions remain at current levels—they're running at about 40 billion tons a year—adds to the eventual misery. Had the United States started to lead by example three decades ago, the situation today would be very different. It's still not too late to try—indeed, it's imperative to try—but, to quote Boris Johnson, 'humanity has long since run down the clock on climate change'" (Kolbert, 2021, 22).

3.4.68 "We Are Still Knocking on the Door of Climate Catastrophe"

Back at the World Climate Summit in Glasgow, Scotland, the *New York Times* reported on November 13: "Negotiators Strike a Climate Deal, but World Remains Far from Limiting Warming...Some activists called the agreement in Glasgow disappointing, but it established a clear consensus that all countries need to do much more." In other words, everyone came out of the summit knowing what was obvious before it started: that everyone needed to act on the problem, and soon, to preserve a liveable Earth. What came out was short on specifics and enforcement and so, beyond a wish for good behavior, rather useless. More than 100,000 activists marched in the streets around the meeting venue urging more decisive action.

Seth Borenstein, a veteran science reporter at the Associated Press, quoted UN Secretary General Antonio Guterres as saying that "We are still knocking on the door of climate catastrophe" (Borenstein & Jordans, 2021, A-12). Alok Sharma, a chief organizer of the United Nations COP 26 climate summit in Glasgow, called the meeting's final pact between nearly 200 nations to escalate efforts to combat climate change "a fragile win" (Plumer & Friedman, 2021b). Others called it "mejor canada" (Spanish: "better than nothing").

To the end, delegates were arguing over such issues as whether the final document should advise countries to "phase down" or "phase out" the use of coal. The lukewarm phrase "phase down" won. The document said *nothing* about oil or coal, although for the first time in COP history, it *did* use the term "fossil fuels." Delegates were advised to aim at keeping their emissions increased to 1.5 °C (2.7 °F), the same target set at the climate summit in Paris (2016). Some specialists asserted in the meantime that 4.0 °F would be a more realistic target by the year 2100.

The final statement also requested that nations phase out "inefficient" subsidies for fossil fuels. "Inefficient" was defined as subsidies that encourage wasteful consumption. Worldwide, about $423 billion are spent per year on subsidies that range from those that help the poor pay fuel bills to direct grants for fossil fuel companies that do such things as stimulate production. John Kerry, a US special envoy for climate change, called such subsidies "the definition of insanity" (Sengupta et al., 2021, A-1).

The statement also urged wealthy nations to "at least double" funding to protect poor nations from the hazards of a hotter planet. Poorer nations asked wealthy nations for billions of dollars to protect themselves from the hazards of a hotter planet. Nothing, as usual, was required. The final document meekly urged richer and poorer nations to have a dialogue on the question. And so the meeting broke up with a commitment to meet again before the end of 2022.

3.4.69 November 17, 2021: New Delhi—Smothered in Coal Pollution, Car Exhaust

According to the Associated Press, as published in the *New York Post*, "Authorities in New Delhi closed schools indefinitely and shut down some coal-burning power plants Wednesday to reduce air pollution in India's smog-shrouded capital and neighboring states, as the country weighs an unprecedented and more far-reaching step: a lockdown in New Delhi. The dirty-air crisis in the city of more than 20 million people has underscored India's heavy dependence on coal, which accounts for 70 per cent of the country's power. The New Delhi state government said it is open to the idea of a weekend lockdown to reduce automobile traffic and potentially other air-polluting activity in the city, and it is awaiting the go-ahead from India's Supreme Court" (Associated Press, 2017). As in years past, according to the Earth Observatory, "Sensors in Delhi and elsewhere in northern India have recorded soaring levels of air pollution. ▶ Sensors in the capital area—including ▶ one at the U.S. Embassy—recorded concentrations of fine particulate matter (PM2.5) and coarse particulate matter (PM10) well above 400 $\mu g/m^3$ on several occasions in November. Since particulate matter is linked to a range of respiratory, cardiovascular, and other ▶ health problems, World Health Organization ▶ guidelines recommend that 24-h mean PM2.5 concentrations be kept below 15 $\mu g/m^3$. The high pollution levels led to ▶ partial lockdowns, school closures, and halts in construction in Delhi and ▶ other cities" (A Shot of Smoke for Delhi, 2021).

3.4.70 November 26, 2021: Soaking the Smithsonian

Officials at Washington DC's Smithsonian Institution, the largest public and research complex of museums in the world, have been calculating how much addi-

3

tional water will be necessary to inundate it. The museums run to west to east from near the White House (which is on a low hill) to the foot of Capitol Hill. Many of the buildings were constructed on fill land above extensive swamps that before their construction drew water from the Tidal Basin (near the Jefferson and Lincoln memorials, which feed off the Potomac River).

Many of the institution's research collections (as opposed to public displays) are kept in basements, some of which have already begun to leak. Even now, a stagnant tropical storm could flood many of the buildings on the National Mall between basement and first floor levels. Only a few of the newer buildings, such as the African American Museum of History and Culture, have much stronger pumping equipment. Other buildings containing priceless collections will be at the mercy of garbage cans and sandbags as weather warms and storms become more intense. Floods also could ruin electrical and ventilation systems that "keep the humidity at the right level that protect[s] priceless art, textiles, documents, and specimens on display" (Flavelle, 2021d, A-14).

3.4.71 December 2019: Devastating Out-of-Season Tornadoes in the US Midwest

Parts of the US Midwest sustained a night of apocalyptic storminess. Omaha's high was 74 °F, about 35 °F above average, before an evening of very unusual tornadoes and thunderstorms pummeled the area.

The previous Friday night, about 200 miles south and east of Omaha, a devastating line of tornadoes formed in Missouri and Arkansas and then moved east-northeast into Kentucky, reducing whole towns to kindling. One of the storms had a continuous path of between 200 and 250 miles, the longest path since observations of such things have been recorded in the 1970s. These tornadoes raced across towns in western Kentucky at 100 miles an hour, with winds equal to those of EF-4 hurricanes. The number of recorded tornadoes was about 40; the death toll was at least 100. All in all, this was one of the worst tornadoes in North America in recorded history—and the entire event occurred in *December*, usually the least likely month of the year for such storms (◧ Fig. 3.22).

These lines of storms had climate change written all over them:
1. The time of the year was very unusual (mid-December), acting like May.
2. The intensity of the storms: up to F-4 (among the most severe in history, at *any* time of the year).
3. The size and endurance of the storms: one of them 200–250 miles long and a mile wide.

Temperatures in the upper 80s °F in Mississippi plus atmospheric instability fed these monstrous tornadoes. The damage (human and material) was very difficult to comprehend. Nature can be vicious. And if nature can do this in mid-December, we wondered what it might do in *May*.

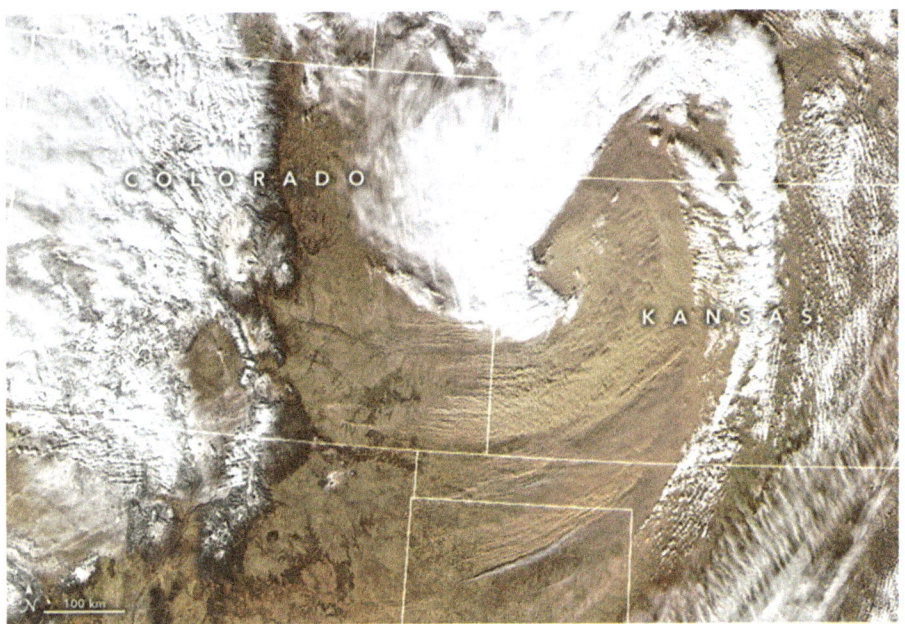

◘ Fig. 3.22   A derecho, a sort of dryland hurricane, roared out of the Rocky Mountains in mid-December 2021, provoking several tornadoes in eastern Nebraska and western Iowa, very out-of-season for twisters. This was only one day after a devastating line of tornadoes formed in Missouri and Arkansas and then moved east-northeastward into western Kentucky, reducing whole towns to kindling. Source: NASA

In addition to tornadoes, straight-line hurricane-force wind gusts as strong as 81 miles per hour struck nine states. Immense dust storms in Colorado and wind-fanned wildfires ravaged Kansas. The first-ever December tornadoes were observed in Minnesota, South Dakota, and western Iowa, following some of the most damaging tornadoes on record in Kentucky and nearby states. Temperatures of 74° in Omaha and Des Moines, both daily records, combining with upper-air moisture, produced winds that drove storm fronts eastward at 100 miles an hour in western Kansas and eastern Colorado.

"We do expect an increase in favorable conditions for severe storms," said John Allen, an associate professor of meteorology at Central Michigan University. "And that means we have to be aware that we can have these extreme events in places or at times that we haven't necessarily thought of before" (Fountain & White, 2021).

Much of the wildly stormy week was unseasonably warm—"It felt like spring; there were people wearing shorts," said Zach Sharpe, who heads the Iowa Storm Chasing Network—"but once the storm front approached, a blast of cold air brought instantly freezing temperatures and 80-miles-per-hour winds. "It was eerie to be chasing tornadoes 10 days before Christmas," Sharpe said. "We were out in our vehicles, listening to Jingle Bells, while tornado sirens were going off".

3

Many scientists held their judgments as to whether this week of weather weirdness was an example of global warming. However, Victor Gensini, a professor of meteorology at Northern Illinois University, "pointed out that all extreme weather events are now taking place against the backdrop of an atmosphere that has been profoundly shaped by humans burning fossil fuels. Assume that all extreme weather events are impacted by climate change," he said....The better question is to what extent climate change might have enhanced the extreme weather event or how likely was the event to occur in the absence of climate change.

Background Information

So what comes with a high wind warning, a red flag warning, a tornado warning, a severe thunderstorm warning, and (the most immediate crisis) the family dog starting to pee on the couch in your basement office, which is also your tornado shelter?

Answer: Is there any contest here, fool? Take out the busting dog!

So I did, then the sky went dark with that puke-green hue that anyone who has witnessed a tornado knows very well.

The sky is very dark, and trees are swaying back and forth. The Weather Service tells us that a severe thunderstorm with enmeshed funnel clouds was moving in with 80 mph winds. And that is not the tornado itself. That is the speed at which it is moving. The sirens are going off. Tornado sighted 20 miles west of our home and shelter.

Don't we love the adventure of truly wacky weather? Is it global warming? The experts are still having a debate, but 74 °F on a mid-December afternoon (after several lethal tornadoes in Kentucky, etc., the previous night) seems convincing.

There were more sirens. The radio says funnel on the ground six miles to our west-southwest, moving east northeast. The clouds are racing as they darken, flashing lightning. The radio warns us of winds on the ground at 100 mph.

Wind-whipped dust clouds (wind: 80 mph) have hit dry areas in Colorado and western Kansas. Funnels hop-scotched all over eastern Nebraska and Iowa. A wind gust of 81 mph hit Offutt Air Force Base near Omaha. A man in Minnesota stepped out of his back door to have a smoke. His family and neighbors later found him dead; his body pinned under a 40-foot fallen tree.

South Dakota also experienced its first December tornado. "The storm system is unprecedented," said Andrew Ansorge of the National Weather Service in Des Moines. "Local people didn't have anything to compare it to. Straight-line winds reached 93 m.p.h in Lincoln, Nebraska" That was similar to a Category 2 hurricane.

The previous week's monster tornado that earlier hâd obliterated Mayfield and other towns in southwestern Kentucky was classified as an E-4, with winds around its funnel of 190 mph.

"It's surreal," said Weather Service Veteran Michael Moritz. "This close to Christmas, to have this this kind of severe weather over such an expansive area." Meteorologists and climatologists in the area said they never had seen anything like it (Gaarder, 2021, A-1). The straight-line winds killed three people in windstorms with sharped-edged, abrasive wind-driven grit that drove cars off roads and ignited wildfires in bone-dry areas west of the tornado zone. One man died when his semi-trailer overturned.

NASA sent satellite photos over the Internet, with headlines, heading, in all caps: "Hurricane-force winds, dust storms, tornadoes, wildfires, snow squalls, heavy rain, and record-breaking heat accompanied an unusual December storm system." "Unusual" just about said it.

So much wild weather had not been experienced by weather watchers in Omaha since that fictional day about 1890 when a tornado picked up a house containing Dorothy and Toto and then dropped them off at the head of the Yellow Brick Road toward the Wizard's place near Omaha. No one had a clue at the time that *The Wizard of Oz* was actually an allegory for a weather forecast for a weird December evening in our little town. (Omaha also has experienced two very violent tornadoes that ripped through the city in March 1913 and May 1975.)

3.4.72 December 2021: Windstorm, Sandstorm, Firestorm, and Snowstorm

On December 30, 2021, high winds roared out of the west and down the front slope of the Rocky Mountains in Colorado. Northwest of Denver, peak gusts reached 115 miles (185 km) per hour—the equivalent of a Category 3 hurricane. Those winds whipped up intense grass and brush fires in south Boulder and blew them east toward the towns of Superior and Louisville, igniting a firestorm.

Tens of thousands of residents were evacuated as flames were blown down streets and through cul-de-sacs. The fire was carried by what climate scientist and Boulder resident Daniel Swain called an ▶ urban firestorm. Blown by hurricane-force winds, the embers leapt from house to house, burning many from the inside out, while torching trees, igniting commercial buildings, and jumping a highway. The next day, survivors endured several inches of snow.

Strong winds and wildfires are not uncommon on the Front Range, but a December wildfire is very unusual. The usual fire season lasts from May to September. One recent ▶ study found that increases in extreme fire weather are being driven by decreases in atmospheric humidity and increasing temperatures. The fact that the worst fire in the state's history occurred in December is a direct indicator of climate change.

In 2021, Colorado saw an ▶ unseasonably warm summer and fall, coupled with ▶ record dryness. The warm, dry spell followed an ▶ unusually wet spring, which reduced wildfires through the summer and fueled the growth of vegetation— which then dried out and provided ample tinder for the December fire (Pratt, 2021).

Many residents said that they felt newly vulnerable as fires that usually were isolated in the mountains blasted through some houses in a minute or two, "vulnerable to the devastating effects of a warmer, drier climate...." "This is a new world we're living in," said Jennifer Balch, the director of the Earth Lab at the University of Colorado—Boulder. "We have to completely rethink where homes are at risk,'" said Korina Bersentes, whose family home was destroyed. She looked at its ruins and said: "'I do fear that this is going to be the new norm in the West. It's not going to be wildfires in the mountains. It's going to be wildfires everywhere'" (Brennan et al., 2022).

The same day, a blizzard in the Central Plains pulled down wind chills to the minus 30 range with up to a foot of snow, plus more of those once-unusual deep-winter tornadoes southeast of the blizzard. The world came into the year 2021 with a weather map displaying minus 8 °F in Omaha and plus 80 °F in Jackson, Mississippi, eastward and northward, the perfect atmospheric brew for tornadoes inland in the South, which formed the next day.

On December 28, 2021, the weather had a touch of the Biblical: In Texas, fish rained from the sky, one event that no one seemed to be able to blame on climate change. Otherwise, one part of the Puget Sound area recorded more than 2 inches of rain in just 12 h. Washington, DC, received close to a foot of snow. The downpours and heavy snow continued well into January in the Pacific Northwest. For awhile, all of Western Washington's highways across the Cascades were blocked by excessive snow and ice, and Interstate 5, running north-south near Puget Sound shut down because of rain-induced mudslides. At least 38 avalanches rumbled down the snow-and-ice-caked mountainsides on both sides of Snoqualmie Pass, about 60 miles east of Seattle's urban core. Workers clearing snow through Stevens Pass, in the Cascades, said that after they had cleared the cross-pass highway, the next day it was covered by a four-inch-thick slab of ice. Some areas just east of the passes inn such towns as Leavenworth had 3–4 feet of snow. Hoquiam, on Washington's Pacific coast, received 6 inches of rain in one day (Delkic, 2022).

3.4.73 The Olympics' Brown Slopes: February 2022

Viewers who watched the 2022 Winter Olympics in Beijing who know something about the geophysical facts may have noted, when watching outdoor events that, beyond the manufactured snow, everything was very brown and very dry. The area between Beijing and the Great Wall has very little plant life. Part of that is the season (winter), but the area also has had several years of nearly continuous drought. The Chinese have been trying to plant trees (as the "Great Green Wall") and failing, for several years. When the wind blows from the Northwest, grainy sandstorms something obscure Beijing. This year, the sky above the skiing venues until the last few days remained mainly blue. To ensure enough water to make artificial snow, several hundred farmers were moved out of the area, and the water supply of Beijing was tapped. The Olympics' snow-making was among the largest creations of snow and ice in the history of the world. The Beijing Winter games "were, for the first time….tak[ing] place almost entirely on artificial snow…a water management operation of enormous scale…forshadowing the reality of snow sports everywhere as the planet warms," wrote Matthew Futterman and Raymond Zhong in the *New York Times* (Futterman & Zhong, 2022).

At the same time, a report from the University of Maine indicated that ice and snow on Mt. Everest that fell over several decades were melting 80 times as fast as it took to accumulate. In the short term, this melting will provide a wealth of water for a billion people, mainly in China and India. Eventually, however, the same people will face water starvation in a warming world.

3.4.74 Violent Rain and Wind Storms Now Routine Worldwide

The week of February 13, 2022, brought news of violent wind and rain and wind in Britain and Brazil. A 122-mile-per-hour gust was recorded in England, where one weather official said the storm would be the worst the country had seen in 30 years. At least seven deaths were reported. At least 78 people died in torrential rain and mudslides near Rio de Janeiro in Brazil and 7 died in a similar event in Wales, where a 122-mile-per-hour wind gust was reported. Both episodes were widely attributed to climate change. The deluge dumped in a month's worth of rain on Wales in one day and night. And in Australia, a land is beset by drought and deluge on a regular basis; a "rain bomb" hit in the northeast during the last week of February, with 9 feet of rushing water described as waves and sheets, killing at least nine people.

Results

Paired with previous years' heat and droughts, the cold rain, floods, and snows which close this chapter provide a textbook example of the drought/deluge duality that is characteristic of a rapidly changing climate. Climate subject to increasing levels of fossil fuel combustion will be hotter but sometimes colder and wetter but sometimes drier, as thermal inertia (see ► Chap. 2) affects meteorological cycles. The many examples described in the chapter above show just how wild the weather (which becomes climate over time) can be. Overall, weather and climate will become hotter, but not in a ceaseless cycle, as well as drier and wetter, again in an inconsistent fashion. The chapters to come describe how Native Americans have developed an environmentally based worldview, based on beliefs that have been incorporated into non-Native environmental movements. This chapter is followed by two individual case studies based on Native peoples' reactions to heat and drought, the Navajos (Dene) who not only have faced climate-based disasters long before most people were aware of global warming.

The IPCC on February 28, 2022, issued its most harrowing report to date, stating without qualification that "The dangers of climate change are mounting so rapidly that they could soon overwhelm the ability of both nature and humanity to adapt, creating a harrowing future in which floods, fires and famine displace millions, species disappear and the planet is irreversibly damaged." Written by 270 researchers from 67 countries, the report is an atlas of human suffering and a damning indictment of failed climate leadership, said António Guterres, the United Nations secretary general. With fact upon fact, this report revealed how people and the planet are getting clobbered by climate change.

The IPCC sketched the toll: hundreds of millions of people could struggle against floods, deadly heat waves, and water scarcity from severe drought. Mosquitoes carrying diseases like dengue and malaria will spread to new parts of the globe. Crop failures could become more widespread, putting families in places like Africa and Asia at far greater risk of hunger and malnutrition. People unable to adapt to the enormous environmental shifts will end up suffering unavoidable loss or fleeing their homes, creating dislocation on a global scale.

3

The report hit news desks in the midst of a deadly invasion of Ukraine by Russia, as most of the world rose in revulsion against horrific war crimes. Thus, the IPCC's report gathered dust on many news desks around the world, except those of scientists, some of whom said they thought as if they ought to go on strike until climate change was taken more seriously. Bruce C. Glavovic, a professor at Massey University in New Zealand, and two colleagues called upon climate scientists to stage a mass walkout, stopping their research until nations en masse take action on global warming. "We've had 26 Conference of the Parties (COP) meetings, for heaven's sake," he said, referring to United Nations global warming summits. More scientific reports, another set of charts. "I mean, seriously, what difference is that going to make?" (Zhong, 2022). Naomi Oreskes, a Harvard historian of science, agreed with Glavovic's assessment. "Every time the I.P.C.C. comes out with yet another report, [it says] yet again, that the science is unequivocal. Well, then, why do we need another report?" (Zhong, 2022).

3.5 Questions and Exercises

1. The point is often made in the study of climate change that one event does not make a trend. That is the difference between weather and climate. The confluence of many events over wide areas during a relatively short time does, indeed, change long-term trends. That is how weather becomes climate and how it is now rapidly changing.

    Please discuss and debate this statement. How much of a weather trend is necessary to establish climate? Can you work from local examples at all seasons?

2. *What role do United Nations' reports on climate play in combatting global warming?*

3. **The 2021 International Panel on Climate Change's (IPCC) 3000-page report is relevant not only because of the specific ways in which it states the relationship between humankind, greenhouse gases, global warming, and the damage that this rough climate regime can cause but also the ways in which this information can be used to support legal arguments which may, in years to come, be used to formulate deliberations which can be used to define environmental law and penalties for violations of it.**

    Discuss the use of legal means to combat climate change. What practices are now legal that could become illegal under an Earth-centered legal system?

4. The secretary general of the United Nations, Antonio Guterres, has called the most recent IPCC's report a "'code red'" for humanity. Essentially all of the Earth's increase in temperatures after the mid-nineteenth century (i e , since widespread use of coal and oil began) has been driven by humanity's increase in the use of fossil fuels, the report said.

What did Mr. Guterres mean by "code red"? Please be specific.

5. "The most important takeaway [from the most recent UN report on climate] is that climate change is now certain, and it's here," said Diana Bernstein, a climate scientist at the University of Southern Mississippi. "If we don't address it immediately ... it's going to make everybody's lives miserable" (Chrobak, 2021).

    Point out and discuss specific examples of how a changing climate is making life more miserable. Include plants as well as animals other than humans if you wish. Also feel free to counterargue. In what ways has climate change made life better for some plants, animals, and humans?

6. Almost all the world's glaciers are retreating, and ice loss is causing polar regions to warm very quickly due to the loss of heat-reflecting frozen cover. Heat waves are hotter and more frequent, and the hottest days in a decade are 1.2 °C hotter than they were between 1850 and 1900. Ocean heat waves have doubled in frequency since the 1950s.

    Discuss ways in which melting of ice compounds (accelerates) further melting. Bring in the role of "albedo," the whiteness of ice vis-a-vis ocean water.

7. An upper-level ridge of high pressure that slid over Alaska in June 2019 unleashed a heat wave of astonishing intensity. With temperatures soaring into the 80s and even 90s °F in some parts of Alaska, several all-time and daily temperature records fell. Anchorage, Kenai, and King Salmon broke all-time records on July 4, 2019. In Anchorage, the record was not just broken; it was obliterated. The city reached 90 °F (32 °C) on Independence Day.

    What sort of atmospheric pattern produced a high of 90 °F in Anchorage on a summer day, when the average mid-summer high is 62 °F? Where did this heat come from?

8. *Using the chapter headings in this chapter, discuss some of them in their climatic (longer-range) context. How could some of them be used in light of your personal experience?*

References

A Shot of Smoke for Delhi. (2021, November 11). *Earth Observatory*. Retrieved from https://earthobservatory.nasa.gov/images/149086/a-shot-of-smoke-for-delhi.

Adam, K. (2021, September 23). Boris Johnson tells world leaders to 'grow up' on climate change, takes aim at Kermit the Frog. *Washington Post in Seattle Times*. Retrieved September 30, 2021, from https://www.seattletimes.com/nation-world/boris-johnson-tells-world-leaders-to-grow-up-on-climate-change-takes-aim-at-kermit-the-frog/?utm_source=marketingcloud&utm_medium=email&utm_campaign=Morning+Brief+9-23-21_9_23_2021&utm_term=Registered%20User.

3

Andreoni, M. (2021, November 3). Brazil, facing skeptics, seeks role, as climate leader. *New York Times*, A-8.

Associated Press. (2021, November 1). Schools close as smog-laden India capital considers lockdown. Retrieved from https://nypost.com/2021/11/17/schools-close-as-smog-laden-india-capital-considers-lockdown/.

Barry, C. (2021, October 21). Venice Copes with Higher Tides and Climate Change. Associated Press MorningWire. Retrieved October 23, 2021, from https://apnews.com/article/floods-climate-science-lifestyle-business-7e87c573c7ad9129a8d1a4fcaaa3f9e1?user_email_address=ff355df48f607b11ff93cb7e60dbb14d&utm_source=Sailthru&utm_medium=email&utm_campaign=MorningWire_Oct20&utm_term=Morning%20Wire%20Subscribers.

Borenstein, S. (2021a, July 11). *Summer trend: West Gets Hotter Days, East Hotter Nights* (p. A-5). Associated Press in Omaha World-Herald.

Borenstein, S. (2021b, July 25). *Numbers show climate change impact* (p. A-15). Associated Press in Omaha World-Herald.

Borenstein, S., & Jordans, F. (2021). *Compromise seals climate deal* (p. A-12). Associated Press in Omaha World-Herald.

Brennan, C., Healy, J., & Kasakove, S. (2022, January 2). After catastrophic fire, Colorado fights a new hazard: 10 inches of snow. *New York Times*. Retrieved June 5, 2022, from https://www.nytimes.com/2022/01/01/us/colorado-fires-snow-boulder-superior-louisville.html.

Brown, M. (2021, December 8). *Wildlife smoke carries risks* (p. A-15). Associated Press in Omaha World-Herald.

Caldwell, C. (2021, November 26). Bankers took over the climate change summit. *New York Times*, A-21.

Carlowicz, M. (2021, August 23). Henri Soaks the Northeast. *NASA Earth Observatory*. Retrieved August 27, 2021, from https://earthobservatory.nasa.gov/images/148742/henri-soaks-the-northeast?src=eoa-iotd.

Chea, T. (2021, August 29). *California drought devastates almond producers* (p. D-7). Associated Press in Omaha World-Herald.

Chrobak, U. (2021, August 9). To prevent catastrophic global warming, we need to leave fossil fuels in the ground. *Popular Science*. Retrieved August 16, 2021, from https://www.popsci.com/authors/ula-chrobak.

Cvijanovic, I., Santer, B. D., Bonfils, C., Lucas, D. D., Chiang, J. C. H., & Zimmerman, S. (2017). Future loss of arctic sea-ice cover could drive a substantial decrease in California's Rainfall. *Nature Communications, 8*, 1947. Retrieved November 30, 2017, from https://www.nature.com/articles/s41467-017-01907-4.

Davenport, C. (2021, October 14). U.S. plans to build wind farms along most of coastline in climate push. *New York Times*, A-18.

Delkic, M. (2022, January 10). Record rain and snow shut down roads and schools in Washington State. *New York Times*. Retrieved January 12, 2022, from https://www.nytimes.com/2022/01/09/us/seattle-record-snow-flooding.html.

Fassihi, F. (2021, July 22). Severe water shortages add a volatile element to challenges in Iran. *New York Times*, A-10.

Flash Floods from Ida Swamp the Northeast. (2021, September 6). *NASA Earth Observatory*. Retrieved September 12, 2021, from https://earthobservatory.nasa.gov/images/148792/flash-floods-from-ida-swamp-the-northeast?src=eoa-iotd.

Flavelle, C. (2019, August 8). The food supply is at dire risk, U.N. Experts Say. *New York Times*, A-1, A-8.

Flavelle, C. (2021a, August 5). In bill, parties recognize climate change crisis. *New York Times*, A 5.

Flavelle, C. (2021b, September 24). The cost of insuring expensive waterfront homes is about to Skyrocket. *New York Times*. Retrieved September 29, 2021, from https://www.nytimes.com/2021/09/24/climate/federal-flood-insurance-cost.html.

Flavelle, C, (2021c, October 21). Julian E. Barnes, Ellen Sullivan, and Jennifer Steinhauer. Reports Lay Out Climate's Threat to U.S. Security. *New York Times*, A-1, A-13.

Flavelle, C. (2021d, November 26). Saving history with wet-vacs in Washington. *New York Times*, A-1, A-14.

Fountain, H. (2021, August 17). Dwindling water levels in lake mead force cuts in supply in the west. *New York Times*, A-13.

Fountain, H., & White, J. (2021, December 18). Rising from the Antarctic, A Climate Alarm: Wilder winds are altering currents. The sea is altering carbon dioxide. Ice is melting from below. *New York Times*. Retrieved December 19, 2021, from https://www.nytimes.com/interactive/2021/12/13/climate/antarctic-climate-change.html.

Friedman, L. (2022, January 28). Kerry tells top polluters 'we all must move faster to fight climate change.' *New York Times*, A-18.

Friedman, T. (2018, January 23). The tweet trump could never send Tehran. *New York Times*. Retrieved from https://www.nytimes.com/2018/01/23/opinion/trump-iran-climate-change.html.

Fuller, T., & Shawn H. (2021, August 27). Wildlife smoke chokes lake Tahoe, once an oasis of fresh air. *New York Times*, A-1, A-13.

Futterman, M., & Zhong, R. (2022, February 7). Turning China's peaks white comes at a price. *New York Times*, A-1, D-9.

Gaarder, N. (2013, March 30). *Arctic warming blamed for last year's heat and this year's chill.* Omaha World-Herald. Retrieved April 1, 2013, from http://www.omaha.com/article/20130330/NEWS/130339998/1685#artic-warming-blamed-for-last-year-s-heat-and-this-year-s-chill.

Gall, C. (2021, August 5). Turkey's forest fires fuel a political inferno online. *New York Times*, A-6.

Green, J. (2021). *The apocalypse reviewed: Essays on a human-centered planet*. Dutton.

Healy, J. (2021, September 7). In fight against Caldor fire, hotel workers are a backbone. *New York Times*, A-12.

Heating Up in Tokyo. (2021, July 27). *NASA World Observatory*. Retrieved July 29, 2021, from https://earthobservatory.nasa.gov/images/148650/fires-rage-in-turkey?src=eoa-iotd; https://earthobservatory.nasa.gov/images/148616/heating-up-in-tokyo?src=eoa-iotd.

Horowitz, J. (2021, December 12). Greek Island burns in a sign of crises to come. *New York Times*, A-1, A-6.

Hubbard, B. (2021, October 21). 24 in Syria are executed over series of wildfires. *New York Times*, A-9.

Hubler, S. (2021, July 30). The year summer came with dread. *New York Times*, A-10.

Hurricane Ida Batters Louisiana. (2021, August 30). *NASA Earth Observatory*, NASA. Retrieved September 1, 2021, from https://earthobservatory.nasa.gov/images/148767/hurricane-ida-batters-louisiana.

Innis, M. (2016, April 9). Climate-related death of coral around world alarms scientists. *New York Times*. Retrieved April 10, 2016, from http://www.nytimes.com/2016/04/10/world/asia/climate-related-death-of-coral-around-world-alarms-scientists.html.

Isai, V. (2021, August 11). Heat wave could worsen a devastating wildfire season in Canada. *New York Times*, A-6.

Isai, V., & Gundlock, B. (2021, August 5). Canadian farmers race to save cattle from drought. *New York Times*, A-4.

Jimenez, J. (2021, August 16). July was earth's hottest month on record, NOAA says. *New York Times*, A-15.

Jourbert, L. (2021, August). The edge of survival. *National Geographic*, 110–137.

King, R. (2021, September 1). Harmful algal blooms are impacting lake Eire. *Spectrum News*. Quoting NASA. Retrieved September 20, 2021, from https://spectrumnews1.com/oh/columbus/weather/2021/09/11/harmful-algal-blooms-impact-lake-erie.

Kintisch, E., & Stokstad, E. (2008). Ocean CO2 studies look beyond coral. *Science, 319*, 1029.

Kitsantonis, N. (2021, August 11). Fires ravage Greek towns as countries send support. *New York Times*, A-6.

Kolbert, E. (2021, November 15). Running out of time at the U.N. climate conference. *The New Yorker*, 21, 22.

Kwai, I. (2021, July 27). Between heat and floods, England endures extremes. *New York Times*, A-10.

Levin, R. (2001, December 6). Can California Tourism Survive climate change? *New York Times*, B-7.

McDonald, B. (2021, October 12). The large and costly battle to contain a megafire. *New York Times*, A-10.

Melley, B. (2021, September 17). *Fighting fire with fire to protect sequoia trees.* Associated Press. Retrieved September 18, 2021, from https://apnews.com/article/fires-environment-and-nature-forests-trees-wildfires-16c2303399942a4015d3a780d2eec4e3?utm_source=Sailthru&utm_medium=email&utm_campaign=Sept17_MorningWire&utm_term=Morning%20Wire%20 Subscribers.

Melley, B., & Metz, S. (2021, August 25). *Crews struggle to stop fire bearing down on Lake Tahoe.* Associated Press in the Seattle Times. Retrieved August 26, 2021, from https://www.seattletimes. com/nation-world/nation/utm.

Miller, R. W., Rodriguez, D., Randazzo, R., & Wilkins, T. (2012, August 31). Caldor fire jumps highway amid evacuation order; Scorches its way toward Pristine Lake Tahoe USA Today. Retrieved September 2, 2021, from https://www.usatoday.com/story/news/nation/2021/08/31/caldor-fire-lake-tahoe-evacuations-highway-89/5661531001.

Miller, S. (2012, August 26). Blaze approaches Lake Tahoe. *USA Today*, 2-A.

Murphy, P. (2012, September 2). *NBC Evening News.*

Myers, S. L. (2021, July 22). A somber toll as record rain swamps China. *New York Times*, A-1, A-6.

Myers, S. L., Bradsher, K., & Buckley, C. (2021, September 1). Extreme weather challenges city life in China. *New York Times*, A-1, A-8.

Nguyen, D. (2021, August 2). *Warming rivers in west killing fish* (p. A-6). Associated Press in Omaha World-Herald.

Nir, S. M. (2021, September 3). Cornered in rooms or cars by torrents of water, dozens are found dead. *New York Times*, A-12, A-13.

O'Connor, M. R. (2021, November 15). Towering infernos: What is it like to fight a megafire? *The New Yorker*, 30–36.

Olmos, S., Fountain, H., & Romero, S. (2021, July 23). Heat wave, poor policies, and erratic flames create a catastrophe. *New York Times*, A-11.

Pendell, R. (2021, December 17). *Climate action* (p. A-7). Public Pulse, Omaha World-Herald.

Piangigini, G. (2021, July 27). Residents evacuated as wildfires ravage Sardinia in 'Disaster Without Precedent.' *New York Times*, A-10.

Pianigiani, G. (2021, August 12). Sicily registers record-high temperature as heat wave sweeps [an] Italian Island. *New York Times*. Retrieved August 14, 2021, from https://www.nytimes.com/2021/08/12/ world/europe/sicily-record-high-temperature-119-degrees.html.

Plumer, B. (2021, October 13). Fossil-fuel use may peak soon, but perils of climate change still loom. *New York Times*, A-7.

Plumer, B., & Friedman, L. (2021a, November 5). Dozens of countries pledge to stop using coal power. *New York Times*, A-10.

Plumer, B., & Friedman, L. (2021b, November 13). Negotiators strike a climate deal, but world remains far from limiting warming. *New York Times*. Retrieved November 14, 2021, from https://www. nytimes.com/2021/11/13/climate/cop26-glasgow-climate-agreement.html?.

Pratt, S. (2021, January 4). Colorado faces winter urban firestorm. *NASA Earth Observatory*, 20–22.

Pyro Cumulous Clouds From Wildfires. (2021, July 20). *NASA Earth Observatory*. Retrieved July 22, 2021, from https://earthobservatory.nasa.gov/images/148678/an unusually-smoky-fire-season-in-sakha?orc=coa-iotd.

Research Shows More People Living in Floodplains. (2021, September 28). *NASA Earth Observatory*. Retrieved September 30, 2021, from https://earthobservatory.nasa.gov/images/148866/research-shows-more-people-living-in-floodplains.

Romero, S. (2021a, August 4). Saguaros like it hot. But maybe not quite this hot. *New York Times*, A-13.

Romero, S. (2021c, September 8). Paddling, skiing, hiking, and coughing. *New York Times*, A-16.

Santora, M. (2021, October 20). Africa's last remaining glaciers are melting away. *New York Times*, A-9.

Scales, H. (2021, November). An icy world in Meltdown. *National Geographic*, 101–121.

Schmall, E. (2021, October 21). Heavy rains kill dozens in India and Nepal. *New York Times*, A-10.

Sengupta, S. (2018, January 18). Warming, water crisis, then unrest: How Iran fits an alarming pattern. *New York Times*. Retrieved from https://www.nytimes.com/2018/01/18/climate/water-iran.html?

Sengupta, S., Friedman, L., & Plumer, B. (2021, November 12). Glasgow climate talks are down to the wire on money, ambition and fossil fuels. *New York Times*, A-1.

Smialek, J. (2021, October 8). Fed's Brainard signals focuses on climate change. *New York Times*, B-2.

Tabuchi, H. (2021a, September 2). Lack of power hinders assessment of toxic pollution. *New York Times*, A-14.

Tabuchi, H. (2021b, September 14). Cleaner jet fuels may help climate, but experts worry. *New York Times*, B-7.

Tarter, N. (2021, December 13). The age of megafires, the mail. *The New Yorker*, 3.

Troiasnovski, A. (2021, July 19). As Frozen Land Burns, Siberia Trembles: In Russia's far northeast, people take arctic temperatures in stride, but 100-degree days are another matter entirely. *New York Times*, A-6.

Tunkersley, J., Rogers, K., & Friedman, L. (2021, November 3). Nations cut deals to slash methane and save forests. *New York Times*, A-1, A-8.

Turner, N. (2022, February 17). *A surge of squid amid a heat wave*. Seattle Times in Omaha World-Herald, A-8.

Unusually Smoky Fire Season in Sakha. (2021, August 8). *NASA Earth Observatory*. Retrieved August 11, 2021, from https://earthobservatory.nasa.gov/images/148678/an-unusually-smoky-fire-season-in-sakha?src=eoa-iotd.

Vigdor, N., & Lukpat, A. (2021, October 25). 2 unusual storms converge and pummel California coast. *New York Times*, A-18.

Voiland, A. (2012, October 12). *A record-breaking year for fire in Sakha*. NASA Earth Observatory. Retrieved October 16, 2012, from https://earthobservatory.nasa.gov/images/148943/a-record-breaking-year-for-fire-in-sakha?src=eoa-iotd.

Wakabayashi, D., & Hsu, T. (2011, October 8). Google says it will ban false claims on warming. *New York Times*, B-2.

Weiss, K. R. (2018). Drying lakes: Warming climates, drought, an overuse are draining some of the world's biggest lakes. Threatening habitats and cultures. *National Geographic*, 108–131.

Wilson, S., Palley, S., & Brulliard, K. (2021, September 1). *Caldor fire continues to rage outside South Lake Tahoe*. Washington Post. Retrieved September 5, 2012, from https://www.washingtonpost.com/nation/2021/09/01/caldor-fire-south-lake tahoe/

Winfield, N., McHugh, D., & Ritter, K. (2021, November 1). *Leaders make mild pledges on climate* (pp. A-1, A-2). Associated Press in Omaha World-Herald.

Zalasiewicz, J., & Williams, M. (2016). *Ocean worlds: The story of seas on earth and other planets*. Oxford University Press.

Zhong, R. (2022, March 1). These climate scientists are fed up and ready to go on strike. *New York Times*. Retrieved March 1, 2020, from https://www.nytimes.com/2022/03/01/climate/ipcc-climate-scientists-strike.html.

References and Resources (NASA Earth Observatory)

Colorado State University. (2021, June 30). A pair of pyrocumulonimbus plumes in British Columbia.

Fromm, M., et al. (2010). The untold story of pyrocumulonimbus. *Bulletin of the American Meteorological Society, 91*(9).

Fromm, M., et al. (2000). Observations of boreal forest fire smoke in the stratosphere by POAM III, SAGE II, and lidar in 1998. *Geophysical Research Letters, 91*(9).

3

NASA. (2010, October 19). Fire-breathing storm systems.

NASA Earth Observatory. (2019, August 13). Flying through a fire cloud.

NASA Earth Observatory. (2014, August 6). Evolution of Pyrocumulus over California.

NASA Earth Observatory. (2010, August 31). Russian firestorm: Finding a fire cloud from space.

Peterson, D., et al. (2018). Wildfire-driven thunderstorms cause a volcano-like stratospheric injection of smoke. *Climate and Atmospheric Science, 1*(30).

Peterson, D., et al. (2021). Australia's Black Summer pyrocumulonimbus super outbreak reveals potential for increasingly extreme stratospheric smoke events. *Climate and Atmospheric Science, 4*(38).

U.S. Naval Research Laboratory. (2021, July 23). US Naval Research Laboratory PyroCB Press Conference, July 16, 2021.

U.S. Naval Research Laboratory. (2021, July 22). US Naval Research Laboratory PyroCB Press Conference, July 21, 2021.

U.S. Naval Research Laboratory. (2021). NRL PyroCb.

Wired. (2021, July 27). Oh good, now there's an outbreak of wildfire thunderclouds.

World Weather Attribution. (2021, July 7). Western North American extreme heat virtually impossible without human-caused climate change.

Further Reading

Albright, R., Caldeira, L., Hosfelt, J., Kwiatkowski, L., Maclaren, J. K., Mason, B. M., Nebuchina, Y., Ninokawa, A., Pongratz, J., Ricke. K. L., Rivlin, T., Schneider, K., Sesboüé, M., Shamberger, K., Silverman, J., Wolfe, K., & Zhu, K. (2016). Reversal of ocean acidification enhances net coral reef calcification. *Nature, 531*, 362–365.

Caldeira, K. (2016, February 24). Reversal of ocean acidification enhances net coral reef calcification. *Nature*, 530. Retrieved March 1, 2016, from http://www.nature.com/nature/journal/vaop/ncurrent/full/nature17155.html.

Einhorn, C. (2021, October 6). Coral reefs in decline worldwide, study finds. *New York Times*, A-11.

Fedorov, A. (2018). Permafrost-Landscape Map of the Republic of Sakha (Yakutia) on a Scale 1:1,500,000. *Geosciences, 8*(12), 465.

Fountain, H. (2017, December 7). In a warming California, a future of more fire. *New York Times*. Retrieved from https://www.nytimes.com/2017/12/07/climate/california-fires-warming.html.

Fountain, H. (2021, July 20). A wildfire so overwhelming that it controls the weather. *New York Times*, A-1, A-11.

Fountain, H. (2021, August 13). It rained at the summit of Greenland. The showers are another troubling sign of a changing arctic, which is warming faster than any other region on earth. That's never happened before. *New York Times*, A-8.

Hauser, C. (2021, December 16). 'Off the charts' storm rolls across Midwest. *New York Times*, A-21.

Kolbert, E. (2006, November 20). The darkening sea: What carbon emissions are doing to the oceans. *The New Yorker*, 66–75.

Murphy, H. (2021, August 6). System of currents is slowing, study finds. *New York Times*, A-5.

National Geographic. (2021, October 8). Siberia's massive wildfires are unlocking extreme carbon pollution.

NASA Earth Observatory. (2021, August 8). An unusually smoky fire season in Sakha.

Parfenova, E., et al. (2019). Assessing landscape potential for human sustainability and 'attractiveness' across Asian Russia in a warmer 21st century. *Environmental Research Letters, 14*(6), 065004.

Plumber, B., Zhong, R., & Friedman, L. (2022, February 28). Time is running out to avert harrowing future, climate panel warns. *New York Times*. Retrieved March 1, 2022, from https://www.nytimes.com/2022/02/28/climate/climate-change-ipcc-un-report.html?campaign_id=2&emc=edit_th_20220301&instance_id=54529&nl=todaysheadlines&regi_id=35795487&segment_id=8426.

Ponomarev, E. (2016). Wildfires dynamics in Siberian Larch Forests. *Forests, 7*(6), 125.

Royte, E. (2021, July). Too hot to live. *National Geographic*, 40–65.

Soja, A., et al. (2007). Climate-induced boreal forest change: Predictions versus current observations. *Global and Planetary Change, 56*(3–4), 274–296.

Soja, A., et al. (2020). Space-based observations for understanding changes in the arctic-boreal zone. *Reviews of Geophysics, 58*, e2019RG000652.

The Washington Post. (2021, August 10). Siberia endures 'nasty burning' amid worst fire season in decades.

Voiland, A. (2012, October 12) A record-breaking year for fire in Sakha. *NASA Earth Observatory*. Retrieved from https://earthobservatory.nasa.gov/images/148943/a-record-breaking-year-for-fire-in-sakha?src=eoa-iotd.

York, A., et al. (2021). Wildland fire in high northern latitudes.

4

Mother Earth vs. Mother Lode

A Native American Environmental Ethos

Contents

B. E. Johansen, *Global Warming and the Climate Crisis*,
https://doi.org/10.1007/978-3-031-12354-2_4

Learning Objectives

1. Students will learn two points of view toward exploitation of natural resources, one characterized as regarding Earth's resources as a source of material wealth and use for economic development (Mother Lode), and the others that view these resources as part of natural creation (Mother Earth), to be utilized only in context with natural values.

2. Students will study the critical remarks of Native American authors and philosophical figures such as Lakotas (Sioux) Standing Bear and Black Elk, along with Anglo-American attitudes toward lands and resources, such uranium, gold, and silver.

3. Students will study and discuss how each of these points of view impacts our relationships toward the Earth and, more particularly, to problems associated with climate change.

4. Why discuss Native American philosophy in a class on climate change? Students will discuss why such discussion is a necessary addendum to a fuller understanding of attitudes about our relationship to the land and living things, as well as well as Native Americans' central role in our culture as well.

5. Students will compare and contrast native and non-native points of view from Standing Bear: "Our altars were built on the ground and were altars of thankfulness and gratefulness. They were made of sacred earth and placed upon the holiest of all places—the lap of Mother Earth."

6. Students will appreciate that Native American philosophy often combines spiritual and environmental themes in ways that appeal to many non-Indian environmental activists today. A lively scholarly debate has flared regarding how Native Americans generally conceived the Earth.

7. Students will discuss statements such as the following: Discovery goes both ways in any encounter. So while Europe did not "discover" America, it *was* quite a discovery for Europe. For roughly three centuries before the American Revolution, the ideas that made it possible were being discovered, nurtured, and embellished in the growing English and French colonies of North America, as images of America became a staple of European literature and philosophy. America provided a counterpoint for European convention and assumption. America became, for Europe and Europeans in America, at once a dream and a reality, a fact and a fantasy, the real, and the ideal.

8. Students will discuss ideas that may be unfamiliar to them, such as Standing Bear's: "There is a great difference in the attitude taken by the Indian and the Caucasian toward nature, and this difference made of one a conservationist and the other a non-conservationist of life. The Indian, as well as other creatures that were given birth, were sustained by the common mother—Earth. He was therefore kin to all living things and he gave to all creatures equal rights with himself.... The... Caucasian.... Bestowing upon himself the position and title of a superior creature, others in the scheme were, in the natural order of things, of inferior position and title; and this attitude dominated his actions toward all

things. The worth and right to live were his, thus he heartlessly destroyed. Forests were mowed down, the buffalo exterminated, the beaver driven to extinction and his wonderfully constructed dams dynamited.... The white man has come to be the symbol of extinction for all things natural to this continent" (Standing Bear, 1978, 166).

4

Overview

A Native American hunter thanks the buffalo before taking its life to sustain his people's survival. Examples of a Native American environmental ethos are examined, such as the conflict over the Dakota oil pipeline, during which hundreds of mainly Native Americans with non-Indian support braved winter cold and water hoses and ultimately prevailed.

From these two poles of experience are derived two points of view toward an environmental ethos, quoting historical points of view from notable Native American authors and philosophical figures such as Chief Standing Bear and Black Elk. These two points of view influence how we define ourselves and our surroundings. This chapter provides a point and counterpoint regarding these two ways of perceiving America's living environmental philosophy often combines spiritual and environmental themes in ways that appeal to many non-Indian environmental activists today. A lively scholarly debate has flared regarding how Native Americans generally conceived of the Earth. Some ethnohistorians maintain that Native Americans possessed little or no environmental philosophy and that any attempt to assemble evidence to sustain a Native American ecological paradigm is doomed to failure because the entire argument is an exercise in wishful thinking by environmental activists seeking sentimental support for their own views.

On the other hand, appreciation is due such statements as Standing Bear's: "The Indian[s] were sustained by the common mother—Earth. He was therefore kin to all living things and he gave to all creatures equal rights with himself.... The... Caucasian.... Bestowing upon himself the position and title of a superior creature, others in the scheme were, in the natural order of things, of inferior position and this attitude dominated his actions toward all things. The worth and right to live were his, thus he heartlessly destroyed. Forests were mowed down, the buffalo exterminated, the beaver driven to extinction and his wonderfully constructed dams dynamited.... The white man has come to be the symbol of extinction for all things natural to this continent" (Standing Bear, 1978, 166).

On the *New York Times* editorial page, I read a piece about Frank Herbert's environmental themes in "Dune." "Native peoples were at the cutting edge of environmentalism in Mr. Herbert's day, and they still are [but]...the scale has enlarged. It's not only wilderness that needs defending, but also the delicate balance of gases in our shared atmosphere. Here, Indigenous activists have been indispensable." The author of the article, Daniel Immerwahr, a professor of history at Northwestern University, mentioned the enormous protests against the Dakota Access Pipeline that was canceled after many people realized that it would be used to carry slurry from tar sands mining from Alberta to the US Gulf of Mexico coastline to be refined into gasoline and other products.

Growing up on the Washington coast with the Quileute and Hoh peoples, Herbert saw how logging companies had ruined the Washington State coast between the Olympic mountains, the Pacific Ocean, and the Strait of Juan de Fuca. Native peoples told Herbert that "White men are eating the Earth…. They're going to turn this whole planet into a wasteland just like North Africa." After some hesitation, wrote Immerwahr, Herbert agreed with his Quileute and Hoh hosts. According to Herbert's son, Brian, Herbert said that "The world would become a 'big dune'" (Immerwahr, 2021, A-20). And so, the elders whom Herber used to construct his science fiction classic "Dune," about a sand-besieged planet with no water (that was made into a movie a few years after the book's initial publication in the early 1960s and remade in 2021). Non-Indians still hunger for an alternative to the rip-and-run ethic of human relationship with the Earth and its atmosphere.

Eye Catcher

Students will discuss ideas that may be unfamiliar to them, such as Standing Bear's: "There is a great difference in the attitude taken by the Indian and the Caucasian toward nature, and this difference made of one a conservationist and the other a non-conservationist of life. The Indian, as well as other creatures that were given birth, were sustained by the common mother—Earth."

How We Define Ourselves, and Our Surroundings

We will not emerge from a future of a malign climate crisis until we not only change our technology, as necessary as that is, but also until we reform how we think about the Earth, redefining our relationship to the nature, sphere, and all the plants and animals with whom we share it. We also need to recompose how we define ourselves and our surroundings. New technology alone will not do this for us. Come along, please, and explore some climate science leavened with Native American Studies and papal theology.

4.1 Native American Philosophy: Spiritual and Environmental Themes Combined

4.1.1 Native American Examples of Earth as Mother

Native American philosophy often combines spiritual and environmental themes in ways that appeal to many non-Indian environmental activists today. A lively scholarly debate has flared regarding how Native Americans generally conceive the Earth, past and present. Some ethnohistorians maintain that Native Americans possessed little or no environmental philosophy and that any attempt to assemble evidence to sustain a Native American ecological paradigm is doomed to failure because the entire argument is an exercise in wishful thinking by environmental activists seeking sentimental support for their own views.

Assertions that a Native American concept of Mother Earth used no concept of "Mother Earth" before immigrant peoples fantasized it in their own name during the twentieth-century environmental movement misses an astonishing amount of the historical record, not to mention Native American oral histories. References to "Mother Earth" in Native American cosmology are not scarce. They are abundant, according to George Cornell: "Native peoples almost universally view the Earth as a feminine figure. The Mother provides for the sustenance and well-being of her children: it is from her that all subsistence is drawn. The relationship of native peoples to the Earth, their Mother, is a sacred bond with the creation."

Those who dismiss a native ecological ethic as an invention of modern-day hippies and pan-Indianists are missing something much deeper than mere mentions of "Mother Earth" in nineteenth-century primary sources. They are missing the fundamental nature of many Native American traditions, the terms in which native thought conceptualizes the land and the life it nurtures. To Roger Dunsmore (1997) Western thought creates hierarchies and categories that do not exist in Native American thought. The very cognitive map for conceptualizing life is different, as illustrated in recent time by the example of "…A Wasco Indian logger (a faller), who quit logging and sold his chainsaw because he couldn't stand hearing the trees scream as he cut into them." This is a worldview in which "[T]he whole world is perceived and valued. Even the flies." A sense of a web of life connecting all things framed the cognitive map of Chief Joseph when he said, "The Earth and I are of one mind."

References to indigenous affection for nature permeated the thoughts of the Lakota (Sioux) Luther Standing Bear, who wrote: "The Lakota was a true naturalist, a lover of Nature. He loved the earth and all things of the Earth, the attachment growing with age. The old people came literally to love the soil and they sat or reclined on the ground with a feeling of being close to a mothering power…. In talking to children, the old Lakota would place a hand on the ground and explain: 'We sit in the lap of our mother. From her we, and all other living things, come. We shall soon pass, but the place where we now rest will last forever….' Our altars were built on the ground and were altars of thankfulness and gratefulness. They were made of sacred earth and placed upon the holiest of all places—the lap of Mother Earth" (Standing Bear, 1978, 192, 194, 200) (◘ Figs. 4.1 and 4.2).

Like Black Elk, Tecumseh, Black Hawk, and others, Standing Bear invoked the image of Mother Earth in his writing. Luther Standing Bear's use of the Mother Earth image is particularly striking when placed next to similar language used by Black Elk which has been passed to us through accounts by Neihardt, Epes Brown,

◘ **Fig. 4.1** Offering at the waterfall. See the text depicting Standing Bear's concept of Mother Earth from late nineteenth century (Standing Bear, 1978, 192, 194, 200). (Source: National Museum of the American Indian)

◘ **Fig. 4.2** Offering to the sun. San Ildefonso. See the text depicting Standing Bear's concept of Mother Earth from late nineteenth century (Standing Bear, 1978, 192, 194, 200). (Source: National Museum of the American Indian, Smithsonian, 1927)

and others. Unlike accounts attributed to Black Elk, Tecumseh, and Sea'th'l, however, the use of the image by Standing Bear raises no questions of interpretation, because he wrote in English acquired at the Carlisle Indian School.

"There is a great difference in the attitude taken by the Indian and the Caucasian toward nature," Standing Bear wrote, "and this difference made of one a conservationist and the other a non-conservationist of life. The Indian, as well as other creatures that were given birth, were sustained by the common mother—Earth. He was therefore kin to all living things and he gave to all creatures equal rights with himself.... The...Caucasian.... Bestowing upon himself the position and title of a superior creature, others in the scheme were, in the natural order of things, of inferior position and title; and this attitude dominated his actions toward all things. The worth and right to live were his, thus he heartlessly destroyed. Forests were mowed down, the buffalo exterminated, the beaver driven to extinction and his wonderfully constructed dams dynamited.... The white man has come to be the symbol of extinction for all things natural to this continent" (Standing Bear, 1978, 166).

Standing Bear also was a critic of European-American society generally in words appreciating nature and his forefathers and foremothers that recall those of Sea'th'l, and many others named above. He also evoked the same sacred tree of life that was familiar to Black Elk: "The white man does not understand the Indian for the reason that he does not understand America. He is too far removed from its formative processes. The roots of the tree of his life have not yet grasped the rock and soil. The white man is still troubled with primitive fears; he still has in his consciousness the perils of this frontier continent.... He shudders still with the memory of the loss of his forefathers upon its scorching deserts and forbidding mountain-tops. The man from Europe is still a foreigner and an alien. And he still hates the man who questioned his path across the continent" (Armstrong, xi–xii).

Daniel Wildcat, a Yuchi and a member of the Muscogee Nation of Oklahoma and a professor at Haskell Indian Nations University, Lawrence, Kansas, speaking at the University of Colorado Boulder's Center of the American West on September 29, 2011, said that, until recently, "Native people were members of tribes, not nation-states, with a relationship to nature that defined species on which people depended for survival in familial terms, as relatives, not as exploitable resources. The bison on the Great Plains, salmon among the Coast Salish, and corn across much of today's North America (Turtle Island) were 'the central relative we acknowledged'" (Berry, 2011).

Indigenous environmental activism (including visceral opposition to development of extractive industry) stems from long historical experience with resource colonization, which works in synthesis with an ethos that invests animus in everything natural (◘ Figs. 4.3 and 4.4).

❏ Fig. 4.3 Portrait of Standing Bear. While European religions often restrict their blessings to humanity alone, Native Americans interpret "all my relations" to mean all of nature—animate (in European conception) and not. The rocks under our feet are regarded as animated. This respect for nature is fundamental and enduring and at the root of traditional Native American responses to economic development. Definitions of "balance" are couched in this context, counterpoising protection of "Mother Earth" with an immigrants' ethos that seeks a "Mother Lode," rooted in Genesis 1:28 ("Be fruitful, multiply, fill the Earth, and subdue it"). Naomi Klein, in "This Changes Everything: Capitalism and the Climate" (2014:183), writes of industrialists who "view nature as a bottomless vending machine". (Source: National Museum of the American Indian, Smithsonian)

4.1.2 "Welcome to My Moonscape"

Tar Sands Mining as Violation of Mother Earth Relationship

One of the greatest contrasts between Mother Earth and Mother Lode thinking is the opening, in recent years, of tar sands mining in Alberta, Canada. Exploitation of tar sands at the surface requires a form of strip mining that scars the Earth in ways that will not be quickly nor easily repaired. It's the last thing the Earth needs when carbon dioxide levels in the atmosphere already have risen to more than 420 parts per million, more than 40% above peak preindustrial levels, with damage to climate, as well as rising seas and oceanic acidity.

❏ Fig. 4.4   Portrait of Standing Bear. While European religions often restrict their blessings to humanity alone, Native Americans interpret "all my relations" to mean all of nature—animate (in European conception) and not. The rocks under our feet are regarded as animated. This respect for nature is fundamental and enduring and at the root of traditional Native American responses to economic development. Definitions of "balance" are couched in this context, counterpoising protection of "Mother Earth" with an immigrants' ethos that seeks a "Mother Lode," rooted in Genesis 1:28 ("Be fruitful, multiply, fill the Earth, and subdue it"). Naomi Klein, in "This Changes Everything: Capitalism and the Climate" (2014:183), writes of industrialists who "view nature as a bottomless vending machine." (Source: National Museum of the American Indian, Smithsonian)

On a local level, many people worry about oil spills in fragile areas such as the Nebraska Sand Hills that could contaminate the Ogallala aquifer in an area where water is scarce. This aquifer supplies 78% of public water and 83% of irrigation water in Nebraska, almost a third of the irrigation water used in the United States. One look at the accompanying photograph will be enough to make the point that this new source of fossil fuels ruins everything it touches, replacing the homes of ancient forests and Indigenous peoples with a toxic wasteland.

4.1.3   A Native American Ecological Ethos Shapes Non-Native Thinking

Awareness of economic development's costs also animates the non-Indian environmental movement. "This, then, is the nemesis that modern Western man, together with his imitators, has brought upon himself by following the directive given in the first book of *Genesis*," wrote the great English historian Arnold Toynbee. "That directive has turned out to be bad advice, and we are beginning, wisely, to recoil from it" (Toynbee, 1973, n.p.).

The development of an environmental ethos in modern times reflects native wisdom and values, as well as recognition that traditional capitalism, practiced absolutely, is a suicide pact, as Edward Abbey said decades ago, "the ideology of the cancer cell," that is, random, all-consuming, and eventually lethal to its host. I will trace the development of this ethos and its adaptation in non-native society as necessary for human survival. My primary exhibit of this adaptation will be increasing awareness of global warming and its intensifying perils for coming generations. Specifically, I will recognize the geophysical principle of thermal inertia, which describes how the system delivers evidence of temperature rise 50 years (in the air) and about 150 years (in the oceans) after the carbon dioxide emissions that cause them. The geophysical system thus requires that our industrial and diplomatic systems respond according to the needs of the seventh generation, an accordance with native traditional ecological ethos.

The idea that Native peoples have something valuable to teach majority society has become quite popular. When I proposed to trace an eighteenth-century tutorial by the Haudenosaunee (Iroquois) for Benjamin Franklin about 50 years ago, most of my PhD dissertation supervisors thought I was, if not crazy, at least out of paradigm. Four decades after I started my dissertation journey, during mid-December, 2015, I woke up in South India, having been invited to give a plenary address: "What Has Been Will Be: Native American Contributions to Democracy, Feminism, Gender Fluidity, and Environmentalism" at a global seminar on "Celebrating the Ancient/Contemporary Wisdom of Fourth World" hosted by the Department of English, Acharya Nagarjuna University, Guntur, India.

I was arguing that the "history of westward movement," as it was parsed at that time, was much more nuanced than that. Discovery goes both ways in any encounter. So while Europe did not "discover" America, it *was quite a discovery* for Europe. For roughly three centuries before the American Revolution, the ideas that made it possible were being discovered, nurtured, and embellished in the growing English and French colonies of North America, as images of America became a staple of European literature and philosophy. America provided a counterpoint for European convention and assumption. America became, for Europe and Europeans in America, at once a dream and a reality, a fact and a fantasy, the real, and the ideal.

A detailed case has now been made that native consensual democracy helped shape the thoughts of some of the US founders. Decades after that, the founding mothers of American feminism learned from native matrilineal cultures. Today (witness the recent recognition of same-sex marriage by the US Supreme Court) indigenous American acceptance of gender fluidity has become accepted as well. Environmental points of view that all things are connected or restricted to the human race, especially the male gender—"All men are created equal and endowed by their creator with certain inalienable rights"—also have entered mainstream thought, but only after the perils of industrial pollution and alteration of the atmosphere (climate change) have become manifest. Even today, many Native American homelands remain resource colonies in the United States. One-third of the country's "Superfund" sites (the worst pollution locations) are on American Indian reservations. Human domination of the Earth has been so thorough that even Pope Francis has called into question humanity's subjugation of the Earth commanded by *The Bible's* book of *Genesis*.

4

4.1.4 Environmental Ethos: Mother Earth or Mother Lode?

Native American philosophy often combines spiritual and environmental themes in ways that appeal to many non-Indian environmental activists today. A lively scholarly debate has flared regarding how Native Americans generally conceived of the Earth. Some ethnohistorians maintain that Native Americans possessed little or no environmental philosophy and that any attempt to assemble evidence to sustain a Native American ecological paradigm is doomed to failure because the entire argument is an exercise in wishful thinking by environmental activists seeking sentimental support for their own views. William A. Starna, professor of anthropology at the State University of New York (Oneonta), called the argument that Native Americans had an environmental ethic part of a recently concocted pan-Indian mythology. As he does in the face of evidence that the Iroquois helped inspire democracy, Starna asserts that modern Indian activists created the idea of native environmentalism out of a simplistic desire to emulate Whites.

One wishes to be intellectually charitable and not to unfairly alienate Starna and others who share their beliefs about the recent, non-Indian genesis of the "Mother Earth" image. However, their assertions that Native Americans had no concept of Mother Earth before immigrant peoples fantasized it in their name during the twentieth-century environmental movement misses an astonishing amount of the historical record, not to mention Native American oral histories. References to "Mother Earth" in Native American cosmologies are not scarce. They are abundant, according to George Cornell: "Native peoples almost universally view the earth as a feminine figure. The Mother provides for the sustenance and well-being of her children: it is from her that all subsistence is drawn. The relationship of native peoples to the Earth, their Mother, is a sacred bond with the creation" (Cornell, 1990, 3).

Many native cosmologies conceive of the sky (including the sun) as a masculine counterpart to Mother Earth, as a loving couple who are sometimes prone to many of the failings of human relationships between men and women. While the Christian *Bible* commands subordination of the Earth (in *Genesis*), humankind in many native cosmologies places human beings in a web of interdependent relationships with all facets of the creation. In this web, all things are animate, even objects, such as the pebbles under one's feet, which European languages characterize as lifeless. In the web of Native American experience, the landscape of life envelops *all* Whites such as Starna seems to recoil at the thought that ways of thinking exist other than his own. Some European-Americans recoil at the idea that no way of thinking except his vans have been hearing Native Americans characterize the Earth as mother since shortly after the Mayflower landed. Massasoit, who invited the Pilgrims to the first intercultural Thanksgiving dinner, faced European ideas of land tenure with a few questions of his own: "What is this you call property? It cannot be the Earth, for the land is our mother, nourishing all her children, beasts, birds, fish, and all men. The woods, the streams, everything on it belongs to everybody and is for the use of all. How can one man say it belongs only to him?" (Weaver, 1996, 10).

4.1.5 Tracing Earth as Mother Through Languages

Sometimes a belief in Earth as mother is reflected in Native American languages. In the Algonkian/Ojibwe language, for example, "The words for Earth and the vagina, respectively, *aki* and *akitun*, 'share the same root'" (Paper, 1990, 14). Each Native people in the Americas has its own origin story, but many share common elements. The characterization of the Earth in the feminine, using kin terminology, is one of these Native American perspectives on the environment that were often virtually opposite those of many early immigrants, who sought to "tame" the "wilderness." Most Many Native Americans saw themselves as enmeshed in a web of mutually complementary life. As Black Elk said: With all beings and all things, we shall be as relatives.

References to indigenous affection for nature permeated the thoughts of Luther Standing Bear, who wrote: "The Lakota was a true naturalist—a lover of Nature. He loved the Earth and all things of the Earth, the attachment growing with age. The old people came literally to love the soil and they sat or reclined on the ground with a feeling of being close to a mothering power…. In talking to children, the old Lakota would place a hand on the ground and explain: 'We sit in the lap of our mother. From her, we, and all other living things, come. We shall soon pass, but the place where we now rest will last forever….' Our altars were built on the ground and were altars of thankfulness and gratefulness. They were made of sacred Earth and placed upon the holiest of all places—the lap of Mother Earth" (Standing Bear, 1978, 192, 194, 200).

4.1.6 Everything Is Regarded as Having Personality

Standing Bear defined his people's relationship to everything else on Earth, writing that in the native view everything is animate, as noted above, but also "possessed of personality." He compared the world to a library, with "the stones, leaves, grass, brooks…birds, and animals as its books" (Hughes, 1983, 80). Many times, wrote Standing Bear, the Indian is embarrassed and baffled by the white man's alienation from nature, as reflected in allusions to nature in such terms as "crude, primitive, wild, rude, untamed, and savage" [retain "savage"] (Standing Bear, 1978, 196). To Standing Bear, many Whites imagined Native Americans as savages to "salve… [their] sore and troubled conscience[s] now hardened through the habitual practice of injustice" (Standing Bear, 1978, 251).

Like Black Elk, Tecumseh, Black Hawk, and others, Standing Bear invoked the image of Mother Earth in his writing and speaking. Luther Standing Bear's use of the Mother Earth image is particularly striking when placed next to similar language used by Black Elk which has been passed to us through accounts by Neihardt, Epes Brown, and others. Unlike accounts attributed to Tecumseh and Sea'th'l, however, the use of the image by Standing Bear raises no questions of interpretation, because he wrote in English acquired at the Carlisle Indian School.

4

As Standing Bear said, "There is a great difference in the attitude taken by the Indian and the Caucasian toward nature, and this difference made of one a conservationist and the other a non-conservationist of life. The Indians, as well as other creatures that were given birth, were sustained by the common mother— Earth. He was therefore kin to all living things and he gave to all creatures equal rights with himself…. The…Caucasian…. Bestowing upon himself the position and title of a superior creature, others in the scheme were, in the natural order of things, of inferior position and title; and this attitude dominated his actions toward all things. The worth and right to live were his, thus he heartlessly destroyed. Forests were mowed down, the buffalo exterminated, the beaver driven to extinction and his wonderfully constructed dams dynamited…. The white man has come to be the symbol of extinction for all things natural to this continent" (Standing Bear, 1978, 166).

Standing Bear also was a critic of European-American society generally, in words appreciating nature and his forefathers and foremothers that recall those of Sea'th'l (Seattle), half the width of a continent away. Standing Bear also evoked the same sacred tree of life that was familiar to Black Elk: "The white man does not understand the Indian for the reason that he does not understand America. He is too far removed from its formative processes. The roots of the tree of his life have not yet grasped the rock and soil. The white man is still troubled with primitive fears; he still has in his consciousness the perils of this frontier continent…. He shudders still with the memory of the loss of his forefathers upon its scorching deserts and forbidding mountain-tops. The man from Europe is still a foreigner and an alien. And he still hates the man who questioned his path across the continent" (Armstrong, xi–xii).

4.1.7 An Environmental Ethos: Through the Eyes of Standing Bear

In addition to being rooted in their homelands, Native Americans maintain historical, spiritual bonds to the land that fosters attention to environmental threats. Longtime fishing-rights activist Billy Frank, Jr. said that this connection places protection of environment and its role in sustaining human and all other life "at the top of our priority list" (Russo et al., 2013, 235). "When we say the Okanagan [Native] word for 'ourselves,'" said Jeanette Armstrong, "We are actually saying the ones [that] are 'dream' and 'land' together" (Grossman & Parker, 2012, 38). Chief Willie Charlie has said: "Mother Earth is crying, and we need to pay attention to what she is saying" (Grossman & Parker, 2012, 45–46).

Standing Bear, who watched large-scale Anglo-American immigration change the face of the Great Plains, contrasted European-American and Native American conceptions of the natural world of North America: "We did not think of the great open plains, the beautiful rolling hills, and winding streams with tangled brush, as 'wild.' Only to the white man was nature 'a wilderness' and only to him was the land 'infested' with 'wild' animals and 'savage' people. To us it was tame. Earth was bountiful, and we are surrounded with the blessings of the Great Mystery. Not until the hairy man from the east came and with brutal frenzy heaped injustices upon us and the families we loved was it 'wild' for us. When the very animals of the forest began fleeing from his approach, then it was for us that the 'Wild West' began" (Standing Bear, 1978, 38)

Standing Bear was a severe critic of the Whites' attitudes toward nature. He said he knew of no species of plant, bird, or animal that had been exterminated in America until the coming of the white man. For some years after the buffalo disappeared, there still remained huge herds of antelope, but the hunter's work was no sooner done in the destruction of the buffalo than his attention was attracted toward the deer. "The white man considered natural animal life just as he did natural [Native American] life upon this continent, as 'pests,'" wrote Standing Bear. "Plants which the Indian found beneficial were also 'pests.' There is no word in the Lakota vocabulary with the English meaning of this word" (Standing Bear, 1978, 165).

Activist and author Kurt Russo, who has worked with the Native American Land Conservancy, commented: "Our courage to acknowledge this crisis, our conviction to stand up for unborn generations, our connection to nature and, through nature, to each other, and our resilience—as a family, as a species, as peoples—will determine whether we hear her cry for the agony of extinction or stand idly by and bear witness to a great dying" (Russo et al., 2013, 236).

"Indigenous communities are, in general, in a unique position given their history and knowledge to understand and respond to the crisis," commented Russo. "[They] are, in general, more informed and engaged than the majority of Americans or their political and corporate leaders.... As place-based communities of inter-related families with historical consciousness, indigenous peoples are also more resilient, and thus able to face, rather than deflect or deny, the true magnitude of the crisis [that] will have long-lasting and potentially catastrophic consequences for every life-form and human community" (Russo et al., 2013, 234).

4.1.8 Native Wisdom's Value for Today

Battles between Mother Earth and Mother Lode philosophies are not solely an historical exercise or a library study. They are occasionally used on the evening news, in the *New York Times*, and elsewhere in our daily lives. Consider proposals for the Pebble Mine, one of the world's largest gold and copper strip mines versus one of the planet's most prominent sockeye salmon runs. This battle is taking place in and near Bristol Bay, Alaska, about 200 miles from Anchorage, with a switch in federal government commitment every time the president comes from a different party. President Obama opposed the mine, after which President Trump loved the idea of golden waste ponds that he insisted would not harm the fish. President Joe Biden then turned course again, on behalf of the fish.

To hear the mining companies tell it, there will be no problems for the migrating salmon if the strip mine is built at Pebble Mine. Never mind the surging rivers of toxic mine waste close to the salmon streams. The fishermen whose families are fed by the fish catch objected. In September, 2020, lawyers for President Biden filed suit

4

with what was said will be permanent protections for the fish. We shall see how permanent these are the next time someone who visualizes the Earth as vending machine (such as Donald Trump) comes into power in the White House.

Embracing the Mother Lode way of doing things in this case involves a mine more than a mile square and one-third of a mile deep, with an estimated total yield of at least $300 billion worth of gold and copper. What will be lost, according to the mine's opponents, is habitat for half of the world's sockeye salmon, along with associated streams and other water sources. Plans also anticipate construction of a 270-mW power plant and 82 miles of new roads, plus large ponds for dumping of mining "tailings," (waste) produced by separating gold and copper from unsalable rock, some of which is toxic. Thus is a natural habitat turned into an industrial wasteland. The proponents of all of this says that its effects will be no big deal for its $300 billion return in gold and copper, including 130 miles of streams, more than 2800 acres of wetlands, and 130 acres of open water. US President Joe Biden's succinct evaluation of the whole plan is: "This is no place for a mine." A great majority of native and non-native fishermen agree with him.

This is basic survival behavior for a sustainable world and, it goes without saying, one very important example of how indigenous ideas inform the thoughts and actions of many people, indigenous and not. Chief Standing Bear said a century and a half ago: Let us put our minds together and see what life we can make for our children.

Native wisdom (traditional ecological knowledge in this case) reflects what scientists refer to as the laws of nature, illustrated here by the behavior of thermal inertia in climate change science, which is discussed elsewhere in this book. It's very important to understand that planning for the seventh generation reflects the same kind of thinking required to solve global climate problems. The world must anticipate effects *before* they become obvious. That is what ties the "Mother Earth" philosophy together with climate science.

4.1.9 Pope Francis' Sense of Environmentalism

The keepers of religious doctrine have realized just how effective humankind has been at subduing the Earth and just how damaging that subjugation has become. During the summer of 2015, Pope Francis, who has become well-known for directly tackling many controversial issues, made climate change a Vatican priority by issuing an encyclical (essentially a policy statement) detailing how the burdens of global warming worldwide fall disproportionally on the poor. Indigenous peoples around the world bear a disproportionate burden of environmental damage. In the United States today, Native peoples often live on ruined, exhausted land, suffering toxic consequences. Fully one-third of the Superfund sites declared by the US Environmental Protection Agency are on Native American lands.

Let us ask the tough questions. When it comes to sustainability, what *really* works? What *really* matters? In the long run, can a capitalistic system change its character to embrace standards of performance not predicated on growth, ones that improve the quality of life rather than sheer production, and ones that improve the quality of life rather than sheer production solely in the service of profit?

Can a system predicated on growth adapt to a sustainable world in which having less "stuff" will be preferable? How can we adjust our desires to fit a new world in which more is not always better? Will our basic values change along with our energy sources?

A major—perhaps *the* major question facing an Earth and its human denizens in a time of worldwide environmental crisis is: can capitalism change its character? A sustainable environment can make good business. Witness the growth of alternative forms of energy. Can capitalism factor respect for the Earth that sustains us all into its calculus of development? If so, it may be a positive force in a new, sustainable world. If not—if it retains attributes of the cancer cell—then ultimately, our progeny will inherit an exhausted, poisoned world.

Can capitalism, with its appetite for pell-mell (and often environmentally destructive) growth, survive in a new world in which geophysical reality demands that we restrain our demands upon the Earth? Are we ready to operate with an accounting system that brings everyone to account for the toll that our activities exact on the Earth and its atmosphere? Can we fashion a system in which polluting the atmospheric commons is defined as a criminal act for which sizable fines are levied and people serve time in prison? Such a system would redefine some present-day free choices (e.g., to trash the commons) as illegal acts. At that point, with indigenous advice, we will be truly respecting Mother Earth.

Because thermal inertia serves us with the impact of today's carbon dioxide emissions 50–150 years in the future, the geophysical system *requires* that we heed traditional ecological knowledge to anticipate the effects of our actions on the seventh generation. To do otherwise virtually guarantees that future generations will inherit a scorched, desolate world. The survival of human peoples, as well as the plants and animals upon which we depend, *requires* planning seven generations hence. This is *not* an optional luxury.

Let us explore the ideas that combine matters of the spirit with economy, for both the spirit and the body require nourishment. These ideas will provide sustenance to all peoples, and they will be provided it in an absence of war. War kills spirit and consumes resources that peoples need to live. War must be made obsolete. The other option is annihilation of our habitat and each other. These ideas are based in in beliefs that that will make tomorrow work. The stakes have never been so high, nor time so short. Solving the problems that face humankind cannot be solved by quibbling. Exercise of nationalistic pride condemns all of to a desiccated home planet.

Even as surviving Indigenous peoples and their cultures have been pulverized by the industrial machine—and even as the ecologically unsustainable nature of this ferocious juggernaut becomes more obvious—the colonizers (who flatter themselves with the descriptor "first world") continue to learn, absorb, and change, even in the act of conquest. The important question eventually may become whether these dominant forces can change fundamentally enough, and quickly enough, to avoid a climatic apocalypse.

4.2 Summary

4.2.1 Can Capitalism Change Its Character?

James E. Hansen, longtime director of NASA's Goddard Institute for Space Studies and the first person to discuss global warming in a scientific context, observed natural limits as early as 1981. "The bottom line is this: business-as-usual, if it continues for even another decade, will be disastrous for the planet. We can have a stable climate, clean air, and an unpolluted ocean. And clean energies yield good jobs. It is up to the public to make sure that we get onto a path that stabilizes climate and allows all the creatures of Creation to continue to thrive on this planet" (Johansen, 2007, 7). The effects of climate change are not theoretical, and they are not speculative problems that can be handed off to future generations. Economic activity around the world, as well as the lives of animals and plants, are being affected today by rising temperatures. This is not merely a matter of a few degrees on the thermometer but of alterations in the environment that sustains all of us.

Greenhouse gases have no morals, loyalty, nor party affiliation. Carbon dioxide is not having a debate with us. It merely retains heat. Thus, in 50 years, when our children are grandparents, the planetary emergency in which we are now tasting the first course will be a dominant theme in everyone's life, unless we act now. Within a decade or two, thermal inertia will take off on its own, portending a hot, miserable future for coming generations.

A major—perhaps *the* major question facing an Earth and its human denizens in a time of worldwide environmental crisis is: can capitalism change its character? A sustainable environment can make good business. Witness the growth of alternative forms of energy. Can capitalism factor respect for the Earth that sustains us all into its calculus of development? If so, it may be a positive force in a new, sustainable world. If not—if it retains attributes of the cancer cell—then ultimately, our progeny will inherit an exhausted, poisoned world.

Are we ready to operate with an accounting system that brings us all to account for the toll that our activities exact on the Earth and its atmosphere? Can we fashion a system in which polluting the atmospheric commons is defined as a criminal act for which sizable fines are levied? Such a system would redefine some present-day free choices (e.g., to trash the commons) as illegal acts. At that point, with indigenous advice, we will be truly respecting Mother Earth.

Because thermal inertia provides us with an estimate of today's carbon dioxide emissions 50–150 years in the future, the geophysical system *requires* that we heed traditional ecological knowledge to anticipate the effects of our actions on the seventh generation. To do otherwise virtually guarantees that future generations will inherit a scorched, desolate world. The survival of human peoples, as well as the plants and animals upon which we depend, *requires* planning seven generations hence. This is *not* an optional luxury.

This is basic survival behavior for a sustainable world and, it goes without saying, one very important example of how indigenous ideas inform the thoughts and actions of everyone. Native wisdom (traditional ecological knowledge in this case) reflects what scientists refer to as the laws of nature, illustrated here by the behavior of thermal inertia. It's very important to understand that planning for the seventh generation reflects the same kind of thinking required to solve the global climate problem. The world must anticipate effects before they seem obvious. That is what ties the "Mother Earth" philosophy together with climate science.

Regarding the famous passage in Genesis 1:28, the text of the Hebrew *Bible* (the most basic text for biblical scholars) does not justify absolute dominion of humankind over nature (as, unfortunately, it has been often interpreted since the seventeenth century); it emphasizes stewardship. See, for instance, the official documentation of the European Ecumenical Assembly in May 1989 (Basel, Switzerland): "We are to be stewards in God's world. Stewardship is not ownership. God the creator remains the sole owner, in the full sense of the term, of the entire creation" (p. 42).

4.2.2 Pope Francis' Words: *"A New Sense of the Human Family"*

Pride in language and culture is salubrious to individuals' sense of history and identity. Excess nationalism that prevents international cooperation to solve harmful worldwide maladies is quite another thing. As Pope Francis has pointed out: For all of our connectivity due to expansion of social media, ability to communicate can breed contempt as well as mutual trust. For all our hyper-connectivity, Francis has said, we witness a fragmentation that has made it more difficult to resolve problems that affect us all. The Pope's encyclical, titled "Brothers All," also said: "The forces of myopic, extremist, resentful, and aggressive nationalism are on the rise." The Pope's document also advocates support for migrants, as well as resistance to nationalist and tribal populism.

Francis broadened his critique to the role of market capitalism, as well as nationalism has failed the peoples of the world when they need cooperation and solidarity in the face of the worldwide corona virus pandemic. Humankind needs to unite into "a new sense of the human family [Fratelli Tutti, "Brothers All"], that rejects war at all costs" (Pope, 2020, 6-A).

Results
Nationalism will pose problems for the sustainability of the Earth as our little blue and green orb becomes more crowded over time. The old ways, in which nationalistic arguments often end in devastating wars, are obsolete, given that the Earth and all the people, plants, and other animals that it sustains are faced with the existential threat of a climate crisis that within two centuries, more or less, will flood large parts of coastal cities and endanger many species of plants and animals. To survive, we must listen to the Earth and observe her travails, because they are increasingly our own.

Let us explore the ideas that will make tomorrow work.

4

4.3 Questions and Exercises

 1. Frank Herbert's environmental themes in "Dune" pose some interesting issues. "Native peoples were at the cutting edge of environmentalism in Mr. Herbert's day, and they still are [but]…the scale has enlarged. It's not only wilderness that needs defending, but also the delicate balance of gases in our shared atmosphere. Here, Indigenous activists have been indispensable." **For those who have read "Dune" or seen the movie: did you notice the environmental or Native American themes?**

   Did you notice the strong Native American presence in the enormous protests against the Dakota Access Pipeline that was canceled after many people realized that it would be used to carry slurry from tar sands mining from Alberta to the US Gulf of Mexico coastline to be refined into gasoline and other products? If Herbert was alive today, he might have recalled that, growing up on the Washington State coast with the Quileute and Hoh peoples, he saw how logging companies had ruined the Washington State coast between the Olympic mountains, the Pacific Ocean, and the Strait of Juan de Fuca. Native peoples told Herbert that "White men are eating the Earth. They're going to turn this whole planet into a wasteland…." Reflect, please, on the different cultures compared in the title of this chapter.

2. **Native American philosophy often combines spiritual and environmental themes in ways that appeal to many non-Indian environmental activists today.** A lively scholarly debate has flared regarding how Native Americans generally conceived the Earth. Some ethnohistorians maintain that Native Americans possessed little or no environmental philosophy and that any attempt to assemble evidence to sustain a Native American ecological paradigm is doomed to failure because the entire argument is an exercise in wishful thinking by environmental activists seeking sentimental support for their own views.

   Reflect on what you know, or who you know, to begin a discussion of various ways to regard environmental issues. Bear in mind that such points of view are not *a preserve of any single ethnic group. However, lessons of history and historical perception inform everyone. Comments?*

3. Furthermore, those who dismiss a native ecological ethic as an invention of modern-day hippies and pan-Indianists are missing something much deeper than mere mentions of "Mother Earth" in nineteenth-century primary sources. They are missing the fundamental nature of many Native American traditions, the terms in which native thought conceptualizes the land and the life it nurtures. *Further comments? Without engaging in unfair stereotypes, discuss environmental themes that affect everyone who shares the Earth.*

4. Western thought creates hierarchies and categories that do not exist in Native American thought. The very cognitive map for conceptualizing life is different, as illustrated in recent time by the example of "…A Wasco Indian logger (a faller), who quit logging and sold his chainsaw because he couldn't stand hearing the trees scream as he cut into them" (Dunsmore, 1997, 7). *See below for an expanded discussion of this theme.*

5. Western thought creates hierarchies and categories that do not exist in Native American thought. The very cognitive map for conceptualizing life is different, as illustrated in recent time by the example of "…A Wasco Indian logger (a faller), who quit logging and sold his chainsaw because he couldn't stand hearing the trees scream as he cut into them." This is a worldview in which "[T]he whole world is perceived and valued. Even the flies" (Dunsmore, 1997, 7).

   Here we are engaging in a discussion of two sharply divergent ways of relating to environmental issues. Is it fair, in our time, to attribute one stance to one group, or should we concentrate on the message inherent in the example? Discuss and honor all points of view.

6. **Like Black Elk, Tecumseh, Black Hawk, and others, Standing Bear invoked the image of Mother Earth in his writing and speaking.** Luther Standing Bear's use of the Mother Earth image is particularly striking when placed next to similar language used by Black Elk which has been passed to us through accounts by Neihardt, Epes Brown, and others. Unlike accounts attributed to Black Elk, Tecumseh, and Sea'th'l (Seattle), however, the use of the image by Standing Bear raises no questions of interpretation, because he wrote in English acquired at the Carlisle Indian School. There is a great difference in the attitude taken by the Indian and the Caucasian toward nature, and this difference has made one a conservationist and the other a non-conservationist of life. The Indian, as well as other creatures that were given birth, were sustained by the common mother—Earth.

   Discuss the image of Earth as mother by Standing Bear. Some of you may even wish to read one or more of his books. They are in libraries and otherwise available. Some of you may also wish to read about the same theme being used by other tribal peoples around the world or in the history of other countries. Try, for example, the writings of Henry David Thoreau.

7. One of the greatest contrasts between Mother Earth and Mother Lode thinking is the opening, in recent years, of tar sands mining in Alberta, Canada. Exploitation of tar sands at the surface requires a form of strip mining that scars the Earth in ways that will not be quickly nor easily repaired. The opponents assert that oil sands are a relatively new form of fossil fuel—the last thing the Earth needs when carbon dioxide levels in the atmosphere have risen to more than 420 parts per million, more than 40% above peak preindustrial levels, with damage to climate, as well as rising seas and oceanic acidity.

   Debate and discuss: how necessary are tar sands mines for expansion of the oil supply at a time when use of fossil fuels, eventually, will be dangerous to survival of all living things? Or is this point of view simply a bias or a "hoax," as some well-known political figures have contended? We already have addressed the difference between a "Mother Earth" and a "Mother Lode" point of view when it comes to treatment of natural resources. Have you any further comments now that we are further along on this subject?

8. **Furthermore, two points of view vis-à-vis the ecological world have been compared: The characterization of the Earth in the feminine, using kin terminology, is [a] Native American perspective on the environment that is often virtually op-**

posite those of many early immigrants, who sought to "tame" the "wilderness." Most many Native Americans saw themselves as enmeshed in a web of mutually complementary life. As Black Elk said: "With all beings and all things, we shall be as relatives" (Black Elk, 1953, 105). Standing Bear defined his people's relationship to everything else on Earth, writing in the native view that everything is animate—"possessed of personality," he said. He compared the world to a library, with "the stones, leaves, grass, brooks…birds, and animals as its books."

9. *Here is another point of view for evaluation, specially addressing Indigenous peoples vis-à-vis Mother Earth vs. Mother Lode points of view. It is worth discussing.*

During the summer of 2015, Pope Francis, who has become well-known for directly tackling many controversial issues, made climate change a Vatican priority by issuing an encyclical (essentially a policy statement) detailing how the burdens of global warming worldwide fall disproportionally on the poor. Indigenous peoples around the world bear a disproportionate burden of environmental damage. In the United States today, Native peoples often live on ruined, exhausted land, suffering toxic consequences. Fully one-third of the Superfund sites declared by the US Environmental Protection Agency are on Native American lands.

10. *Furthermore, Pope Francis here addresses the existential question: The Pope Francis' encyclical was part of a broader campaign by the Pope to advocate protection of the Earth and all of creation. The Pope prompted Catholic theologians to reinterpret Genesis to emphasize stewardship over subjugation.*

This course is a place for serious thinking about searingly important issues no matter which values you profess. Pope Francis has a seat at this table, and so do Indigenous peoples, as does everyone else on the planet Earth. Let us ask the tough questions. When it comes to sustainability, what *really* works? What *really* matters?

Question for summary discussion (maybe even a final examination essay test). Bite into this:

The effects of climate change are not theoretical, and they are not speculative problems that can be handed off to future generations. Economic activity around the world, as well as the lives of animals and plants, are being affected today by rising temperatures. This is not merely a matter of a few degrees on the thermometer but of alterations in the environment that sustains all of us.

Greenhouse gases have no morals, loyalty, nor party affiliation. Carbon dioxide is not having a debate with us. It merely retains heat. Thus, in 50 years, when our children are grandparents, the planetary emergency in which we are now tasting the first course will be a dominant theme in everyone's life, unless we act now. Within a decade or two, thermal inertia will take off on its own, portending a hot, miserable future for coming generations.

11. Can a system predicated on growth adapt to a sustainable world in which having less "stuff" will be preferable? How can we adjust our desires to fit a new world in which more is not always better? Will our basic values change along with our energy sources?

A major—perhaps *the* major question facing an Earth and its human denizens in a time of worldwide environmental crisis is can capitalism change its character? A sustainable environment can make good business. Witness the growth of alternative forms of energy. Can capitalism factor respect for the Earth that sustains us all into its calculus of development? If so, it may be a positive force in a new, sustainable world. If not—if it retains attributes of the cancer cell—then ultimately, our progeny will inherit an exhausted, poisoned world.

12. **Here's a summary question and debate exercise or an essay test that will offer some brain exercise.**

    (a) **Can capitalism, with its appetite for pell-mell (and often environmentally destructive) growth, survive in a new world in which geophysical reality demands that we restrain our demands upon the Earth? Are we ready to operate with an accounting system that brings us all to account for the toll that our activities exact on the Earth and its atmosphere? Can we fashion a system in which polluting the atmospheric commons is defined as a criminal act for which sizable fines are levied? Such a system would redefine some present-day free choices (e.g., to trash the commons) as illegal acts. At that point, with indigenous advice, we will be truly respecting Mother Earth.**

References

Berry, C. (2011, October 1). Scholar Daniel Wildcat in discussing the environment: It's relatives, not 'resources.'. *Indian Country Today Media Network*. Retrieved February 23, 2021.

Grossman, Z., & Parker, A. (2012). *Asserting native resilience: Pacific rim indigenous nations face the climate crisis*. Oregon State University Press.

Johansen, B. E. (2007, October). Global warming, 'thermal inertia,' and tomorrow's news. *Nebraska Report, 7*.

Paper, J. (1990). Through the earth darkly: The female spirit in native American religions. In C. Vecsey (Ed.), *Religion in native North America* (pp. 3–19). University of Idaho Press.

Russo, K., Grossman, Z., & Parker, A. (2013). Asserting native resilience: Pacific Rim indigenous nations face the climate crisis. Corvallis: Oregon State University Press, 2012. *American Indian Culture and Research Journal, 37*(2), 234.

Weaver, J. (Ed.). (1996). *Defending mother earth: Native American perspectives on environmental justice*. Orbis Books.

Winkelmann, R., Levermann, A., Ridgwell, A., & Caldeira, K. (2015). Combustion of available fossil fuel resources sufficient to eliminate the Antarctic ice sheet. *Science Advances, 1*(8), e1500589. https://doi.org/10.1126/sciadv.1500589. Retrieved October 26, 2015, from http://advances.sciencemag.org/content/1/8/e1500589

World economic forum feels climate change pressure. (2013, January 25). *Environment News Service*. Retrieved February 27, 2015, from http://ens-newswire.com/2013/01/25/world-economic-forum-feels-climate-change-pressure/

Further Reading

Gillis, J. (2016, March 22). Scientists warn of perilous climate shift within decades, not centuries. *New York Times*. Retrieved April 7, 2016, from http://www.nytimes.com/2016/03/23/science/global-warming-sea-level-carbon-dioxide-emissions.html

Hansen, J., Sato, M., Hearty, P., Ruedy, R., Kelley, M., Masson-Delmotte, V., Russell, G., Tselioudis, G., Cao, J., Rignot, E., Velicogna, I., Tormey, B., Donovan, B., Kandiano, E., von Schuckmann,

K., Kharecha, P., LeGrande, A. N., Bauer, A. M., & Lo, K. W. (2016). Ice melt, sea level rise and superstorms: Evidence from paleoclimate data, climate modeling, and modern observations that 2 °C global warming could be dangerous. *Atmospheric Chemistry and Physics, 16*(3), 761–3812. https://doi.org/10.5194/acp-16-3761-2016. Retrieved April 22, 2016.

Harvey, C. (2015, September 11). Scientists confirm there's enough fossil fuel on earth to entirely melt Antarctica. *Washington Post*. Retrieved October 22, 2015, from http://www.washingtonpost.com/news/energy-environment/wp/2015/09/11/scientists-confirm-theres-enough-fossil-fuel-on-earth-to-melt-all-of-antarctica/?wpmm=1&wpisrc=nl_headlines

Hughes, J. D. (1983). *American Indian ecology*. University of Texas Press.

Standing Bear, L. (1978). *Land of the spotted eagle [1933]*. University of Nebraska Press.

Toynbee, A. (1973, September 16). The Genesis of pollution. *New York Times*.

Waters, C. N., Zalasiewicz, J., Summerhayes, C., Barnosky, A. D., Poirier, C., Gałuszka, A., Cearreta, A., Edgeworth, M., Ellis, E. C., Ellis, M., Jeandel, C., Leinfelder, R., McNeill, J. R., Richter, D. d B, Steffen, W., Syvitski, J., Vidas, D., Wagreich, M., Williams, M., Zhisheng, A., ..., Wolfe, A. P. (2016, January 8). The Anthropocene is functionally and stratigraphically distinct from the Holocene. *Science, 351*. https://doi.org/10.1126/science.aad2622. Retrieved March 11, 2016.

4

A Struggle to Survive

The Navajos' and Hopis' Sea of Sand Dunes

Contents

© The Author(s), under exclusive license to Springer Nature
Switzerland AG 2023
B. E. Johansen, *Global Warming and the Climate Crisis*,
https://doi.org/10.1007/978-3-031-12354-2_5

⊜ Learning Objectives

1. Students will become acquainted with methods of dryland farming which have sustained the Navajo and Hopis for several centuries.

2. Students also will come to know the difference between a little rain and snow and none at all. What plants and animals are first to go in unprecedented drought?

3. Students will learn the global meteorological patterns that cause frequent droughts in the homelands of the Navajo and Hopi including Hadley cells (which govern atmospheric air circulation), location of mountain ranges, relationships to air flow and wind patterns, etc.

4. Students will learn why sand dunes are migrating faster across the landscape at speeds heretofore unknown in Navajo country. During 2009, the USGS measured dune migration as fast as 112–157 ft per year. Some dunes moved more than 3.3 ft in a single windstorm.

5. Students will be able to analyze why conditions described below are happening in this place at this time: Streams that once were sources of precious water have dried up, feeding the wind with plumes of gritty, irritating sand. The USGS found that "The formation and movement of active dunes on the downwind side of stream-bed sand sources is presently endangering housing and transportation, potentially jeopardizing native plants and grazing lands, increasing health hazards to humans and animals, and affecting regional air quality" (Redsteer et al., 2011b, c).

6. Students will be able to discuss strategies that may help to stabilize moving bodies of sand in the future.

Overview

The Hopis and Navajos in the American Southwest long ago learned how to deal with intermittent drought. Drought during the last half century has been almost without relief, however. The drought has been particularly intense during the last two decades. Navajos who have no indoor plumbing or tap water have been forced to travel several miles for water as their wells run dry while also forced early sale of livestock as former scanty pastures turn to naked dirt. "Perhaps among the worst of those impacts," wrote Terri Hansen in the Indian Country Today Media Network (May 6, 2014). During the 1996–2009 drought period, the extent of dune fields increased by about 70% (also see ◼ Fig. 5.1). These dunes are moving at rates of approximately 35 m per year, covering houses, burying cars and snarling traffic, degrading grazing and agricultural lands, contributing to the loss of rare and endangered native plants, and when they occur contributing to poor air quality, a serious health concern for many of the reservation's 173,667 residents. The Navajo Nation occupies the driest one-third of the Navajos' traditional homeland (Redsteer et al., 2013).

◘ Fig. 5.1 Sandstorm over Clovis, NM. Persistent drought in the American Southwest is forcing Navajos who have no indoor plumbing or tap water to travel several miles for water as their wells run dry while also forced early sale of livestock as former scanty pastures turn to naked dirt. "Perhaps among the worst of those impacts," wrote Terri Hansen in the Indian Country Today Media Network (May 6, 2014), "are the runaway sand dunes it has unleashed, which extend over one-third of the 27,000-square-mile reservation." (Source: Library of Congress. Public Domain)

Eye Catcher

"When you're living in the desert, you don't expect it to get even worse," Russell Begaye, a Navajo Nation Tribal Council delegate from Shiprock, NM, told the Indian Country Today Media Network. Laura Paskus reported (Paskus, 2015a, b) that "he pointed out that reservoir levels are dropping, farming plots are becoming sandier, and the rain and snowfall have declined" during a drought that, punctuated by a few flooding rain and snow events, has now stretched for 20 years. "We know what the long-term effects are going to be: We're going to be out of water. That has to be everybody's concern," said Navajo Department of Emergency Management's Rosalita Whitehair (Paskus, 2015a, b).

Definition

The 25–40% of Navajos who haul their own water pay 20 times per volume that of non-Navajos who have piped-in supplies on per capita income that is less than half of the US average—before adding the expense of round trips that average 28 miles (Cozzetto et al., 2013, 569). During droughts, which are becoming more frequent, both the cost of water and the distances required to acquire it increase.

5.1 Effects of Acute Drought on the Navajos and Hopis

Native peoples already are often intensely affected by climate change because many reservations were established in areas with extreme, challenging environments that immigrating European Americans did not want. In the Southwest, this generally meant dry land (also often with less access to water supplies in streams, rivers, and lakes). The Navajo Nation occupies the driest one-third of the Navajos' traditional homeland (Redsteer et al., 2013, 389). Helen Hunt Jackson described Navajos' eviction from the best land about a century and a half ago: "From tract after tract of [ancestral] lands they have been driven out year by year by the white settlers of the country until they can retreat no further, some of their villages being literally on the last tillable spot on the desert's edge or in mountains' far recesses... In southern California today are many fertile valleys which thirty years ago were the garden spots of these same Indians."

Context: The Expansion of Drought in the United States
Drought also has become pervasive in other parts of the American Southwest. "Our 30,000-acre reservation is pretty dry because of drought," said Lawrence Snow, Land Resources Manager for Utah's Shivwits Band of Paiutes. "Wildfires in the last decade have burned half our acreage and changed the landscape. We've got fewer trees, and bark beetles are trying to kill off the ones we do have. Once the fires happened and took out the ground cover, major storms brought big flooding" (Allen, 2012). When native areas in the American Southwest receive rain and snowfall, it increasingly comes in flooding deluges. Arizona's Havasupai Tribe between 2008 and 2010 endured several damaging floods.

Context: Expanding Deserts World-Wide Due to Expansion of Hadley Cells
Some of the Navajos' enduring drought stems from changes in worldwide atmospheric circulation compelled by a warming climate. Even though warmer air generally holds more moisture, not everyone sees more precipitation in a globally warmed world. Many deserts already are expanding, in a worldwide pattern influenced by atmospheric circulation patterns that meteorologists call Hadley cells.

Most deserts range between 20° and 40° north and south latitude. While precipitation patterns are also influenced by other factors (such as ready access, or lack thereof, to ocean-borne moisture), rainfall is strongly influenced by Hadley cell circulation. Rising air portends instability, low pressure, and storminess; descending air generally produces high pressure and clear skies. In a warmer world, Hadley cells expand northward in the Northern Hemisphere and southward below the equator, which causes deserts to expand. Near the equator, warm, moist air rises, cools, and unleashes downpours. In the upper troposphere, the air spreads north and southward toward both poles, descending at about 30° north and south latitude, and creating deserts. For reasons that are not yet fully understood, as temperatures rise, the Hadley cells then expand further north and south of the equator. (The Navajo and Hopi homelands, at about 32–33° north latitude, are square inside of the worldwide zone of descending air.)

Thus, as sand dunes marched across the Navajo Nation and California suffered its worst drought on record, deserts crept northward into Spain from Africa, Tehran, and other parts of Iran suffered water shortages; Sao Paulo, Brazil, nearly ran out of water in 2014 and 2015; and Australia's multi-year drought scorched the agricultural valleys of the Darling River, meanwhile provoking wildfires that reached the suburbs of Sydney. All of these areas (among others) were within reach of expanding Hadley cells (Johansen, 2017). Droughts in regions where Hadley cells favor descending air now span the globe, from Australia to Spain, Iraq, Afghanistan, parts of China, and the American Southwest, including California, Nevada, New Mexico, Arizona, and Texas. In China, the Gobi Desert, also within the northern reaches of Hadley cell range, has been expanding, sending occasional dust storms into Beijing (when the area is not receiving downpours), aggravating air pollution from coal-fired power plants that gives that city and others in China and India the dirtiest air on Earth. In Iran, Lake Urmia, once plied by cruise ships, has lost nearly all of its water, and water rationing has been proposed for Tehran. As in Southern Russia, major rivers in Iran also have run dry. Groundwater levels also have declined as increasing numbers of wells tap finite aquifers. Rising temperatures also have accelerated evaporation everywhere (Johansen, 2017).

The drought that has been afflicting the Navajos, Hopis, and Pueblos is part of the same multi-year "mega drought" that has been ravaging California. According to bioclimatologist Park Williams, a professor at the Lamont-Doherty Earth Observatory at Columbia University (Rice, 2014, A-1), this is developing into the worst drought in the area since the 500-year event that devastated classical Pueblo civilization between about 800 and 1300 CE. "When considering Western North America as a whole, we are currently in the midst of a historically significant mega-drought," Williams said. Nearly a millennium ago, the same drought also played a major role in devastating much of classic Mayan civilization in present-day Central America. Large areas experienced warm weather without humans burning fossil fuels (except firewood); that warm epoch also enticed the Vikings to settle in Greenland and to give it a name that, 300 years later, during the Little Ice Age, proved misleading (Johansen, 2017).

5.1.1 From Little Water to Nearly None

Navajo and Hopi lands in Arizona have always been relatively dry, but climate change in recent years often has made matters worse. The Navajo and other Southwestern native peoples have made a fine art of surviving on little water for centuries. The ways they farm and the animals they herd are used to it. There is a difference, however, between little rain and nearly none, and that's what they've been dealing with for 20 years. Cindy Dixon's sheep, for example, used to forage scrub on the desert near Farmington, New Mexico. By 2014, however, even that had died, as Dixon turned to expensive bales of hay. "The landscape around her Navajo Reservation homestead," wrote Bobby Magill in Climate Central, "was as brown and bleak as the open-pit coal mine a few miles to the west and well within earshot."

Dixon lives without electricity or running water, but her sheep cannot eat sand. "Since it's all dry and bare and deserted—no vegetation—I have to constantly buy hay and grain to keep the sheep fed," Dixon said, looking at the land around her trailer. "This is a bad, bad area for livestock" (Magill, 2014). The lack of forage is compounded by coal dust blowing in from the mine on stiff winds that are now pushing growing sand dunes across the brown, desiccated land. Sometimes, Dixon cuts her own grocery spending so that she can buy hay for the sheep.

By 2014, sand dunes were "covering housing, causing transportation problems, and contributing to loss of endangered native plants and grazing land" (Cozzetto et al., 2013, 569). Rainfall in some parts of the Navajo Nation fell to 3 in. a year during the latest drought (Redsteer et al., 2011b, c). Because of the enduring drought, "More than one-third of Native lands on the Colorado Plateau (Navajo Nation and Hopi tribal lands) are covered with sand dunes and sand sheets," according to the US Geological Survey (Redsteer et al., 2011b, c). Lands that once were marginally productive for grazing of sheep and dryland agriculture (a long-time practice among the Navajo and Hopi) are becoming true, water-starved deserts. Or, as the USGS phrases it: "Reactivation of inactive dunes could have serious consequences on human and animal populations, agriculture, grazing, and infrastructure on the Navajo Nation and similar areas in the Southwest" (Redsteer et al., 2011b, c). Wind and drought have been worst in the spring.

Dunes are migrating faster across the landscape at speeds heretofore unknown in Navajo country. During 2009, the USGS measured dune migration as fast as 112–157 ft per year. Some dunes moved more than 3.3 ft in a single windstorm (◘ Fig. 5.2).

The Grand Falls dune field has grown by 70% (laterally and downwind) in 15 years (from 1992 to 2007). The drought has continued since then, punctuated by

◘ **Fig. 5.2**   The Grand Falls dune field has grown by 70%. (Source: Library of Congress, copyright released by ES Curtis. Public Domain)

very occasional deluges that quickly run off the cracked, parched, and increasingly sandy earth.

Streams that once were sources of precious water have dried up, feeding the wind with plumes of gritty, irritating sand. The USGS found that "The formation and movement of active dunes on the downwind side of stream-bed sand sources is presently endangering housing and transportation, potentially jeopardizing native plants and grazing lands, increasing health hazards to humans and animals, and affecting regional air quality" (Redsteer et al., 2011b, c).

The Navajo Nation has experienced several decades of rising temperatures, declining snowfall, decreased (or, in some cases, eliminated) streamflow, resulting in water scarcity that has "magnified the impacts of drought that began in 1996 and continues today" (Redsteer et al., 2013, 390). Streamflow data and historic information on surface-water features (such as springs, lakes, and streams) show significant changes over the past century (Redsteer et al., 2010). Historical sources as well as elders' accounts describe many watercourses that are dry today. Some began to disappear in the early to middle twentieth century. "Moreover," wrote Redsteer et al. (2013, 390), "significant reductions in the number and length of stream reaches with perennial flow have occurred since 1920, and for some historic ephemeral streams, no flow during spring run-off and summer rains occurs today." Many surface-water features are now dry year-round or ephemeral and began to disappear in the early to mid-1900s.

5.1.2 Navajos and Hopis Share Drought Histories

Navajo elders shared observations of the drought's impacts with scientists, including "declines in snowfall, surface water features, and water availability" (Redsteer et al., 2011c), along with lack of available water and changing socio-economic conditions as leading causes for the decline in their ability to grow corn and other crops. "Other noticeable changes reported in these accounts include the disappearance of springs and the plants and animals found near water sources or in high elevations, such as certain medicinal plants, cottonwood trees, beavers, and eagles" (Redsteer et al., 2010). The elders also noticed changes in wind speed and endurance that were changing the behavior of sand and dust storms (see ◘ Fig. 5.3).

Drought also has inhibited the Navajos' use of corn pollen, which has a central role in many Navajo ceremonies. When ceremonies are interrupted, this disruption demonstrates a paradigm shift in climate going back hundreds if not thousands of years (Johansen, 2017).

Navajo traditional livelihoods have always been tied to the land, and with more heat, less rain, and expanding sand dunes, basic survival is becoming more difficult. Unemployment on the Navajo reservation was 47% in 2021, and the poverty rate was 37% or more. At least a third of Navajo families haul their water 10 miles or more. The median household income was $24,000 in 2014 (Magill, 2014). Amenities that are taken for granted in many US and Canadian urban areas do not exist. Coal is mined to power electricity-generating plants that supply Tucson, Las

Fig. 5.3 Before the storm, by Edward Curtis. (Source: Library of Congress, copyright released by Curtis. Public Domain)

Vegas, and Phoenix, their lines running above Navajo hogans with no power, not even enough to power a single lightbulb.

By 2014, sand dunes were "covering housing, causing transportation problems, and contributing to loss of endangered native plants and grazing land" (Cozzetto et al., 2013, 569). Rainfall in some parts of the Navajo Nation fell to 3 in. a year during the latest drought (Redsteer et al., 2011b, c). Because of the enduring drought, "More than one-third of Native lands on the Colorado Plateau (Navajo Nation and Hopi tribal lands) are covered with sand dunes and sand sheets," according to the US Geological Survey (Redsteer et al., 2011b, c). Some of this power generation has since been shut down to avoid expensive pollution controls. The methane "hot spot," which covers about 2500 square miles, roughly half the area of Connecticut, near the intersecting borders of Colorado, Arizona, New Mexico, and Utah, was described in the October 16, 2014, issue of *Geophysical Research Letters* (Kort et al., 2014). The Four Corners methane plume by itself comprises 10% of US methane emissions, according to US Environmental Protection Agency estimates (Minard, 2014).

5.1.3   Personal Accounts of Rampaging Sand Dunes

Lester and Louise Williams, who live near Tuba City, have a sand dune as a neighbor, and it's moving in. Despite it, the Williamses say they have no plans to leave their house. Kathy Ritchie (2014) described the sand dune astride the Williams' house: "Lester Williams—aka 'Chee Willie'—and his wife, Louise, live just a few feet from a massive sand dune. The gigantic pile looks to be around 20 feet high. This is the Williamses' fifth house, and they share it with their children and grandchildren. The very same dune that looms right outside swallowed their four previous homes, along with their sheep corrals. Inside the tiny home, family photos hang from the wall, along with a calendar from Hank's Trading Post and a framed poster of the Canadian Rockies.... Chee Willie doesn't speak English, but he's incredibly animated when he speaks Navajo. Through a translator, he talks about the difficulties of living in this kind of environment, where the wind whips the sand so furiously that the family can't leave the house. He says he once tried to remove the sand himself, but it came back. The sand always comes back" (Johansen, 2017).

Louise, Chee Willie's wife, said that blowing sand makes breathing very difficult. The wind howls, and the sand swirls. Yet, they stay. Home is a sacred place among Navajos. Home is a sacred bond—over and above the practical difficulty of obtaining a new lease, moving livestock, and the fact that more and more Navajo land is laced with migrating sand dunes (Johansen, 2017).

Like heavy snowdrifts, dunes move across roads and block them with increasing regularity. "As we're packing up and preparing to leave Chee Willie's house, Tohannie said that yesterday's windstorm shifted a nearby dune, causing it to cover part of a road used by the handful of families in the area, including Tohannie's parents," Ritchie wrote. "I got stuck with my son," he said. "We had to shovel our way out." These are not small drifts, and they can't be moved, like snow, with a plow. And, of course, they never melt. Ritchie described one dune as "[a] magnificent sculpture shaped from mostly eroded Navajo and Entrada sandstone" (Ritchie, 2014). "Many of the dunes in the area are unexpectedly tall, some measuring anywhere from 30 feet to 40 feet high. Begay tells me that near Preston Mountain, some 45 minutes north of where I'm standing, the dune field is even higher, possibly 60 feet to 80 feet in places" (Ritchie, 2014). The dominant plant species in some areas, where any survive, has become tumbleweed, which has evolved to move with the wind, anchoring nothing.

Across the reservation, persistent winds have been driving sand into homes and across roads. Most of the roads are not paved, and their surfaces can become parts of the moving dunes. Even major paved roads have been blocked. On April 16, 2013, driven by winds gusting to 60 miles an hour, drifts of sand closed parts of Interstate 40, as traffic backed up 12 miles. A NASA satellite photographed the dust plume from space (◘ Fig. 5.4).

Changes in Navajo Topography Associated with Climate Change

Margaret Hiza Redsteer, who is of Crow descent, was raised in their homeland near the Montana-Wyoming border, but during the 1970s, having married a Navajo, moved to his homeland, and mothered three children. Many people told her how much plant life in the area had changed over the years. Only later did she begin to associate these changes with climate change. In the meantime, Redsteer and her family had moved to Flagstaff, Arizona, when she was 29 years of age in 1986, where she studied for a PhD. Shortly after 2000, Redsteer, now employed by the US Geological Survey, switched her focus of study from volcanic deposits near Yellowstone National Park to the effects of climate change on the Navajo Nation as intensifying drought contributed to growth and migration of sand dunes there. She wrote several academic papers and had a key role in the National Climate Assessment, released by President Barack Obama in 2014. She has become known for linking elders' recollections with weather records to trace the evolution of climate change.

Interviews of elders fill gaps in USGS data. "We have aerial photographic surveys of the study area from 1934 and from 1954, but between those years there were big changes. Our interviewing not only provides another line of evidence, but it also fills in a lot of the data gaps," Hiza-Redsteer said (Navajo Memory, 2011).

As Bobby Magill wrote in *Climate Central* (2014): "Navajo elders remember wetter times, when winter snows were knee-deep, water always ran in springs and arroyos, and the rangeland among the canyons, mesas, and volcanic hills could support large herds of livestock, a mainstay of the Navajo economy." Some elders recalled a time when they were children, with moist ground until the Fourth of July, climate data on the Navajo Nation indicates a marked drying trend since the middle 1940s, and a warming of 4 °F in many areas since the 1960s. The decrease in snowfall has long-term implications. "Snow is like water in the bank," Redsteer said. "It takes a long time to melt. It soaks into the ground slowly."

During the past 60 years, the Southwest has experienced swings between very wet and very dry, but the current drought has dominated the past 20 years, with brief wet periods in 2004, 2005, and 2010 doing little to alleviate the long-term drought (Johansen, 2017).

◘ Fig. 5.4 Wise Navajo drivers carry shovels: Kee Tohannie, Begay's grandfather and Huskie Tohannie's father, always carries a shovel, tire chains, and sometimes a hatchet in case he or one of his neighbors is marooned in the sand. "I've been stuck in the sand many times—it's a lot of digging," he said. A Navajo driver said: "You just have to know how to drive in sand. Like you learn to drive in snow" (Ritchie, 2014). ("Before the Storm," by Edward Curtis. Source: Library of Congress, copyright released by ES Curtis. Public Domain)

5.1.4 Elders Recall a Wetter Time

Navajo farmer Jonathan Yazzie said that for many years, he grew squash, corn, zucchini, chilis, and even water-hungry cantaloupe. The drought has put him out of business. "The water is just not there no more," he said. "We're down to 15 sheep. No cattle. Two horses. That's our kids' future" (Magill, 2014). He has no access to irrigation that he has sought from the Navajo and Hopi governments. Without it, continued drought will force his family to move. Traditional herding is dying. "I don't know a single young Navajo person today who's thinking about having their own sheep herd," Redsteer said. "Part of that is due to their own market economy. The feasibility of doing that is just impossible now. "There is still a holdout group of elderly people who really don't have a choice. Their language is Navajo. Their culture is Navajo. They really don't have any other place to go" (Magill, 2014). An account in *The Navajo Times* remarked that "Hardly anyone is making a profit from their livestock anymore; they've become expensive pets" (The Navajo, 2011).

"The elders often talk about the difference in the quality and quantity of grasses— how tall, how thick, how much of it there used to be. Some people say when they were young and herding sheep they had to stay right with the herd. If they didn't the sheep would get lost in the grass. It's not like that now," she told *High Country News* (Margaret Hiza, 2012). Elders' recollections provide details that statistics do not. For example, they describe streams used for irrigation that no longer exist, indicating intensifying drought spanning several years. Decades ago, the annual cycle usually included winter rains, a dry, windy spring, and a summer monsoon. Rain and snow have decreased at all seasons, and springs are warmer now. "We've learned from the elders that the soil stayed moist all through the spring until the summer monsoon arrived," she said. "Now, if you were to go out in the springtime during the dry windy season, you could dig a very big trench and not run into any wet sand or soil. The ecological effects are huge because shallow rooted plants aren't going to do as well.... The elders have told us that when there were cottonwoods in the Little Colorado River there were lots of beavers. They used to see cranes migrate through the area in the spring, stopping in the marshes around lakes that aren't there now" (Margaret Hiza, 2012). Redsteer has conducted more than 100 interviews that can be correlated to check for vagaries of memory. Medicine men's memories are very helpful because they use plants in ceremonies and remember what species have changed over time. People also relate changes in vegetation and animal life that change with elevation.

Some of the elders take the blame for changes in the weather, saying that it is spiritual retribution for their unwillingness or inability to follow traditional ways, the Original Instructions. Redsteer tells them that it's geophysics, not their fault, a world changing due to alterations in atmospheric carbon dioxide and other greenhouse gases that are beyond their power.

5

5.1.5 Droughts Expected to Intensify

A technical report for the National Climate Assessment, issued in 2013, said that the Four Corners area probably will continue to endure warmer weather on average during coming decades, as soil continues to dry with droughts becoming more intense and frequent. The drought—and spread of sand dunes—is worst in the southwestern quarter of the Navajo reservation, where many families may be forced to move, according to (Redsteer, 2014), who has used the recall of elders as well as weather data to trace climatic changes. Average annual snowfall across the Navajo Nation declined from about 31 in. in 1930 to about 11 in. by 2010, according to a United Nations case study: "Every tribal elder mentioned the lack of snowfall," Redsteer said. "They describe winters where the snow was 'chest high on horses.' The snowfall declined significantly during the 20th century, and is still declining in recent years" (Magill, 2014). Elders' memories have been especially important in recent years, as heat and drought have accelerated, because many US government weather stations were shut down during the early 1980s to save money.

By 2014, Margaret Redsteer had studied the spreading Navajo sand dunes for 14 years, in collaboration with the Navajo Nation. She also worked with Northern Arizona University's Tribal Environmental Education Outreach Program in a continuing effort to stabilize dunes with native vegetation. The battle has never been easy, and the changing climate grants no favors. While native plants require time to become established after rare rains, invasive tumbleweeds explode nearly overnight. Tumbleweeds suck up moisture before native plants get it, according to Redsteer. "Tumbleweed is a major blow to rangeland conditions," Redsteer said. "It is amazing how huge the areas are that are affected by tumbleweeds" (Ritchie, 2014).

5.1.6 Skeptics Differ with the Majority View on Drought and Heat

Not everyone believes the science, however, according to an account in *The Navajo Times*. After a full day of talking about climate change, it was evident that not everybody is a convert. Jones Begay, a rancher on Black Mesa, said he hasn't seen much evidence of warming up in the mountains, where the springs still flow, the grass still grows, and snowstorms strand people for weeks at a time. Asked if he believes the desert could start creeping higher, he conceded it could happen—but he doesn't believe that because of anything he heard at the meeting. "It's in the *Bible*," said Begay, a regular at the Friends Church. "I don't believe the professors, but I believe God". Yurth said.

"We're seeing that people who live in the drier lowlands are seeing a different timing of changes than people who are living higher, among the buttes, ponderosa, pinon and juniper trees. We're trying to understand that difference more clearly," said Redsteer (Margaret Hiza, 2012). Peoples' livelihoods have changed. "A lot of people have already moved away from having livestock. There is just no water for them; there is no feed. And to haul hay to the reservation all the time is really expensive. You're often making a poor living or losing money in the deal. People have some livestock now, just not very many, and mostly for ceremonial purposes" (Margaret Hiza, 2012). Redsteer has extended her work to other indigenous cultures and has noted similarities, such as increasing climatic volatility. "A lot of them say that they can't predict the weather anymore. Things have changed so much that their traditional calendars don't work. From people in the Amazon, in Africa, in Asia, that's a worldwide unified statement" (Margaret Hiza, 2012).

The Navajos and Hopis live in areas that are "just on the edge of being habitable," Redsteer said. Climate change (especially persistent drought) is taking much of this area past a dangerous threshold. "The annual moisture here has historically been just enough to get by. When there is even a small change, there is a huge effect," she said. John Leeper, director of the Navajo Water Management Branch of the Navajo Nation in Fort Defiance, Arizona, said that "If the current trends that she identifies continue, much of the Navajo Nation will be severely impacted, and much of the Navajo Nation will become uninhabitable" (Navajo Memory, 2011).

"The Navajo Nation is intended to be a permanent homeland for the Navajo people," Leeper said. "However, much of that homeland may be in jeopardy if these trends cannot be successfully mitigated. Not only has Margaret's work identified and documented the current trends, her work also gives us perspective on the steps that can, and must, be taken to reverse many of the most damaging of these trends. Her work will help to ensure that the Navajo people will be able to find their livelihoods here long into the future" (Navajo Memory, 2011).

5.1.7 Attempts to Stabilize Spread of Deserts

Redsteer recommends use of sand barriers to stabilize dunes, as well as seeding areas dominated by dunes to encourage vegetation. "If we're going to do research for people's benefit, we have to try to see what kind of solutions there are," she said. Redsteer has tried sand barriers that are being used as part of China's Green Great Wall to hold back sand dunes advancing across the Gobi Desert, given her by visiting scientists. She admits that the geophysical circumstances make her efforts look very small. "It's like putting a thimble over my head to try to stay dry in a downpour," she said (Yurth, 2011).

Redsteer has addressed Navajo chapter houses (local governments), sometimes convening more than a hundred residents at a time. "The Navajo Nation is in one of the longest droughts in recorded history," she told a sea of grim-faced Central Agency residents sweating in the 90° heat. "It's not a simple matter of overgrazing. There are a number of other factors that make it so critical to address this issue," she told about 120 people who met at the Many Farms Chapter House in September, 2011, "to discuss the alarming desertification that has taken place on the Navajo Nation in the last 30 years" (Yurth, 2011). John Leeper of the Navajo Nation Water Resources Department said that "Every public water system on the rez, not to mention the $20 million livestock industry, are under siege by the drought" (Yurth, 2011).

"First of all, the traditional way of adapting to dry seasons was to move," Redsteer said. These days, "If you have a reservation, and the reservation is established where there are the most limited water resources in the region, the odds of you being able to make it through dry seasons are stacked against you" (Taylor, 2014). "People on the reservation use one-tenth of the water that people in Phoenix use every day. How do you conserve when you are already using so little? They don't have lawns, they don't wash their cars on a regular basis. It's hard to say, 'Well, we really need to conserve now,'" she said with a laugh (Taylor, 2014).

Old habits die hard. "One of the real ironies is that western water law is 'use it or lose it'. Phoenix … to keep its Colorado River allocation, has to use that allocation or it will lose its rights to it. So in some ways there's a disincentive to conserve," Redsteer said (Taylor, 2014). That may change if the water runs out and life becomes impossible.

Even as drought intensifies across the Navajo nation, some residents who haul their own water assert that the government continues to offer large amounts of water for oil drilling (including fracking) as well as coal-fired electric power generation. Laura Paskus reported for the Indian Country Today Media Network on April 11, 2015, that "Lori Goodman, with the nonprofit group Diné CARE, or Diné Citizens Against Ruining Our Environment, has been fighting back against the tribe's use of water for mining and development for more than a decade." "The thing that keeps the Navajo in poverty is [lack of] water," said Goodman. The Navajo tribal leadership doesn't understand the value of water, she said, and has continued to sign away water rights, sell water to industry, and not plan for the future. "The water is always for extraction; they're using it for fracking, for power plants, for mining," she said (Paskus, 2015a, b).

Russell Begaye, a Navajo Nation Tribal Council delegate from Shiprock, New Mexico, is still worried. In particular, he fears that groundwater contamination from fracking and reinjection. The Navajos, he noted, are still dealing with a legacy of groundwater contamination from uranium mining. "We really don't know what is being pumped into our aquifer and how that is being impacted," said Begaye, who ran for president of the Navajo Nation in 2015. "I keep saying, 'I don't think anyone really knows what will happen when you're putting water that's chemically-laced back into the ground, if it mixes with the good clean water being used by our people" (Paskus, 2015a, b).

The Hopis, whose land is surrounded by that of the Navajos, are suffering from the same drought. Their residency in the area is longer than that of the Najavos, and their oral history tells of severe dry spells lasting decades. Without precise measures of time, precipitation, and temperatures, it is impossible to tell which drought, and when, was the most severe. All the Hopis have is a sense of whom among humans lived and who died. Endurance can't be measured on a thermometer or a rain gauge. Intuition tells the Hopis that this one is testing human ability to live through a megadrought even with some latter-day technology (Johansen, 2017).

In dusty fields amidst the mesas, cattle die and rot in the dry heat. Farmers are being urged to use traditional dryland techniques that can raise crops on every little water. At the same time, the Hopis criticize increasing size and water consumption in non-Indian towns and cities in Arizona and New Mexico.

"Why isn't the governor cutting off water resources to southern Arizona?," asked Clark Tenakhongva, vice-chairman of the Hopi Tribe, which is in the north-eastern part of the state. "Cut out the pools. Cut out the water recreation areas. Cut out the golf courses, and you'll start resolving some of the issues the state of Arizona is looking at right now" (Romero, 2021).

In a parched landscape where the Hopi honed water-harvesting methods over centuries, the tribe estimates the reservation has about 2200 head of cattle that consume about 66,000 gallons of water a day. The King Ranch in southern Texas has 30,000 head.

Results

Simon Romero of the *New York Times* described Hopi "dry farming" techniques (2021): "Over the course of centuries, Hopi farmers developed techniques and seeds adapted to the dry climate. Forgoing pesticides, they focused on oasis-like mesas with farmable floodplains, moisture-retaining soils and springs, what Hopi call ▶ qatsi suphelawta, or the 'perfect location for life,'" according to the archaeologists Wesley Bernardini and RJ Sinensky. The area receives an average of 8.5 in. of rain and snow per year (Romero, 2021). During the drought, rain and snow have often been less than that. Springs that used to supply life-sustaining water have been drying up. Sand dunes envelop areas that used to sustain farms. This is what global warming looks like from the ground. Hopi ranchers reduce their herds into the single numbers. Farmers of corn, beans, and squash, foods that have sustained the Hopis for 1200 years or more, watch as the drought consumes their herds.

The Navajos' and Hopis' drought also has intensified because of coal mining at and under Black Mesa (since curtailed) that sucked billions of gallons of water out to provide slurry that provided electricity for large cities such as Los Angeles. An aquifer that was being drawn down provided drinking water to the Hopis. Within about 50 years, the level of the aquifer fell 40 ft (Romero, 2021). What rain that Hopiland now receives comes in bursts that often damages farm fields, before it quickly washes away and dryness returns, again.

5.2 Questions and Exercises

 1. When you're living in the desert, you don't expect it to get even worse," Russell Begaye, a Navajo Nation Tribal Council delegate from Shiprock, NM told the Indian Country Today Media Network. Laura Paskus reported (2015a, b) that "He pointed out that reservoir levels are dropping, farming plots are becoming sandier, and the rain and snowfall have declined" during a drought that, punctuated by a few flooding rain and snow events, has now stretched for 20 years.

Just how deep a drought can a family or a village withstand even if they are accustomed to living on a little water as possible? Take an inventory of your water usage each day and estimate how which you need, absolute minimum. Compare it with the Navajos and Hopis in this chapter. Bear in mind that every drop must be hauled from a well that may be a few miles away. The average one-way trip for water is 14 miles.

2. Perhaps among the worst of those impacts, wrote Terri Hansen in the Indian Country Today Media Network (May 6, 2014), are the runaway sand dunes which extend over one-third of the 27,000-square-mile Navajo reservation. During the 1996–2009 drought period the extent of dune fields increased by some 70%. These dunes are moving at rates of approximately 35 m per year, covering houses, burying cars and snarling traffic, degrading grazing and agricultural lands, contributing to the loss of rare and endangered native plants, and when they occur contributing to poor air quality, a serious health concern for many of the reservation's 173,667 residents.

   Take your own water inventory and cut it in half, and then in half again. What stays? What goes? Bear in mind that you are living in very dry surroundings, which are very hot in the summer, with no options for making ice or showering. Cooking and drinking water must be boiled before it can be used.

3. The drought also has become pervasive in other parts of the American Southwest. "Our 30,000-acre reservation is pretty dry because of drought," said Lawrence Snow, Land Resources Manager for Utah's Shivwits Band of Paiutes. "Wildfires in the last decade have burned half our acreage and changed the landscape. We've got fewer trees, and bark beetles are trying to kill off the ones we do have. Once the fires happened and took out the ground cover, major storms brought big flooding" (Allen, 2012). When native areas in the American Southwest receive rain and snowfall, it increasingly comes in flooding deluges. Arizona's Havasupai Tribe between 2008 and 2010 endured several damaging floods.

   How much annual rain and snowfall where live falls in downpours that run off the land uselessly? Some of that may help raise well levels slightly. Envision an entire new way of life based on water scarcity. And while you are empathizing with the Navajo and Hopi, don't forget water required for your sheep, goats, and crops raised with minimal waterer supplies.

4. The drought that has been afflicting the Navajos, Hopis, and Pueblos is part of the same multi-year "mega drought" that has been ravaging California. According to bioclimatologist Park Williams, a professor at the Lamont-Doherty Earth Observatory at Columbia University (Rice, 2014, A-1), this is developing into the worst drought in the area since the 500-year event that devastated classical Pueblo civilization between about 800 and 1300 CE.

   You are now living in historic times, waterwise. Can you squeeze your water budget any further without going out of business, leaving your farm (Sell it? Who would buy a place that is so dry?) and moving to a city. And what would you do there? Most of your life has between spent on a tiny, nearly waterless farm.

5. By 2014, sand dunes were "covering housing, causing transportation problems, and contributing to loss of endangered native plants and grazing land" (Cozzetto et al., 2013, 569). Rainfall in some parts of the Navajo Nation fell to 3 in. a year during the latest drought (Redsteer et al., 2011b, c). Because of the enduring drought, "More than one-third of Native lands on the Colorado Plateau (Navajo Nation and Hopi tribal lands) are covered with sand dunes and sand sheets," according to the US Geological Survey (Redsteer et al., 2011b, c).

> *This is your new reality, unless climate patterns change quickly. Consult with your fellow "farmers" in a class discussion. You have important decisions to make.*

6. And one more problem; Coal is mined to power electricity-generating plants that supply Tucson, Las Vegas, and Phoenix, their lines running above Navajo hogans with no power.... The methane "hot spot," which covers about 2500 square miles, roughly half the area of Connecticut, near the intersecting borders of Colorado, Arizona, New Mexico, and Utah, was described in the October 16, 2014, issue of *Geophysical Research Letters* (Kort et al., 2014). The Four Corners methane plume by itself comprises 10% of US methane emissions, according to US Environmental Protection Agency estimates (Minard, 2014). Give it a read at your campus library or order it through interlibrary loan. *Search on Google, etc. to find whether this rather enormous irony has been described in the off-reservation media.*

7. **One for discussion only:** Lester Williams—aka "Chee Willie"—and his wife, Louise, live just a few feet from a massive sand dune. The gigantic pile looks to be around 20 ft high. This is the Williamses' fifth house, and they share it with their children and grandchildren. The very same dune that looms right outside swallowed their four previous homes, along with their sheep corrals. Inside the tiny home, family photos hang from the wall, along with a calendar from Hank's Trading Post and a framed poster of the Canadian Rockies.... Chee Willie doesn't speak English, but he's incredibly animated when he speaks Navajo. Through a translator, he talks about the difficulties of living in this kind of environment, where the wind whips the sand so furiously that the family can't leave the house. He says he once tried to remove the sand himself, but it came back. The sand always comes back.

   For further comment—Wise Navajo drivers carry shovels: Kee Tohannie, Begay's grandfather and Huskie Tohannie's father, always carries a shovel, tire chains, and sometimes a hatchet in case he or one of his neighbors is marooned in the sand. "I've been stuck in the sand many times—it's a lot of digging," he said. "You just have to know how to drive in sand. Like you learn to drive in snow" (Ritchie, 2014).

References

Allen, L. (2012, June 14). Southwest tribes struggle with climate change fallout. *Indian Country Today Media Network*. Retrieved July 22, 2012, from http://indiancountrytodaymedianetwork.com/article/southwest-tribes-struggle-with-climate-change-fallout-118386more-beavers-defecating-and-disease

Hansen, T. (2014, May 6). 9 Tribal nations taking a direct hit from climate change. *Indian Country Today Media Network*. Retrieved June 23, 2014, from https://indiancountrytodaymedianetwork.com/2014/05/06/9-tribal-nations-taking-direct-hit-extreme-weather-154746

Johansen, B. E. (2017, May 9). The Navajos' sea of sand dunes. Research & reviews. *Journal of Ecology and Environmental Sciences*. Retrieved from https://www.rroij.com/open-access/the-navajos-sea-of-sand-dunes-.php?aid=86333

Minard, A. (2014, October 28). Methane 'hot spot' seen from space hovers over four corners. *Indian Country Today Media Network*. Retrieved December 10, 2014, from http://

indiancountrytodaymedianetwork.com/2014/10/28/methane-hot-spot-seen-space-hovers-over-four-corners-157560

Navajo memory complements science in study of climate change. (2011, October 21). *USGS Newsroom*. Retrieved July 15, 2012, from http://www.usgs.gov/newsroom/article.asp?ID=3007&from=rss#. VAb2Wmv29vZ

Paskus, L. (2015a, April 9). 'We're going to be out of water': Navajo nation dying of thirst. *Indian Country Today Media Network*. Retrieved May 22, 2015, from http://indiancountrytodaymedianetwork. com/2015/04/09/were-going-be-out-water-navajo-nation-dying-thirst-159948

Paskus, L. (2015b, April 11). Why is the Navajo nation giving water to fracking while its people are parched? *Indian Country Today Media Network*. Retrieved May 2, 2015, from Indiancountrytodaymedianetwork.com/2015/04/11/why-navajo-nation-giving-water-fracking-while-its-people-are-parched-159972

Redsteer, M. H., Bogle, R., Vogel, J., Block, D., Valence, M., & Middleton, B. (2010). The history and growth of a recent dune field at Grand Falls, Navajo Nation, NE [Northeast] Arizona. *Geological Society of America Abstracts with Programs, 42*(5), 416. Paper No. 170-5.

Redsteer, M. H., Bogle, R. C., & Vogel, J. M. (2011b). *Monitoring and analysis of sand dune movement and growth on the Navajo Nation, Southwestern United States*. Fact Sheet Number 3085. U.S. Geological Survey.

Redsteer, M. H., Kelley, K. B., & Francis, H. (2011c). *Increasing vulnerability to drought and climate change on the Navajo Nation*. Paper GC43B-0928, delivered at American Geophysical Union Annual Meeting, 5–9 December 2011, San Francisco.

Redsteer, M. H., Bemis, K., Chief, K., Gautam, M., Middleton, B. R., & Tsosie, R. (2013). Unique challenges facing Southwestern Tribes. In G. Garfin, A. Jardine, R. Merideth, M. Black, & S. LeRoy (Eds.), *Assessment of climate change in the Southwest United States: A report prepared for the National Climate Assessment* (pp. 85–404). Island Press. Retrieved November 3, 2013, from http://www.swcarr.arizona.edu/sites/default/files/ACCSWUS_Ch17.pdf

Rice, D. (2014, September 3). California's 100-year drought. *USA Today*, A-1, A-2.

Ritchie, K. (2014). Dune and gloom. *Arizona Highways*. Retrieved August 25, 2014, from http://www.arizonahighways.com/extras/dune.asp

Romero, S. (2021, October 2). In Arizona, drought ignites tensions and threatens traditions among the Hopi. *New York Times*. Retrieved October 22, 2021, from https://www.nytimes.com/2021/10/02/us/arizona-megadrought.html?campaign_id=2&emc=edit_th_20211003&instance_id=41933&nl=todaysheadlines&regi_id=35795487&segment_id=70560&user_id=8953ac8150496623ee2c782e2065b2e1

Taylor, K. (2014, March 19). Drought hits harder in already parched Indian country. *Aljajeera America*. Retrieved March 24, 2014, from http://america.aljazeera.com/articles/2014/3/19/drought-is-nothingnewinindiancountry.html

Yurth, C. (2011, September 7). Committee hears reports about the impacts of climate change as the Navajo nation suffers from drawn-out drought. New America Media: Indigenous. *The Navajo Times*. Retrieved October 1, 2011, from http://newamericamedia.org/2011/09/the-navajo-nation-suffers-from-drawn-out-drought.php

Further Reading

Cozzetto, K. K., Chief, K., Dittmer, K., Brubaker, M., Gough, R., Souza, K., Ettawageshik, F., Wotkyns, S., Opitz-Stapleton, S., Duren, S., & Chavan, P. (2013). Climate change impacts on the water resources of American Indians and Alaska Natives in the U.S. *Climatic Change, 120*, 569–584.

Hansen, T. (2014a, May 7). Climate disruptions hitting more and more tribal nations. *Indian Country Today Media Network* Retrieved June 2, 2014, from http://indiancountrytodaymedianetwork. com/2014/05/07/climate-disruptions-hitting-more-and-more-tribal-nations-154747

Kort, E. A., Frankenberg, C., Costigan, K. R., Lindenmaier, R., Dubey, M. K., & Wunch, D. (2014, October 16). Four corners: The largest U.S. methane anomaly viewed from space. *Geophysical Research Letters, 41*(19), 6898–6903.

Magill, B. (2014, May 28). The Navajo nation's shifting sands of climate change. *Climate Central*. Retrieved February 15, 2022, from https://www.climatecentral.org/news/navajo-nation-climate-change-17326

Margaret Hiza Redsteer uses Navajo memories to track climate change. (2012, April 4). *High Country News*. Retrieved April 7, 2021, from http://www.hcn.org/articles/geologist-margaret-hiza-redsteer-tracks-climate-change-through-navajo-memories

Specious Solutions and Speculations

Contents

© The Author(s), under exclusive license to Springer Nature Switzerland AG 2023
B. E. Johansen, *Global Warming and the Climate Crisis*,
https://doi.org/10.1007/978-3-031-12354-2_6

We have designed a civilization based on science and technology and at the same time have arranged things so that almost no one understands anything at all about science and technology. This is a clear prescription for disaster. We may for a while get away with this mix of ignorance and power but sooner or later it is bound to blow up in our faces.
Carl Sagan (Powell, 2011)

Nature cannot be fooled.

Richard Feynman (Powell, 2011)

Learning Objectives
1. Students will be examining and learning the advantages and disadvantages of several proposed replacements for fossil fuels as major energy sources for building our homes and offices, propelling our transportation, keeping us properly cooled or warmed, and manufacturing nearly everything we will want or need. The major disqualification for this array of tasks is that our new nominee for this absolutely necessary role is that this new substance, process, or idea cannot mine, or use, or otherwise emit fossil fuels or release the same into the atmosphere such as carbon dioxide, methane, or any other gas or substance that adds any of these materials into our atmosphere in any way, shape, or form. All proposals will be considered, but please bear in mind that 8 billion or more people on Earth as well as many billions more plants and animals are in one hell of a hurry, whether most of us know it, or not. And whether they know for not, this is *the existential question* that living things on Planet Earth have ever faced. A beginning list of proposed solutions follows. This list us not exclusive. Please add your own.

Overview
The race is on to decelerate human emissions of carbon dioxide, methane, and other greenhouse gases that are taking the Planet Earth down a heating, destructive path toward a globally warmed world. Since the need for solutions became more and more manifest about a half-century ago, the crowd of suitors for world energy savior has formed a lengthening line, from those who would lock carbon dioxide into rocks deep underground to others who would maintain a haze of sulfur between the Earth's atmosphere and the sun. Sulfur has cooling properties and also would shroud the surface in a dirty haze. If all of this makes an observer feel as if he or she has a stumbled into a convention of big stakes curiously arrayed circus barkers, rest assured. While a lot of money has been spent on some of these nonstarter "solutions," none has yet heated or air-conditioned a home. For that, try more prosaic approaches, such as wind and solar power, both of each are now less expensive to bring "on line" than oil-powered energy or coal. Nuclear power is still on the stage, although its extremely high construction costs and safety concerns pose problems. Hydropower has a bid in this lottery, but we are running out of damsites. One thing is for certain, however, we need solutions, probably several of them, and quickly, because continued dependence on fossil fuels, sooner rather than later, is going to burn our metaphorical house (our Earth) down.

Escape from 2021's summer of climatic havoc was difficult without hearing a cacophony of official voices exclaiming about the malign results of climate change. Actually, what nearly everyone who is now in this conversation seems to have discovered is the power and violence with which climate around the world can change as excess greenhouse gases have been added to the atmosphere at an accelerating rate. Human-caused increases of greenhouses gases are said to be the primary culprit of deadly weather. Ratchet up the search for solutions.

6.1 The Evidence Becomes More Evident…

For many people, the evidence has had to slap them in the face. Decades of warnings from esteemed scientists have not convinced them—yet. What is a course in atmospheric physics, however, up next to some of the largest wildfires in the history of record keeping leaping from tinder dried by a world-class drought that turns luxury homes to stilts and ashes within a few minutes?

Heads continued to turn during the late summer of 2021, as a strong Category 4 hurricane drowned and flattened much of New Orleans and then turned into a monstrous tropical depression that generates deluges of historic (religiously oriented observers said "Biblical") proportions much of the entire eastern coast of the United States, as millions more people who are afflicted by life-wrecking weather ask…

6.1.1 What Can We Do?

A textbook in Urban Geography is not much help when it comes to containing huge flash floods. Most of any big city's ground is paved with asphalt on roofs and roads and the cement of sidewalks. No one who laid all of that asphalt and concrete was thinking of dealing with 3 in. of rain in 1 h.

We have been building more wind and solar power, both of which have been providing incremental gains in the battle against eventual apocalypse. Within a few years, for example, Iowa will have 100% wind-powered electricity. The chief corporate architect of this clean-energy coup is the billionaire Warren Buffett, head of Berkshire Hathaway, which has bought out Iowa's formerly public utility, MidAmerican Energy. Several other US states are following the same road, but not nearly as quickly. One of the more interesting examples is Texas, where early motivations came from ex-US President George W. Bush. Wind power is also gaining traction in many European countries, most notably Germany. China also has become a major player in renewable energy.

In the rest of the world, solar power on a mass scale is being realized through "solar farms" the size of several football fields. Debate has arisen over what should be done to ensure that we have enough land for other purposes, such as raising food. The usual solution is to build the solar arrays in deserts with very little rain and plenty of sunshine.

The balance of this chapter delves into wish-level solutions that are more desperate, of a kind that may be undertaken by peoples who may later wish they hadn't. Take most of the pitches to come with a huge grain of salt.

6.2 From Nearly Nonexistent to Existential in One Human Lifetime

Ours is a climate crisis that hardly anyone except a few scientists used to care about suddenly rendered very, very real. In a little more than one human lifetime, it has risen to the status of a scientific and political priority, having shot from nonexistent to existential, prompting that aforementioned cacophony of voices proposing solutions. As we debate, however, the carbon dioxide curve continues to arch upward, from about 280 parts per million (ppm) to more than 400. The higher that number, the more heat the lower atmosphere is holding, and the higher surface temperatures will eventually become; the volume of rivers, seas, and oceans will rise, flooding cities, towns, and submerging countryside.

We begin with some of what we can't or shouldn't do. As Robert S. Young, professor of geology at Western Carolina University, said: "No matter what you spend in vulnerable coastal areas, you can't protect everything from every storm. Quite simply, it is impossible to stop land loss and guarantee the safety of people so long as the climate continues to change, the sea level continues to rise, and warming seas continue to create supercharged storms. It is impossible even if you spend lots of money, and do it the right way (Young, 2021, A-23).

It may look dramatic to build gigantic levees around cities that are sinking amidst rising seas, as New Orleans did after Hurricane Katrina and before 2021s Hurricane Ida, or as Italy's Venice has been doing amidst copious debate, as waters rise in its famed canals. Levees built for billions of dollars may even hold back the ocean for a while. However, water has a pernicious way of finding a way over, under, or around walls, especially if someone has built them of porous limestone.

6.2.1 The Problem Is What We Put in the Air

"Ultimately," comments Professor Young (2021, A-23), "We must also recognize that the coastal zone will be in ever-increasing peril until we tackle the changing climate in a meaningful way. All this 'resilience spending' is just a Band-Aid, not a cure, We can build all the seawalls, dunes, beaches, and marshes we want, but *the problem long term is not what we put on the ground, but what we put in the air*" [emphasis added]. Briefly, maybe even a bit simplistically, we must drastically reduce the amount of greenhouse gases (mainly, but not only, carbon dioxide) that we—nearly all of us—have been pouring into Earth's atmosphere, and, bearing in mind the perils of thermal inertia (discussed in detail in ▶ Chap. 2, on science), *do it quickly*. This is not a future peril, and it's not the least bit optional. Every minute we wait, in fact, the less optional it becomes.

Returning to the storms for which the word "unprecedented" was being described as "understated." Journalists nearly burned holes in their keyboards describing "the sheer volume of rain [that] simply overwhelmed the infrastructure of a region built for a different meteorological era" (Newman, 2021, A-1). Officials warned that the unthinkable was quickly becoming the norm. The intensity of the storm (or the fires, etc.) was jaw-dropping. In Central Park, 3 in. of rain fell in 1 h, breaking a record set a few days earlier by Hurricane Henri. The New York Subway became a river, as it shut down. Interstate highways turned into coursing rivers of stalled, wrecked cars backing up against hillsides.

The intensity and number of storms was indicated by a statistic from the US government's *National Climate Assessment:* "In the Northeast, the strongest 1% of storms now produce 55% more rainfall than they did in the middle of the twentieth century, when global warming first became widely associated with rising greenhouse gas levels" (Plumer, 2021a, A-15).

Every 5 or 6 years since the early 1990s, the Intergovernmental Panel on Climate Change (IPCC) has compiled reports on the state of the Earth's climate crisis. At first, the IPCC was a small voice in a conceptual universe dominated by other voices—a tree hugger's oddity, perhaps, with a few forward-looking scientists' warnings standing out. One of these belonged to James Hansen, director, at the time, of NASA's Goddard Institute for Space Studies, on Manhattan Island in New York City. Hansen had the prescient distinction of being one of the first (with colleagues), about 1980, to begin asking basic questions about survival of humanity, as well as other life on Earth, due to steady warming of the atmosphere. Hansen had begun his academic quest not by studying Earth, but by asking why Venus, a planet somewhat like Earth in other ways, was so hot.

Hansen's consideration of this question took him to the climate of Venus, the subject of his PhD dissertation. It was the greenhouse warming of Venus, in fact, that led him, during the 1970s, to devote his scientific life to the study of global warming on Earth. Venus, millions of years ago, may have been more like Earth—a little warmer, perhaps, owing to its position closer to the sun, but not too warm for liquid oceans. The geophysics of Venus differs from Earth's in one important way; however, it has no plate tectonics, which allow subsurface pressures here to discharge piecemeal. Such discharges express themselves in volcanic eruptions and earthquakes that can be deadly close at hand but across the entire Earth maintain a rough equilibrium.

On Venus, subsurface pressures are said to have built up to enormous levels and then exploded in a single, spectacular blast that provoked a runaway greenhouse effect. Temperatures rose and the oceans boiled away. As a result, today's Venus has a surface atmosphere hot enough to melt lead. Most of Venus' carbon is now in the atmosphere, not in its crust (as on Earth), observed by Hansen (April 15, 2013).

So that's Venus—no place for a picnic. What about Earth? Here, with technological ingenuity and a desire for profit, comfort, and convenience, human fossil fuel corporations are supplying us with products manufactured from carbon extracted from the crust, combusted as energy, and released into the atmosphere as carbon dioxide, methane, and other trace gases, some of them manufactured by human beings. We are, in other words, mimicking natural processes on Venus, but in a deceptively slower way.

6.2.2 IPCC Reports Become More Emphatic

The IPCC's first reports today sound rather timid. Global warming had not even broached the debates of US presidential politics in 2000, when Al Gore tried to make an issue of it on this very public of stages. The rest of the field was stumbling along behind him, asking whether humankind had any role at all in the steady warming of the atmosphere that was becoming more evident. Republicans and others already were already rallying their usual forces of persistent denial.

Every half decade or so, the case made by the IPCC's member scientists became more urgent—not only that humankind has a role in the warming of the planet but also whether it was the major influence, with foreseeable consequences portending very identifiable and unpleasant results. The debate over climate change thus became part of mainstream political debate and other discourse.

And into the midst of this weather rumble, the IPCC in 2021 dropped its most strident warning yet. It was, as IPCC reports usually are, several hundred pages long. However, its main message could be boiled down into a few words, such as: get off of fossil fuels quickly or reap the meteorological whirlwind.

At some point—exactly *when* we do not yet really know, although the IPCC has said roughly 2040 on our present emissions path—feedbacks may take off on their own and accelerate the changes (in a "hyperthermal"). By that time, the climate deniers will have been discredited, but it may be too late. We will be on our way to a world where the air in most places will eventually become simply too hot to sustain human (or most other) life. And, as had become usual as time had passed, Jim Hansen was still standing in the midst of the storm almost four decades later saying that, "it was not an exaggeration to suggest," writing in an e-mail to colleagues posted September 26, 2013, having said 8 years earlier that "based on best available scientific evidence, that burning all fossil fuels could result in the planet being not only ice-free, but human-free."

The New York Times reported on its front page August 9, 2021, to summarize the IPCC's latest report: "Nations have delayed curbing their fossil fuel emissions for so long that they can no longer stop global warming from intensifying over the next 30 years, although there is still a short window to prevent the most harrowing future," a major United Nations scientific report concluded in 2021 (Plumer & Fountain, 2021, A-1). That was in 2021, 19 years before the IPCC's prospective "hyperthermal." Today's older teenagers will barely be broaching middle age by then. Solutions are becoming more urgent as the climate clock ticks down.

The report was based on more than 14,000 studies and approved by 195 nations around the world, including (after the departure of Donald Trump) the United States, the world's largest source of greenhouse gases, after China (Plumer & Fountain, 2021, A-1), the most comprehensive summary to date of the physical science of climate change (Plumer & Fountain, 2021, A-1). Contrasting with reports a quarter-century prior, the United Nations summary left no doubt whatsoever that humankind is the primary source of increases in fossil fuel emissions and subsequent warming, violence of storms, and more.

Consider the Words "Thermal Inertia"

This was not a tale for the squeamish by 2022, because, for those who dig into the science, what's to be learned becomes even more ominous as time passes. We have covered this ground before, but it is worth a second look. The concept of thermal inertia is at the basis of the IPCC's more ominous warnings. People as a whole do not act until the hot wind and violent storms warn them that the warnings are real. The political infrastructure likewise marches to the drummer of public opinion. Only then do specialists such as architects and engineers adjust their operations to accommodate the new (and rapidly intensifying) reality. By the time our human apparatus adjusts, we are already 50–150 years behind on climate change's clock.

Why are so many scientists sounding dire warnings that a decade or two of business-as-usual fossil fuel production and consumption will carry the Earth over various "tipping points" beyond which human ability to influence climate change may become irrelevant? Don't we have plenty of time for such a slow-motion crisis to unfold? Once again, please, consider the two words "thermal inertia." You will be hearing them again. Jim Hansen taught me about thermal inertia, the geophysical fact that today's temperatures reflect greenhouse gas emissions of a half-century ago in the atmosphere and a century or two in the oceans. This principle does not get much coverage in most of the press. Here is a temporal analogy: In the oceans, the temperature level at Pearl Harbor is only now broaching December 7, 1941). It is only now beginning to show up in the oceans. On land, the consequences of our lassitude about the climate crisis has just now reached the first Earth Day in 1970.

As has been described in previous chapters, global warming is a deceptively backhanded crisis in which thermal inertia delivers results a half-century or more after our burning of fossil fuels provokes them. Our political and diplomatic debates react *after* we see results. Political inertia plus thermal inertia thus present the humanity and our planet with a challenge to fashion a new energy future *before* raw necessity—the hot wind in our faces—compels action. Global warming is dangerous because it is a sneaky, slow-motion emergency, demanding that we acknowledge a reality century in the future with a system of individual, legal, and diplomatic reaction that reacts in the past tense. Thermal inertia and natural variability allow the climate change deniers among us to argue that geophysical facts do not matter.

Related to thermal inertia is the carbon dioxide level in the atmosphere, which recently passed 420 parts per million. This figure is as high as the Pliocene, 2–3 million years ago, when the oceans were perhaps 100 ft higher. Eventually, the oceans *will* rise, but thermal inertia delays the impact. Today's carbon dioxide level does not instantly melt the glaciers that raise sea level.

People need to understand why their actions today will influence the fate of coastal cities in a century or two. Only if we stop pouring greenhouse gases into the atmosphere *today*, which is, of course, impossible. Today, we are shaping agriculture and the fates of plants and animals centuries well into the future. By that time, today's greenhouse gas levels (which will be higher in 300 years, speeding this process) will have melted much of Greenland's ice, as well as the West Antarctic Ice Sheet. Portions of the larger East Antarctic Ice Sheet will be melting. Most mountain glaciers and the Arctic Ocean's ice cap will exist only in history books and dusty copies of the *National Geographic*.

6.2.3　**This Cake Is Already Being Baked**

NASA satellites have detected a widening rift in Antarctica's Pine Island glacier on the Western Antarctic Ice Sheet (WAIS). This glacier acts as a plug, like a cork in a bottle, between the ocean and the WAIS. Should this ice sheet melt into the sea, the world ocean would rise perhaps 6–8 ft. That "cork" is now breaking up.

At the same time, Greenland's ice is also melting at an accelerating rate, as are mountain glaciers around the world. Should our fossil fuel emissions continue at their present rate—an accelerating, unchecked development of all available reserves, conventional oil and gas, tar sands, fracked shale, and so forth—and if the atmosphere's proportion of greenhouse gases continues rise to levels unknown for many millions of years, we can anticipate an uninhabitable world: not tomorrow, of course, but sooner than many of us believe. Into this equation now ambles the aching need for durable solutions—into this land that is full of honest, educated people who are aware of the problems but are also dealing with the unaware, denying carnival barkers who think it's all a hoax, as well as many others who are out to rip off money in the face of crisis.

6.2.4   Did Anyone Mention "Crisis"?

The need to change our energy and transportation infrastructure today and as quickly as possible will be heightened by a look into a likely future if we do not act, or do so too slowly. The ultimate climatic nightmare becomes possible after the oceans warm enough to convert presently solid methane deposits in the oceans to atmospheric gas, further accelerating warming due to melting permafrost, dissolving ice sheets, and desiccation of rainforests. Once again, this will not be tomorrow's news, although today's inactions will eventually make it more possible. Geophysical evidence suggests that the Earth has suffered bouts of severe warming in the distant past from natural causes that were intensified by release of the planet's stores of greenhouse gases that reside in solid form, beginning with peat and permafrost on land, followed by methane hydrates in the oceans.

The geophysical facts do not cease to matter because some politicians do not recognize them. They also do not vanish as surface knowledge about global warming spawn all sorts of hustles and scams that may sound like progress on the surface but really do nothing to provide real progress on a deadly serious issue.

This book is meant to increase understanding of science and possible solutions, with a realization that humankind's survival instincts *can* run congruent with a sustainable future.

6.2.5   The Cost of Coal and Oil as Energy Is Now Above That of Solar and Wind

The above facts outline the horror story, now made more monstrous by the drumbeat of the daily news, the hot, humid breath in our faces, when and where the story is no longer a question of *if,* but *when.* What about solutions? Do we have any left? Yes, we do, in that "narrow window" described by the IPCC's 2021 report. In a world where solutions are mainly rooted in economics, one of the most important moments in recent environmental history was that glorious day when the cost curve of solar and wind energy for general use in everyday human life dipped below that of oil and coal. In a world driven by profit and loss, this is a major, positive tipping point for solutions.

A more visible moment may have been the introduction, in 2021, by several motor companies (Tesla, Ford, General Motors, Honda, Toyota, et al.) of their first mass-production all-electric vehicles, although one must still investigate the source of their electricity. An electric car or truck is no environmental bargain if the power it uses is derived from coal, natural gas, or oil (as gasoline). On the other hand, a hybrid car generates energy as its battery is being charged and delivers about 650 miles from a standard 15-gallon tank of gasoline. By 2022, even before infrastructure for electric cars had been fully developed, sales of them were exploding. In 2021, about 4% of new cars were electric, double the proportion a year earlier, but still a minuscule proportion of the total. The biggest single corporate presence in the market was Elon Musk's Tesla, which delivered almost a million electric cars in 2021, a 90% increase compared to 2020 (Ewing & Boudette, January 8, 2022, A-11).

6.2.6 The "Year-Round Fire Season"

In the search for solutions, how do we deal with the growing toll of wildfires, which are pouring more and more carbon dioxide into the atmosphere? Wildfires have always occurred, but usually within a given time period, late spring through summer, into early fall. By 2022, however, people and agencies that fight fires found that the fire season had jumped its former seasonal limits on both ends of the calendar. Employees of the US Forest Service referred to their employer as a fire department with a forest service attached.

Kate Brown, governor of Oregon, wrote in the *New York Times* of "a fire season that never ends" (Brown, 2021, A-24) 5 weeks after Portland, Oregon, had reached 116 °F, and Salem, Oregon, 117 °F, while Oregon alone was losing a million acres a year to wildfires, only one front in a world aflame. "Over the long term," she wrote, "We must take concerted, nation-wide action to address climate change" (Brown, 2021, A-24). She also endorsed electric cars and hybrids which use gasoline as they are started and then switch to electricity when the vehicle moves. Such vehicles use very little gasoline and do not depend upon finding a charging station every few hundred miles. Governor Brown made her comments as firefighters in Oregon were battling to contain ▶ the Bootleg Fire, which had already burned more than 400,000 acres along Oregon's border with California, the Beaver State's third largest blaze since 1900. "The harsh reality is that we're going to see more of these wildfires. They're hotter, they're more fierce and obviously much more challenging to extinguish. And they are a sign of changing climate impacts," Brown told CNN.

In February 2022, the *New York Times* reported that a new United Nations report "has concluded that the risk of devastating wildfires around the world will surge in the coming decades as climate change further intensifies what the report described as a 'global wildfire crisis'" (Zhong, 2022). "The scientific assessment is the first by the organization's environmental authority to evaluate wildfire risks worldwide," the report said. Zhong wrote: "It was inspired by a string of deadly blazes around the globe in recent years, burning the American West, vast stretches of ▶ Australia and even in ▶ the Arctic. The images from those fires—cities glowing under ▶ orange skies, smoke billowing around ▶ tourist havens and heritage sites, woodland animals badly injured and killed—have become grim icons of this era of unsettled relations between humankind and nature. The heating of the planet is turning landscapes into tinderboxes," said the report (Zhong, 2022).

6.3   The Need for Solutions: The Scientific Basis

About 34 million years ago, Jim Hansen has said, temperatures sank low enough that ice sheets began forming on Antarctica. Three to four million years ago, much the same process began in the Arctic. Both locked up large amounts of carbon dioxide, as levels declined in the atmosphere. About 2 million years ago, glaciations became cyclical, as ice ages alternated with interglacials. Atmospheric levels of CO_2 declined during glacial advances and rose between them. The last ice age ended about 12,000 years ago (Kolbert, 2020, 24).

"What is now happening, Hansen explained, is that climate history is being run in reverse, and at high speed, like a cassette tape on rewind. *Carbon dioxide is being pumped into the air some ten thousand times faster than natural weathering processes can remove it. 'So humans now are in charge of atmospheric composition,'* Hansen said [emphasis added]. Then he corrected himself. 'Well, since we're determining it, whether we're in charge or not'" (Kolbert, 2020). Emphasis is due here because the fact that we are pumping CO_2 into the atmosphere at 10,000 times [the rate] that nature can remove it, bringing its level down even to natural rates is going to take quite bit of work (i.e., energy in some form), even at a time when sources of greenhouse gases (such as increasing fire coverage and others) are increasing.

Among the many risks of running the system backwards is that the ice sheets that formed during the ice ages will disintegrate, and as humankind pumps much more CO_2 (and other greenhouse gases) into the air much faster than the natural system works, the system cannot cope. Or, perhaps more to the point: human civilization began at a given position of disposition, of so much ice vis-à-vis so much water. If ice is melting, and if it is melting much more quickly than natural rates, sea levels rise, and flooding becomes a very large problem. Human beings have an affinity for oceanic water, and before human beings had the technology to move themselves and freight by air, large amounts of it were moved on ships, over oceans from port to port. (Transportation also produces greenhouse gases, more per pound by air, less if by sea.)

Large trading centers formed, often (but not always) at the intersections of oceans and river systems which, so it happens, are the areas most prone to flooding if sea levels rise. Consider New York, New York, Boston, New Orleans, London, Kolkata, Shanghai, and many other long-standing cities. We have been pouring CO_2, etc. into the atmosphere at accelerating rates since about 1850, since coal and oil became major energy sources.

Given that thermal inertia delays effects of ocean ice melting for about 150 years, we have barely begun to see major rises in the seas. Thermal inertia in the ocean takes roughly a century and a half, so temperature rises there began in a significant way about 1950. I say "about" because other processes get in the way and cause wiggles in the temperature and CO_2 graphs. After time has passed, however, the general direction becomes clear, along with effects on weather and other processes, most of which are negative: extremes of precipitation, rising intensity of storms, intensity and number of heat waves, and so forth. By this time, human beings pay attention as nature ups the ante, but scientists realize that nature has sprung a trap—that the sea level licking at the docks on Manhattan Island (and many other places around the world) is roughly 50–150 years behind what we have already placed "in the pipeline." Today's visible conditions are several decades behind what we already have "paid for." This scientific basis raises the intensity of humankind's need for effective and enduring solutions—quickly.

In the meantime, ice continues to melt, and temperatures rise, with various processes reinforcing each other. Jim Hansen estimates that if all of the Earth's ice melts, sea levels will rise about 250 ft. Run your eyes across a map of the Earth (a globe might even be better, for visual proportion) and give some thought to which cities will be inundated, and where insurance premiums will rise most quickly. At big insurance companies, this process has already begun. If you own a beachfront home anywhere between Brownsville, Texas; New Orleans; Tampa; Miami; Charleston, SC; Norfolk, Virginia; and New York City, keep an eye on your flood insurance bill, if you have one.

Elsewhere, look for knots of population between river deltas and oceans. Take an extra hard look at the Bay of Bengal and the Ganges River Delta, or Shanghai and the Yellow River, the North Sea, the Thames River, London, Paris and the Seine, Miami Beach, Buenos Aires, Sydney, Australia, Mumbai, India, and many more. Stop for a moment, catch your breath, and realize what Jim Hansen was thinking when he added all of this and evaluated when and where it is taking us: "So you can't do that without producing a different planet" (Kolbert, 2020, 27). *This is the "cake" that we are now baking. All of it raises the timely need for effective, enduring solutions.*

How High Must Global CO_2 Go Before All of Earth's Ice Melts?

So, how high can atmospheric carbon dioxide go before we reach the point where we are on a downhill slide to a world without lasting ice? This won't happen all at once, Hansen reminds us. The Earth is still large, and so is remaining ice. Humankind's grandest edifices are puny by comparison. However, the general geophysical betting line these days is that carbon dioxide at 350 parts per million is the last semi-safe stop on this bus. Hey, you say: last I saw, the CO_2 proportion is pushing 420 ppm and generally hasn't stopped rising since coal began to be mined in Britain and oil spurted out of the ground in Pennsylvania.

Why was it, for example, that oil and coal developed as resource bases when they did? One reason is that they were required to fuel the industrial revolution. Another reason was that slavery was no longer a legal source of cheap power. Why, along the same line, was the cotton gin invented when it was? One very important reason was that Black hands were no longer legally available to pick cotton under the duress of slavery, even if cotton had a rapidly growing market. Thus, a scientific invention came along in time to converge with several social, political, and legal trends.

6.3.1 **"Coal is My Worst Nightmare"**

Nearly no one at the time (in the middle to late nineteenth century) paid any attention to the geophysical results of raising the proportion of oil and coal, etc. in the atmosphere. The dominant point of view seemed to be that milder winters would not be bad at all. Only by about 1980 did many scientists begun to ask the correct questions. Only by about and after 1980 did a large number of scientists come to agree with Hansen and others. The timbre of questions increased as the effects of raising greenhouse gas levels became more evident. At first, many people, not the least of whom were some environmentalists, came along slowly, and then with a sense of emergency. Later, they came to be regarded as prescient, especially as the daily weather began acting up more than usual. Even corporate leaders took pride in telling the world that they had some solar panels on the company roof. That is where we are today.

Perhaps 20 years ago—hardly the wink of an eyelash in the history of human beings' tenancy on the Earth, some inquiring minds began to ask how long it would take for the West Antarctic ice cap to melt. At first, the estimates often were too far in the future to matter to anyone now living. At nearly the same time, ice shelves began to crack up. The same thing was noticed among glaciers on the highest mountains. Comparative pictures of well-known mountains' ice packs came into view. Look, for example, at the snow mantle of Hemingway's Kilimanjaro about 1900 compared to today's photographs. Closer to home, Washington State's Mount Rainier (native name: Tahoma) has become instructive as well. The size of the Arctic ice cap over a few decades' time delivers some evidence as well, as do temperature readings, especially in the Arctic and Antarctic.

Soon, wild suppositions became contestable facts, and then generally accepted estimates of dangerous sea-level rise advanced uncomfortably close to our own time, or the lifetimes of our children and grandchildren. Climate science and meteorology are malleable, and we have been watching them change. Net loss of ice at the fringes of Antarctica's ice cap has started, now, not in the century or more believed a few decades ago. Greenland's ice is shrinking. "The extent of the Arctic summer ice is now only a little more than half what it was just 40 years ago," Elizabeth Kolbert wrote in *The New Yorker* (2020, 27). Deserts north and south have been growing fast enough to measure in one human lifetime. Pervasive, economy-killing droughts have grown larger and longer in a few short decades in such places as the western one-third of North America.

Steven Chu, a physicist who holds a Nobel Prize who served as US Energy Secretary under Barack Obama, said: "There's enough carbon in the ground to really cook us. Coal is my worst nightmare" (Kolbert, 2020, 28). In his frequent travels, Hansen quotes him or review Bob Dylan's lines in "The Times They Are A-Changing," adding a touch of global hydrology. Or John Holdren, President Barack Obama's science advisor: "Any reasonable comprehensive and up-to-date look at the evidence makes clear that civilization has already generated dangerous anthropogenic interference with the climate system" (Kolbert, 2020, 29).

Writers and editors also are advised to do their math when estimating when the world as we know it now will be solidly in the past tense. Savor, for example, this sentence: "It's [the 'heat dome'] that rewrote the record book on heat in the Pacific Northwest during late June 2021, [which is] being called a once-in-a-millennium event, which means you might have been expected to experience it once during your lifetime…" (Mann & Hassol, 2021, A-22). "Millennium," in this case, means a thousand years. If we take this phrase literally, we also must ask: how many of us will live for a thousand years? This mistake is made with embarrassing regularity by accomplished reporters and *New York Times* copy editors.

With the heat has come extreme drought in the US Southwest (Arizona, New Mexico, Utah, and parts of California, Nevada, and New Mexico). In this area, lack of water ranks with civilization-destroying "mega droughts" of at least two millennia (2000 years), and it is not over yet. All of these droughts cannot be blamed on today's form of climate change, however. Clearly, in the past, droughts have been aggravated by natural forces other than human beings burning fossil fuels. These droughts without fossil fuels occurred in the Mayas' homeland, on and off, between the 800s and 1500s CE. Our time has delivered a new form of ruination for the global environment that is new in humankind's history—as recent as the mass use of the internal combustion engine and the use of coal and oil as major sources of energy, as well as the end of legalized slavery, the widespread use of electricity, and the defeat of George Armstrong Custer by Lakota and Cheyenne warriors—a major battle in its time that was contested on *horses*. Major concern over the destructive power of greenhouse warming is barely a quarter-century old now—*one-third* of a human lifespan today over much of the Earth. The search for mass-scale solutions of this problem which follow—the good, the bad, and the even worse—is younger than that. With that prelude, let us consider a number of really bad ideas for solutions that have been thrown at this issue and done little except cost billions of dollars.

6.3.2 Coal Capture and Sequestration (CCS): A Bad Idea Whose Time Has Come—and Gone

We now have a fair number of leaders in some of the Earth's most economically powerful countries advertising what they believe is magic that will make carbon dioxide vanish. Wow! Double wow! They made an agreement at a meeting in mid-June 2014 to collectively cut carbon emissions *in half* by 2030. That sounded *really* good. Unfortunately, by 2022, we had not even begun to approach such an ambitious goal.

There is, however, always distance between pledges and actually making and enforcing such goals, especially in countries supported by laws and ethics that are frequently used to support industries' powers to actually take actions that may be unpopular with shareholders and others who draw profits from keeping things more or less as they are. Many such people have not received the message that the entire Earth is sleepwalking into an existential crisis. Even those who pledge to do certain things need more persuasion to be convinced that signing onto a pledge is mainly signing some paperwork and smiling for the TV cameras. After that, the *real* work begins.

There's been a lot of talk in lofty political circles (e.g., the White House, et al.) about "carbon capture and sequestration (CCS)."

Thus, at meetings such as the G-7 (or G-8, depending on political alignments), corks are popped and negotiations undertaken over catered food by notable people with constituencies, in buildings with rich histories. A lot of airline fuel is burned to get together for talks to reduce fossil fuel emissions. Then, at a conference, they smile for the cameras and then fly home without having to do more than pledge to do something about coal. Score one important point for shuck and jive, for stumbling before reaching even the first curve on a very long journey. They did pledge to refuse funding for any project that does not have CCS technology, which is a very poor solution, even in theory. However, the G-7 did pledge to achieve "an overwhelmingly decarbonized electricity sector" by 2030 (without dealing with coal other than a puny fig leaf called CCS). This one was made for 60-s news snatches, dessert for the TV cameras.

6.3.3   Old Habits Die Hard: Use of Coal Has Been *Increasing*

Given the shuck and jive that we have been hearing recently, one might believe that coal is on its final run as the world's most-used and dirtiest fossil fuel. We should be so lucky. Coal is still by far the Earth's major fuel, and its major source of energy, as well as carbon pollution. At the end of 2021, the International Energy Agency estimated that coal's use would *increase 9%* worldwide during 2022. The IEA estimated that coal's use would increase by 20% in North America and the European Union. India and China probably would see increases of 9–12%, according to this forecast. With many important political figures calling for such things as a net-zero energy system by 2050, the situation out in the Real World is much different.

"Coal is the single largest source of global carbon emissions, and this year's [2020] historically high level of coal power generation is a worrying sign of how far offtrack the world is in its efforts to put emissions into decline toward net zero," said IEA executive director Fatih Birol (Gillespie, 2021, A-8). Rising prices of natural gas has been a factor, while talk about solar, wind, and other renewables has produced some progress, thus far, but very little in coal's big picture.

A reminder to everyone: if greenhouse gases had a sense of humor, they would be laughing their heads off. Do we need to be reminded? Greenhouse gases have no sense of humor, nor conscience, nor any of our cherished human notions. They do nothing except accumulate in the atmosphere and hold heat. They will hold more heat as long as human beings use them to pursue warmth, comfort, and profit. Period. Full stop.

6.3.4 Environmentalists Infiltrate ExxonMobil's Board of Directors

Now we consider both the serious and superfluous, that there have been some changes in the world energy mix, but who knows what they will amount to over time, of which we have a shrinking supply when it comes to dealing with climate change. How much of such changes and what amounts to comic opera? For example, in 2020 and 2021, environmental activists in a tiny environmental hedge fund called Engine No. 1 secured three seats on the board of directors of ExxonMobil, the world's largest oil company, in an attempt to get the oil giant to reduce its carbon footprint. This was the first time that Exxon management had lost one or more seats that the firm's management had not recommended. Quite an accomplishment this was, but in the world where policy is made, a minority of votes on a corporate board are, as we say in Spanish, "mejor canada" (Spoken usually with a sense of irony, "better than nothing").

For environmentalists, who only very rarely sit on the boards of very large oil companies, this is good news, of course. Such a thing has never before happened, and given some shrewd politicking, it may give birth to some real changes—as long as the rest of the board can be convinced of the existential nature of the climate change crisis. However, after a dollop of excitement about this very partial coup, thus far, we have not heard much about actual accomplishments by this crew.

In the battle of boardroom politics, Engine No. 1, the tiny hedge fund formed an alliance with a corporate giant in the mutual funds industry, Blackrock, whose chief executive officer, Laurence D. Fink, stressed the necessity of climate action in his annual letter to executives: "No issue ranks higher than climate change on our clients' lists of priorities. They ask us about it nearly every day" (Phillips, 2021, B-3). Nice words, but Engine No. 1 held only 0.02% of ExxonMobil's stock at the time. Blackrock, Vanguard, and State Street companies together, which owned about 20% of ExxonMobil's stock, all had signed on to the Net Zero Asset Managers' Initiative, a commitment to guide their companies to net-zero carbon emissions by 2050. Suddenly, this environmentalists' invasion of ExxonMobil's Board of Directors seemed to have acquired something more than hippie joke status. Stay tuned to see what this insurrection can do for the Planet Earth. The jury, as is said in the cliché hall of fame, is still out. The greenies are still woefully outnumbered in ExxonMobil's boardrooms that they will most assuredly lose any battle more serious than what color to paint the boardroom's walls.

6.3.5 Seawalls to the Rescue? Beware Limestone

One proposed solution to oceanfront flooding is construction of seawalls, especially in places such as Florida, which is essentially a very large sandspit, that is, among the largest and most vulnerable sea-level risks on Earth. The US federal

6

government (using the Army Corps of Engineers) and the City of Miami in 2021 advanced a tentative plan to build 6 miles of seawall, only a small part of which would be on or near oceanfront properties.

The rest of the wall would have been inland, perhaps designed by engineers who recognized that much of the coast is limestone, which is very porous and would allow seawater to easily flow around and under it. Even if Miami and nearby areas were to build 6 miles of wall that could impede seawater in a serious way, what about the rest of Florida's thousands of miles of shoreline, which faces essentially the same problems as Miami? Other sea-wall promotors also might consider the proposed cost of the 6 miles: $6 billion (Mazzei, 2021, A-1). Multiply that by the number of 6-mile long segments necessary to fence off all of Florida, how many Florida municipalities could afford even part of it? In addition to sea-level rise on porous soil, Miami and the rest of Florida face another climate change peril: growing intensity, frequency, and heavier rains from hurricanes. Rising seas also may soon invade the aquifer that supplies much of Miami's drinking water. As seas continue to rise (a full moon is now sometimes all that it takes to flood some areas of the city), insurance may become unaffordable. Substantial flooding also could crack old sewers and septic tanks (Mazzei, 2021, A-18). Given all of this, the climate change deck is increasingly stacked against Miami, with gleaming ranks of towering condos and office buildings lining its shores a little higher from sea level at high tide as a basketball hoop.

6.3.6 The Chief Heat Officer in Athens, Greece

With temperatures steadily trending upward in Greece (as high as 111 °F during the summer of 2021), Eleni Myrivili was appointed as Athens' first chief heat officer. At that time, she was Europe's first person paid to keep her eyes open for ways to beat the heat. Myrivili decided to contend for the job on Athens' record-hot day as she stopped hanging laundry on her rooftop behind the Acropolis, nearly unable to breathe from the heat. "I could only take short, kind of burning breaths," she said, recalling that ash from the fires also turned her black clothes white. "It was scary" (Horowitz, 2021). She took the job as roughly 200,000 acres of parched, forested land burned in Greece.

Jason Horowitz (2021) reported from Athens for the *New York Times*: "It is not just Greece. In recent days, a heat wave on the Italian island of Sicily appears to have resulted in the ▶ hottest recorded temperature in European history, and fires have broken out across the Italian south. Europe's summer of natural disasters has included increasingly frequent extreme weather events that have caused fatal flooding in Germany and Belgium, as well as in ▶ Turkey. Every week there is a new nightmare."

Athens' urban heat has been augmented by human density from a building boom after the country's civil war that was meant to assist migration from rural areas. Athens is the second most densely populated city in Europe after Paris. "Heat is an invisible and insidious killer," Ms. Myrivili said. "Heat is one of those climate hazards that you don't really see. It's hard for people to talk about it. You don't see flying roofs and cars flooded. It is really important to get people to understand why it is dangerous" (Horowitz, 2021).

Athens also has built many miles of heat-sucking black-topped roads as its physical size has grown. Tarred roofs also have replaced gardens and parkland. During heat waves, temperatures on asphalt surfaces in Athens have reached 60 °C (140 °F), high enough to make an extended walk painful (Horowitz, 2021). The heat, as well as smoke from nearby fires, has made everyday life in Athens nearly insufferable. Most of Athens' apartments have no air-conditioning, a legacy of cooler summers past.

An anti-heat activist since 2007 and a social anthropology professor, Myrivili was elected to the Athens City Council (2014) and as deputy mayor (2017–2019), as she focused on climate change issues. She also became an international activist through the Rockefeller Foundation of a program meant to place at least one heat officer on every continent. In 2021, Miami-Dade County (Florida) hired the first heat officer in the United States; Freetown, Sierra Leone, became the first to make such an appointment in Africa.

Athens' city government has been working to provide more residential air-conditioning, as well as cooling centers, with plans to make asphalt more reflective (and less common), and to replace black tarred roofs with reflective solar panels and gardens. On the ground, plans call for more shade-bearing trees.

6.3.7 Mitigation of Global Warming: Shopping Until You Drop?

For some business, eco-greenery is mainly old-fashioned avarice pumped up in the Earth Day drag. In the corporate caverns of Manhattan, Christmas, ad-agency style, "tis the season of greensploitation"—greenwash as marketing strategy. Witness a full-page advertisement in the *New York Times* (A-9, November 16, 2007). The New York department store Barney's displayed an otherwise nude model (lower neck upward) with a gold necklace (hanging a Christmas tree pendant) across her forehead, pitching: "Gorgeous **green** gifts, fabulously **fair trade** fashion, sensationally **sustainable** swag, orgasmic **organic** denim and cashmere, **environmentally conscious** tchotchkes & more."

If we are going to shop until we drop, we might as well convince ourselves we are going to **save the earth**, and do it **in large type, all caps**. We now wallow in a sea of green exploitation—so much that we need to ask: Dare we blow a holistic in this fantasy by asking just how "green" all of this really is? Or is it just a joke played on the overtly environmentally correct?

Another example: Jim Hansen dubbed Minnesota Governor Tim Pawlenty of Minnesota "governor greenwash" after he signed a bill in 2007 calling for 25% renewable energy in his state but then failed to take a stand against a large new coal-fired power plant, Big Stone II, proposed for South Dakota, but supplying power to Minnesota. New coal plants not only aggravate global warming, said Hansen, but they also restrain growth in scale (and reduction in cost) for clean alternatives such as wind and solar (Dear Governor, 2008). Again, the talk was green, but the walk was not. Such things are in style these days.

In one advertising technique, negative perceptions of a company or product are directly confronted with counterpropaganda. Witnessing "clean coal" television advertising with its sparkling white and shimmering light-blue background visuals, one could easily confuse these with pitches laundry detergen. Advertising aimed at alleviating odious image problems can be downright Orwellian. Agencies get paid good money to create "spin"—not to lie, exactly, but to create an image that is only marginally on speaking terms with reality.

Witness, for example, BP's advertising campaign regarding the Gulf of Mexico after a huge oil spill there. BP presented itself as eco-champion and as friend of the American public. Anna Beaty, a master's student in communication at the University of Nebraska at Omaha, studied public reaction to BP's post-spill advertising campaign and found that most people expected propaganda, so they discounted the messages. "The reason that these propagandist approaches need to be exposed is because the public has a right to know how they are being misled. In essence BP [was] attempting to buy [its] way to salvation through these campaigns," she wrote (Beaty, 2011, 38). Many viewers were nearly impervious to this kind of advertising appeal. Viewers expected BP (whose advertising agency once told us in its heyday of greenwash that "BP" stood for "Beyond Petroleum") to manage reality in its favor.

6.3.8 Clean Coal: On a Mission to Save the Planet

In reality, of course, there is no such thing as "clean coal," not in its mining, which is dirty and dangerous, nor its power generation, which is fraught with toxins one can see (such as ash effluent), even after it has been—bring on the laundry detergent—"scrubbed," and things we can't easily see, such as mercury. Coal never was and never will be "clean." It's an advertising agency oxymoron. "Clean coal" is an exercise in corporate mass advertising word association on the order of George Orwell's "war is peace," and "hunger is plenty." The technological fixes that go under that name do very little to reduce the atmosphere's carbon dioxide load. Meet British Petroleum (BP, for those who don't like the eco-conscience-cringing sound of "petroleum" as in "Beyond Petroleum," Earth-friendly frackers, and the "gas lady").

To hear them tell it, the propagators of clean coal are on a mission to save the planet. By making coal-fired energy more efficient, they say it will provide more energy, more reliably, at less expense than all the planet's solar panels and wind turbines ever will. They spread their message at climate treaty negotiations. In 2013, the 19th Conference of Parties (COP) was held in Warsaw, capital of Poland, in a country that derives 90% of its electricity from coal. The Polish government used the occasion to cosponsor the International Coal and Climate Summit, organized by the self-same government and the World Coal Association, as an international showcase for "clean coal."

Others set the bar higher. "The coal industry can and must radically transform and diversify to avoid the worst impacts of climate change," Christiana Figueres, then the United Nations' top climate official, told the CEOs of major coal companies at the 2021 Glasgow COP26 conference. "Let me be clear from the outset that my joining you today is neither a tacit approval of coal use, nor is it a call for the immediate disappearance of coal.... I am here to say that coal must change rapidly and dramatically for everyone's sake" (UN Climate, 2013). She also called upon the coal industry to close inefficient plants. She also urged the coal industry to "implement safe carbon capture, use and storage systems on all new plants, even the most efficient, and to leave most existing reserves in the ground" (UN Climate, 2013). Several environmental groups regard "clean coal" as a dangerous a climate change wolf in sheep's clothing. Many have signed a "People's Declaration on Coal," demanding a moratorium on all new coal projects and the end of any role for coal at United Nations climate talks.

Outside the halls of paid advocacy, "clean coal" is something of a joke these days. Even the World Bank has cut financing of coal-fired power plants to "rare circumstances," countries with "no feasible alternatives," as dealing with climate change becomes a live subject in halls of corporate power. The strategy of this change, according to a Reuters report in 2013, is to "send a signal that coal is a risky investment and prompt countries to turn to alternative energy sources" (Yukhananov & Volcovici, 2013). The change was announced under the directorship of Jim Yong Kim, the first scientist to lead the World Bank, as part of a newly aggressive stance on global warming.

However, Godfrey Gomwe, chair of the World Coal Association's energy and climate committee and chief executive of Anglo-American Thermal Coal, responded, saying that the Warsaw Conference of Parties, otherwise known as COP19, "is an important step toward what we all hope will be a comprehensive climate agreement in Paris in 2015. The technology exists to significantly reduce CO_2 emissions from coal. If we were to raise the global average efficiency of coal plants from its current average of 33% up to 40%, we could reduce global carbon emissions by more than two gigatons" (UN Climate, 2013). That is a very large number that Gomwe left dangling without supportive mathematics. If carbon dioxide had a sense of humor and a head for numbers, it would have gotten a laugh out of that one.

6.3.9   The Trouble with Conference Goals

Shortly before the COP 26 meeting from November 1 to 10, 2021, in Glasgow, Scotland, a personal journal kept by Jim Hansen indicated his frustration at the walking in place attitudes of the series of COP (Conference of Parties) held for years to chart goals meant to guide reduction of greenhouse gas levels in the atmosphere: "Prior COPs have been characterized by self-delusion so blatant that one of us (JEH) [James E. Hansen] describes the backslapping congratulations at the end of the COPs as a fraud," Hansen said. He continued: "We cannot blame it all on the political leaders, however. We scientists deserve a large part of the blame. Scientists were slow to realize how low the targets must be for greenhouse gas (GHG) levels and for global warming to achieve a stable, healthy climate for young people and future generations. We also should have made clearer the effects of lags (delayed responses) in the climate system, as well as the time required to replace energy systems that are the largest source of GHGs."

Political leaders were slow to even set a meaningful goal. At last, with the Paris Agreement at COP21 in 2015, they set a goal to limit global warming to 1.5 °C. Global warming had already reached about 1 °C, so 1.5 °C was believed to be the lowest feasible warming limit. However, the leaders did nothing to realize the two essential actions that the target implied. Instead, they went home and took actions and allowed policies that made the goal unachievable (Hansen & Sato, 2021, n.p.).

Hansen favors a rising carbon fee that is collected from fossil fuel companies and distributed to citizens on an equitable basis. The rising fee will force the companies to change their asset base. Also, Hansen favors use of modern nuclear power, which is cleaner than it was during past accidents, with waste that is tiny compared to fossil fuels.

Even energy companies are coming to regard "clean coal" as an oxymoron. By 2013, coal mining was declining in the United States, and cleaner, often cheaper natural gas was taking its place (i.e., until coal use began to increase again in 2021). In November 2013, the Tennessee Valley Authority's board of directors, which supervises the largest public power utility in the United States, voted to close eight coal-fired power plants and partially replace them with natural gas. Between 2010 and 2013, in the United States, more than 150 coal plants were closed or scheduled for shutdown (Frosch, 2013, A16).

With demand for coal falling (at that time) and environmental regulations becoming more stringent, more coal mines were closing than at any previous time. "Coal plants are the single largest source of dangerous carbon pollution in the United States, and we have ready alternatives like wind and solar to replace them," said Bruce Nilles, director of the Sierra Club's Beyond Coal campaign, which aims to reduce US coal-fired power eventually to zero. We have a choice, which in most cases is cheaper and doesn't have any of the pollution" (Frosch, 2013, A-16, A-19).

New York Mayor Michael R. Bloomberg in 2011 donated $50 million to the Sierra Club through Bloomberg Philanthropies to support its nationwide campaign to eliminate coal-fired power plants. Michael Brune, Sierra Club executive director, said that the gift was used in the Sierra Club's "Beyond Coal" campaign, "which helped block the construction of 153 new coal-fired power plants across the country since 2002" (Torres & Eilperin, 2011).

6.3.10 The Climatic Consequences of Coal

So just how "clean" *is* "clean coal"? As the dirtiest and most pervasive fossil fuel, coal is the poster child of the damage that global warming can inflict on Earth and its inhabitants. Continuing to burn coal to generate electricity all but guarantees that we will face catastrophic changes in our climate. In a widely circulated open letter, Bill McKibben and the legendary environmentalist Wendell Berry called for massive protests at the "Capitol Coal Plant" in Washington, DC during March

2009. Driven by the conscience and the urgency of their mission, Berry, Hansen, and McKibben all publicly faced the prospect of arrest for participating in a non-violent protest against coal burning on the capitol grounds.

"There are moments in a nation's—and a planet's—history," they stated in part, "when it may be necessary for some to break the law in order to bear witness to an evil, bring it to wider attention, and push for its correction. We think such a time has arrived... We have our bodies, and we are willing to use them to make our point. We don't come to such a step lightly. We have written and testified and organized politically to make this point for many years, and while in recent months there has been real progress against new coal-fired power plants, the daily business of providing half our electricity from coal continues unabated. It's time to make clear that we can't safely run this planet on coal at all" (Johansen, 2010, 3).

Despite its role as humankind's major source of greenhouse gas pollution, by 2013, consumption of coal, which comprises 90% of Earth's remaining fossil fuel reserves, was increasing in Europe, the heartland of climatic political correctness. Why? It's not just the sweet music of mass advertising. Coal has real appeal. By conventional methods of accounting (which ignore environmental damage), it is "cheap." It's also convenient and familiar. Poland, most especially, is married to coal, as are China, India, and almost half of the United States' electrical generating capacity in 2014 (less than 20% by 2021). Advocates of coals use have long pitched its utility. The sun doesn't have to shine, nor the wind blow, to use it.

Europe's use of coal rose sharply in 2012 after declining for several years stoked by cheap imports from the United States, where the price of coal has declined (in competition with a flood of inexpensive natural gas, much of it from "fracking"), as well as restrictions on nuclear energy. The US coal exports to Europe rose 26% during the first 9 months of 2012 compared to 2011, as exports to China also increased (Birnbaum, 2013). "It's been very welcome that US greenhouse gas emissions have been going down because of the switch to gas," said David Baldock, executive director of the Institute for European Environmental Policy in London. "But if we're simply diverting the coal somewhere else, particularly to Europe, a lot of those benefits are draining away" (Birnbaum, 2013).

In theory, Germany may have the loftiest goals to "go green" in the world. Part of the plan is to phase out nuclear power by 2022. In practice, by 2013, this meant burning sulfur-laced, low-energy lignite (or "brown coal") mined at home at the highest rate since the early 1990s. Lignite supplied one-quarter of German electricity in 2012. Solar panels have been installed at record rates in Germany, but new coal-fired power plants were being built as well. The *New York Times* reported: ["I]n Jäenschwalde, a stone's throw from the Polish border, the forested countryside quickly drops away into a 300-foot-deep pit stretching for miles. Enormous machines slowly eat away at the Earth and shower soft lignite onto a conveyor belt that feeds directly into a nearby power plant. From the precipice of the mine, the 20-foot-tall trucks at the bottom look like Tonka toys" (Birnbaum, 2013).

6.3.11   Coal Worldwide: Easy to Hate, Tough to Ignore

In addition to the figures for coal's continued use cited above, lignite supplied 25.6% of Germany's electricity in 2012, up from 22.7% in 2010. Hard black coal supplied an additional 19.1%, and its consumption also was rising. The revival of coal use has posed a challenge to Germany's lofty environmental goals. By 2050, the country aims to generate 80% of its electricity from renewable sources, allowing steep reductions in greenhouse gas emissions. Green advocates worry that if Germany's extensive—and pricey—support for renewable energy such as wind and solar power diminishes, coal might fill the gap.

Just as environmental advocates tried to dump goal on the ash heap of history, within a few years, China developed a major appetite for worldwide supplies of coal. "At ports in Canada, Australia, Indonesia, Colombia, and South Africa, ships are lining up to load coal for furnaces in China, which has evolved virtually overnight from a coal exporter to one of the world's leading purchasers," wrote Elisabeth Rosenthal in the *New York Times* (2010).

Coal used to be burned mainly close to mining sites. By 2010, China had changed that pattern, adding the carbon emissions of transport to those of coal consumption itself. During the next decade, China closed some coal plants but continued to be the world's leader in both mining and use of the direct coal on the plant. By 2021, China's marriage to coal showed some signs of ending, but it had a long way to go. At the same time, China was building some of the world's largest solar arrays. With 1.4 billion increasingly affluent people, China has one big leg in coal, and hands in wind and solar, all to raise the living standards of 1.4 billion people. Some of these Chinese solar arrays reach the horizon.

China's economy by 2010 was burning half of the 6 trillion tons of coal used on Earth each year. The price of coal doubled from 2005 to 2010, spurred by Chinese demand. "This is a worst-case scenario," said David Graham-Caso, spokesman for the Sierra Club, "We don't want this coal burned here, but we don't want it burned at all. This is undermining everything we've accomplished" (Rosenthal, 2010). In Australia, environmental groups have blocked coal trains en route to Newcastle, as fleets of kayaking protesters delayed coal being loaded for Asia.

6.3.12   More and Larger Cargoes of Coal for China and India

The world's largest private coal company, Peabody Energy, said it was planning to sell larger and larger amounts of coal to China. "Coal is the fastest-growing fuel in the world and will continue to be largely driven by the enormous appetite for energy in Asia," the company said (Rosenthal, 2010). India's coal imports also rose, from 36 million tons in 2008 to 60 million tons in 2009. This trend continued through 2022. Like China, India has been paying lip service to reducing coal combustion, but its availability, low cost, and increasing demand for electricity in India make large reductions a political nonstarter. India does have a few solar and wind power arrays, but they are, at least at present, little but a matter of experimental value and public relations. The government insists that the electrification of the nation by any available means must precede costly conversion to wind and solar. While India has been converting on a small scale, coal is still the workhorse of electricity production. Note the brown haze in New Delhi's sky and elsewhere in India.

6.3.13 Opposition to New Coal Plants Accelerates

Political action has been intensifying over coal-fired power in the United Kingdom, where the "Kingsnorth Six" were acquitted by a Crown Court jury during September 2008. They were among 23 Greenpeace volunteers who attempted to shut down the Kingsnorth coal-fired power plant. The six painted the smokestack with anti-coal slogans before they were interrupted by police. Their defense (upheld by a jury) was "lawful excuse," that they were protecting property of greater value (the Earth) from the impact of climate change. With briefs submitted by James Hansen and others, the defendants argued that coal has a dominant role in a warming climate that poses a clear and present danger.

Environmental groups have tightened their focus on proposed coal-fired power plants. The Environmental Defense and the Natural Resources Defense Council has assembled "strike forces" to mobilize opposition to new plants state by state. These strike forces played a role in obtaining cancelation of plants in Florida and Texas. In New Mexico, for example, the groups intervened in a dispute over whether to construct a new power plant on the Navajo Nation, where the state government, which was opposed to the project, has no direct power to prevent it. The plant's carbon footprint would have equaled 1.5 million average automobiles. At that time, coal-fired electricity contributed more than half of the 57 million tons of annual carbon dioxide emissions in New Mexico (Barringer, 2007).

Thelma Wyatt Cummings Moore, a Fulton County (Georgia) Superior Court judge on June 30, 2008, blocked construction of the first coal-burning power plant proposed in Georgia in more than 20 years, ruling that the government must limit emissions of carbon dioxide. This was the first time that a court had applied an April 2007 ruling of the US Supreme Court recognizing that carbon dioxide is a pollutant under the federal Clean Air Act to an industrial source. The judge overturned the Georgia Environmental Protection Division's approval of an air pollution permit for Dynegy's proposed Longleaf power plant south of Columbus, Georgia. "In a case that is being watched across the country, Judge Moore sent a message that it is not acceptable for the state to put profits over public health," said Justine Thompson, executive director of GreenLaw, the Atlanta public interest law firm that represented the environmental groups (Georgia Judge, 2008).

Between 2006 and 2008, plans for 83 coal-fired power plants in the United States voluntarily withdrew or were denied permits by state regulators. Several utilities were moving away from coal. Xcel Energy, for example, erected 274 wind turbines in northeastern Colorado (Warner, 2009). In early August 2007, Missoula, Montana's Mayor Democrat John Engen won the city council support to buy electricity from a new coal-fired plant starting in 2011, to save the city money. He then was inundated by hundreds of e-mails and phone calls from protesting constituents. Late in July 2011, US Senate Majority Leader Harry M. Reid told chief executives of four power companies that he would "use every means at my disposal" to stop plans to build three coal-fired plants in Nevada. "There's not a coal-fired plant in America that's clean. They're all dirty," Reid said, urging a turn toward wind, solar, and geothermal power (Mufson, 2007, D-1).

All in all, coal-fired plants in the United States have been closing at a rapid rate: 290 plants closed from 2010 to May 2019, 40% of the country's generating capacity. For example, a unanimous vote of Florida's Public Service Commission rejected a Florida Power & Light proposal to build a coal-fired plant near Lake Okeechobee, Florida. Gov. Charlie Crist said approvingly that the Public Service Commission "sent a very powerful message" and that the state "should look to solar, wind, and nuclear as alternatives to the way we've generated power in the Sunshine State" (Mufson, 2007, D-1).

6

During the subsequent years, the erection of new coal-fired power plants was coming to be understood as a political and environmental nonstarter across most of the United States. Be that as it may, the closure of coal-burning power plants paled beside the total of about 8500 in the world. By 2007, Citigroup had downgraded the stocks of all coal companies. "Prophesies of a new wave of coal-fired generation have vaporized, while clean coal technologies remain a decade away, or more," their report said. The Citigroup analysts said that by 2008, "election politics are likely to turn progressively more bestial for coal. Candidates are already stepping up to 'ban coal.'" The Citigroup report said that coal producers' earnings would probably be hurt by "new regulatory mandates applied to a group perceived as landscape-disfiguring global warming bad guys" (Mufson, 2007, D-1).

By 2010, continuing afterward, several new coal-powered generating plants were being canceled or postponed across the United States. However, at about the same time, 645 coal-fired power plants were producing about half the country's electricity; as recently as May 2007, more than 150 new ones had been planned to meet electricity demand that was rising at a 2.7% annualized rate. A private equity deal worth $32 billion involving TXU Corp. canceled 8 of 11 planned coal plants, as similar plants were scuttled in Florida, North Carolina, Oregon, and other states. In the meantime, late in August 2008, Xcel Energy of Minneapolis became the first builder of coal-fired power plants to agree with New York State's Attorney General Andrew Cuomo to provide investors with an analysis of global warming risks posed by its business.

Wind power by 2014 (continuing to at least through 2022) was replacing coal—and quickly. The Omaha Public Power District, heretofore two-thirds coal-generated (the rest at the time was nuclear), by 2016 generated 33% of its electricity from wind. Iowa was already over a 45% wind power by 2020, with 100% planned for 2050. South Dakota, Kansas, Texas—not liberal bastions—were installing wind at a rapid pace. Even Steve King, an ultraright-wing congressman from Western Iowa (since defeated at the polls for political statements that were an embarrassment even to right-wing Republicans), favored wind power over coal for electricity generation.Climate change concerns often were cited as coal plants were canceled, especially in Florida, where rising sea levels from melting ice in the Arctic, Antarctic, and mountain glaciers already has been eroding coastlines. Florida's

Public Service Commission by 2016 was legally required to give preference to alternative energy projects over new fossil fuel generation of electricity. The states of Washington and California have been moving toward similar requirements. Xcel Energy and Public Service of Colorado were allowed to go ahead with a 750-MW coal-fired power plant only after it agreed to obtain 775 MW of wind power. The federal government suspended its loan program for new coal plants in rural communities after 2009 due in part to uncertainty over a suit filed by Earthjustice during July 2007. The suit cited the government's failure to consider the global warming as it financed new coal plants.

Cancelations or delays of coal power plants continued into 2021 and beyond. In February, NV Energy delayed a plant in Eastern Nevada until real "clean coal" technology becomes available; Southern Montana Electric Generation and Cooperative halted work on a plant near Great Falls, in favor of wind turbines and another plant that will burn natural gas. Michigan Governor Jennifer Granholm, a Democrat, told regulators not to approve any of five new coal-fired plants until "all feasible and prudent alternatives" had been considered" (Watson, 2009). Peabody Energy dropped plans for a coal-fired energy plant in Western Kentucky, in favor of a plant that will convert coal to natural gas. In early March 2009, Alliant Energy dropped plans to build a very large coal power plant in central Iowa that would have provided enough energy to supply 500,000 homes. Several of these actions were being challenged by coal power advocates.

6.3.14 Goodbye Toxic Ash: Solar In, Coal Power Out

After decades of suffering the sickening ash clouds of an old coal-fired plant, the Moapa Paiutes, a tiny American Indian tribe who live about 50 miles north of Las Vegas, Nevada, not only retired it but convinced the Interior Department to site two solar power arrays nearby, creating jobs.

When the Reid Gardner Generating Station, adjacent to the Moapa River Reservation, was closed in 2017, its unwilling neighbors faced a new problem: ramping up pressure on the plant's owner, NV Energy, Inc., to clean up its residue of pollution. On August 8, 2013, the Moapa and the Sierra Club filed suit in US District Court, Las Vegas, Nevada, to legally compel the cleanup. MV Energy was bought out in May 2013 by Warren Buffet's MidAmerican Energy Holdings, which has taken an active role in raising Iowa's use of wind power for electrical generation in Iowa to half of the state's total usage by 2016, with full wind power planned, as stated above. Reid Gardner is the last coal-burning power plant in Nevada.

The Sierra Club lawsuit asserted that the federal Resource Conservation and Recovery Act and the Clean Water Act had both been violated over the years by dumping that compromised the health of nearby residents and threatened the drinking water of millions. The lawsuit alleged that for several years the power plant illegally dumped toxins into the Muddy River, flowing into the Lake Mead reservoir behind Hoover Dam, a source of drinking water for more than 2 million people.

6

On Earth Day, 2012, the Moapa Paiute protested the power plant in a 50-mile "Cultural Healing Walk" to Las Vegas with members of the Sierra Club that took 3 days in 100° temperatures. "We are all looking forward to the retirement of the Reid Gardner coal-fired plant that has for decades polluted our reservation," said Vickie Simmons, a leader of the Moapa Paiutes' committee for health and the environment. "And for the sake of our families' health, we must ensure that the toxic waste from the power plant is fully cleaned up. The safety of our community and the future of our children depend on it" (Moapa Paiute March, 2012).

William Anderson, Moapa Band Chairman, said: "The high percentage of thyroid and respiratory problems is a big concern for the tribal members on the reservation. We need a proper study from air monitoring equipment installed on the reservation to study the emissions we're breathing in. That would help determine what needs to be done for our people's health. We also need more stringent storage conditions for coal ash and a study to be conducted to show the health risks associated with breathing in coal ash" (Moapa Paiute Sue, 2013).

"Now, we have to find out what kind of remediation they're going to do—a complete restoration, a conversion to gas or some other type of project," Tribal President William Anderson told the Associated Press. "To us, the ultimate goal would be to remove everything and put the land back the way it was. We'll be able to come to some closure after almost 50 years" (Moapa Paiute Sue, 2013).

The Indian Country Today Media Network reported that air pollution takes the form of toxic coal dust, which tribal members say arrives in giant clouds that send people scurrying indoors, but that's not the only problem they have observed. "There are also several settling ponds for coal ash residue, there are enormous piles of coal that are uncovered, and a huge coal ash landfill that is also uncovered," said Barb Boyle of the Sierra Club, in an article published in the *Huffington Post* (2012), and Mary Ann Hitt, director of the Sierra Club's Beyond Coal campaign. Boyle said that the tribe "has borne this burden for decades. It's time to stop." The plant was built in 1965, and the Moapa Paiute say that they have witnessed their own standard of living plummet over time. "In my era, we were all healthy people," said Aletha Tom, who runs the Moapa schoolhouse. "We didn't have the asthma, thyroid problems, cancer, diabetes, but we have that on our reservation. It's so major now" (Moapa Paiute March, 2012).

In their long campaign, the Moapa Paiutes won allies. Nevada Senator Harry Reid (since deceased) called the Reid Gardner Power Station a "dirty relic" and supported its closure. Soon, due to the tribe's persistence, local television stations and newspapers brought the plague of ash-borne illness to a broader audience. A video "An Ill Wind: The Secret Threat of Coal Ash" reached people from the tribe's website, reporting that the power station "dumps toxin-laden coal ash, a byproduct of combustion, into landfills that lie just a few hundred yards from the reservation. On windy days, coal-ash dust from the plant billows over the reservation, with clouds so thick that you can see and taste them, tribal members say. At such times, residents don't dare let their children play outside. That apparently offers limited protection, though, as the dust seeps into homes, schools, and cars" (Woodard, 2012). The Sierra Club, Earthjustice, and Greenpeace assisted.

Instead of coal-fired energy, the Moapa Paiutes turned to solar power. Tribal Chairman William Anderson told the Indian Country Today Media Network that "the Interior Department gave the Moapa Paiutes fast-tracked approval to build the first-ever utility-scale solar energy project on tribal lands—which seems especially suitable in this region of year-round scorching sun. The 350-MW plant, now finished, built on Moapa Paiute trust land, generates enough power for 100,000 homes, according to the agency, which said that the project provides lease income for the tribe, as well as 400 new jobs, 15–20 of which are permanent.

A "Coal-Free Chicago"

Elsewhere, in Chicago, residents celebrated the closure of two 100-year-old, coal-fired power plants that had blighted their low-income and mainly Black neighborhoods, after a 10-year struggle. The Environment News Service reported March 2, 2012, that about 60 organizations and groups worked with communities affected by air pollution to make Chicago a coal-free city. The Fisk Station at 1111 W. Cermak Road in the Pilsen neighborhood was shut down, as was the Crawford Station at 3601 S. Pulaski Road in the Little Village neighborhood (Decade-Long, 2012).

"This agreement means a cleaner, healthier environment for the communities around these coal plants," said NAACP President and CEO Benjamin Todd Jealous. "Environmental justice is a civil rights issue, and the NAACP is committed to strong regulation and monitoring of toxic coal emissions. For too long, Fisk and Crawford have been literally choking some of Chicago's most diverse neighborhoods, and some of its poorest" (Decade-Long, 2012). About 600,000 Chicago residents are living within 3 miles of the two plants. The areas were regularly exposed to sulfur dioxide, soot, and nitrogen oxide at dangerous levels.

6.3.15 The Debate Regarding "Mountaintop Removal"

People rallied not only to close dirty old coal-fired power plants but also to stop "mountaintop removal" mining that turns hills and mountains into strip mines. In September 2010, opponents of mountaintop removal strip mining converged on Washington, DC to demand that it end. Others gathered at mine sites to obstruct their operations and risk arrest. By 2021, several of these mines had been shut down not only because they were damaging companies' images but also because the markets for coal were shrinking.

"Mountaintop removal" coal mining, in which Appalachian peaks are blasted off, is so environmentally devastating that the federal government should stop it by refusing to issue permits, a group of scientists wrote in *Science* early in 2010. University of Maryland researcher and lead author Margaret Palmer said, "The science is so overwhelming that the only conclusion one can reach is that mountaintop mining needs to be stopped" (Fahrenthold, 2010, A-3). The same week the paper was released, the Environmental Protection Agency approved West Virginia's Hobet 45 mine after its sponsoring company "made changes that would eliminate nearly 50% of the environmental impacts and protect 460 union mining jobs" (Fahrenthold, 2010, A-3).

"Upper elevation forests are cleared and stripped of topsoil, and explosives are used to break up rocks to access buried coal. Excess rock (mine "spoil," or waste products) was pushed into adjacent valleys, where it buried existing streams" (Palmer et al., 2010, 148). "It obliterates stream ecosystems," said Emily Bernhardt, a professor of biology at Duke University and a co-author of the study. She said that 1500 miles of streams had been destroyed so far. "They've been wiped from the landscape." At these sites, "peaks are sheared off with heavy machinery and explosives, exposing the coal seams inside. Excess rock is used to fill steep Appalachian valleys, some with streams at the bottom, to the brim" (Fahrenthold, 2010, A-3).

Rainwater falling on this rock accretes harmful sulfates. "To us, it's like smoking and cancer. It's just so clear-cut that streams below mine sites are left damaged," Palmer said. "The study also linked mountaintop mining to threats to human health, citing potentially toxic dust in the air, well water contaminated with chemicals from mines and fish tainted with toxic metals" (Fahrenthold, 2010, A-3).

At about the same time, June 16, 2010, more than 50 people demonstrated in front of the Union Pacific Railroad's headquarters in Omaha to emphasize its role in hauling coal to power plants across the United States. The Union Pacific earns more than 20% of its gross revenue from hauling coal for electrical generation.

Environmentalist Bill McKibben, founder of ▶ 350.org, an international campaign of activists seeking to reduce the level of carbon dioxide in the atmosphere, opened the protest, stressing that "Coal is the dirtiest of fossil fuels, emitting more carbon dioxide than natural gas or oil," as a small number of Tea Party demonstrators waved signs in favor of coal power (Ruff, 2010). Nineteenth-century political minds were advocating the fuel that started the industrial revolution. How antique—and destructive.

6.3.16 A Carbon Neutral Airline? When Pigs Fly?

Just a few years ago, the idea of flying a multiton jet airliner thousands of miles on corn stubble and pig fat caused some responsible people to groan "When Pigs Fly!" However, we sometimes forget the wonders of technological evolution. Benjamin Franklin thought it would take immigrants from Europe at least a thousand years to reach the Pacific shore. He was off by about 900 years.

A few more than 100 years ago, the state of aviation technology meant Orville Wright in a small, wobbly by plane with one simple rotating engine. In one sense, in aviation by 2022, some "pigs" had already flown. Airliners flying on bacon grease? Some interesting things had happened. Bear in mind that 250 years ago, Franklin never had seen (nor even contemplated) a smartphone, a personal computer, nor a commercial jet, much less one that flew on biofuel. Bear in mind what people may see in another 250 years—if we are not suffocated by the effluvia of fossil fuels.

Test fights had already been taken by the end of 2021—Seattle to San Francisco, on biofuel. This one was undertaken by Delta Air Lines, with others by United and Alaska. Drivers in the San Francisco Bay Area have become the first motorists in the nation to fill their gas tanks with an algae-based biofuel. Alaska used a combination of aviation fuel and field corn-derived biofuel to power a flight from Seattle to Washington, DC. Test flights with no passengers or freight have flown on biofuel from Seattle to Los Angeles and Washington, DC, as well as other routes.

So let us take a mental flight. Let's ask how long it may take and how much mental muscle power will be required to take thousands of biofueled "jets" (we will need another name) from place to place on Earth and how many "pigs" we will need. What will be the technical requirements? How much biofuel will we need, and how will it be manufactured?

Another side sees such strategies as wishful thinking. Sarah Shifley, a lawyer who volunteers on a team for the climate activist group ▶ 350.org in Seattle, which seeks carbon neutrality by 2050, believes that an airline industry which wants to go carbon neutral while doubling its number of flights by the same year is engaged in a classic have-your-cake-and-eat-it-too strategy.

Shifley, said: "After the summer we've had [in 2021], of heat domes and hurricanes and floods and fires, it's unfathomable to me to be considering doubling air traffic" (Gates, 2021, 12). "Elsewhere, particularly in Europe, flying is already being curbed by government policy. France in April 2021 banned domestic flights between cities with a railroad connection of less than 2.5 h. Various government agencies and organizations around Europe have imposed similar bans on short-haul flights for employee business travel" (Gates, 2021, A-1).

6.4 Geo-Engineering: Sulfur as Savior?

6.4.1 "Bombing" the Atmosphere with Sulfur?

A wrenchingly bad idea to "save" the planet from rising levels of carbon dioxide in the atmosphere is "geo-engineering," which involves a wretchedly gallery of grand plans which assume that the human race is a junkie that never will learn to live without its carbon dioxide "fix." From bombing the atmosphere with sulfur (which temporarily cools it) to building gigantic space mirrors to deflect sunlight, all of these ideas are expensive, short-term, and would do little, at great cost, except postpone the eventual climate change reckoning in a sulfurous haze.

The idea that sulfur cools the atmosphere may first have been noticed by Benjamin Franklin, who was on diplomatic assignment for the aborning United States in France during 1783 when several volcanoes erupted in Iceland, spewing a trail of sulfur in fits and starts for 8 months. Temperatures in France and eastern North America sank. Franklin wrote that "There existed a constant fog all over Europe and [a] great part of North America".

A few decades after Franklin made his supposition, nature illustrated just how abruptly a quick dose of sulfur could cool the Earth for a relatively short period. On April 5, 1815, the largest volcanic eruption in recorded history (before or since) exploded from Mt. Tambora on the island of Sumbawa, now part of Indonesia. A witness said that the entire mountain "appeared like a body of liquid fire" (Kintisch, 2010, 61). The huge eruption ejected a plume of sulfur into the stratosphere world-wide and in 1816 became known as "the year without a summer." Snow fell in Maine during June, and farmers in Upstate New York lost their corn crops as the period between frosts shrank to 70 days from the usual 130. Mohawks at Akwesasne, in far northern New York State, reported frosts into June.

Sulfur *does* cool the atmosphere. Suspended particulates caused by emissions of sulfur dioxide and some other urban air pollutants (aerosols) do increase the net albedo (reflectivity) of the Earth, thus usually exerting a cooling influence on tem-peratures. James E. Hansen has estimated that aerosols cool the climate by about 1 W/m^2, "which has substantially offset greenhouse warming" (Hansen et al., 1997, 231).

Mount Pinatubo in the Philippines erupted in 2001, ejecting about 10 million tons of sulfur into the atmosphere, enough to cool the Earth's near-surface atmo-sphere about 0.5 °C for a year or two, or roughly the increase in temperature attrib-utable to global warming during the previous century (had an El Niño not occurred at the same time, the temperature drop may have been on the order of 0.7 °C) (Crutzen, 2006, 212; Morton, 2007, 132).

Sulfur's effects are temporary. Unlike carbon dioxide, which often remains in the atmosphere for several hundred years, sulfur dioxide washes out within 2 weeks. If sulfur gets into your lungs, however, it can obstruct breathing. Because of its brief residency in the atmosphere, any attempt to cool the Earth using sulfur would have to be persistent. The physical requirements required to lift that much sulfur into the atmosphere year after year would be quite a challenge, especially as increases in greenhouse gas levels over time require more of it. The sulfur dioxide would have to be refreshed at least twice a month. If the requisite sulfur was raised as high as required on aircraft, imagine the amount of fossil fuels *that* would require.

Acid rain also would be created by that many injections of sulfur into the stratosphere. As Eli Kintisch wrote in *Hack the Planet* (2010, 65): "Sulfur dioxide is a pollutant that comes out of smokestacks and forms acid rain in the lower atmosphere…[G]as…released in the upper atmosphere, even in aggressive doses… would add only slightly to the global atmospheric sulfur load." However, a con-tinual sulfurous haze would inhibit solar power from concentrated solar power (CSP) facilities, which gather direct sunlight. Solar panels, which use diffuse sun-light, would not be affected.

Elizabeth Kolbert, who writes frequently about climate change in *The New Yorker*, provided a witty send-up of the idea that geo-engineers can solve global warming by pumping the stratosphere full of sulfur dioxide. In a 2009 review of Steven D. Levitt and Stephen J. Dubner's book *SuperFreakonomics: Global Cooling, Patriotic Prostitutes, and Why Suicide Bombers Should Buy Life Insurance*, she described the book's use of the "Parable of Horseshit"—the fact that urban planners during the late nineteenth century feared that the rising volume of horse excrement would inundate large cities such as New York until a technological fix (the automobile) remedied the problem with another pollutant, the effluvia of fossil fuels. The authors of *SuperFreakonomics* ardently believed that extending an 18-mile long tube into the stratosphere and pumping it full of sulfur dioxide would mimic an ongoing volcanic eruption and cool the Earth. The authors ignore climate science completely, not to mention the technological problems of keeping a hose aloft—"all of which," Kolbert concludes, "goes to show that while some forms of horseshit are no longer a problem, others will always be with us" (Kolbert, 2009, 77). This project, fetchingly named "Stratoshield" by Nathan Myhrvold's Intellectual Ventures, proposed to hold the hose above the Earth with helium balloons, adding a Jules Verne touch (Klein, 2014, 262, 264). The question of how the aforementioned hose would be kept in place against stratospheric winds has been left unaddressed. Perhaps, a decade plus later, the epic zaniness of the entire plan has sunk in.

6.4.2 Geo-Engineering: One More Try

Zaniness aside, for a period shortly after the turn of the millennium, the idea of counteracting human warming with stratospheric sulfur enjoyed a spasm of serious scientific review. A meeting of top scientists was convened at Harvard to consider it. The British Royal Society devoted an entire issue of its journal, *Philosophical Transactions*, to the idea, and the US National Academy of Sciences devoted a workshop to it. The American Meteorological Society called for serious study to "offer strategies of last resort if abrupt, catastrophic, or otherwise unacceptable climate change impacts become unavoidable" (Kintisch, 2010, 12).

Tim Kruger, who manages the Oxford University geo-engineering program, defended robust research into the subject because the threat of global warming, left unaddressed, may require its use, even with problems, as the lesser of several evils. Critics, he argues, compare the downsides with "the climate of today, rather than that of a climate-changed future. This is the equivalent of condemning a drug for having side effects in healthy people before even considering whether the benefits would outweigh any side effects on the ill. Humanity may yet find itself in the position of having to decide which option is the least worse" (Kruger, 2014, 457). After that, the debate cooled, and serious consideration of sulfur to cool the atmosphere waned, as greenhouse gas levels continued to rise. I am haunted by an e-mail message from James Hansen in 2014 that he was losing sleep over humanity's inability to address the existential nature of the climate change issue and peoples' tendency to throw nonsensical "solutions" at it. Hansen had always acknowledged the utter seriousness of the issue and also expressed optimism that the problem is solvable. The cost to his sleep may indicate that human foolishness had begun to erode even Hansen's steadfast hope that humankind has the mental fortitude to address such an important issue.

Paul Crutzen of the Max Planck Institute for Chemistry in Germany, who earned his *bona fides* on this subject partially by telling the world the dangers of stratospheric ozone depletion, prepared a paper advancing the sulfuric solution for *Climatic Change* that was widely circulated before its publication (2006, 211–219). Crutzen acknowledged that bombing the atmosphere with sulfur should be a strategy of last resort but argued that humankind was not mounting an adequate response to global warming. He said that 5.3 million tons of sulfur per year, delivered by airplane (or otherwise) at an annual cost of $50 billion, would compensate for the greenhouse gas warming expected during the twenty-first century.

The idea of a sulfuric solution to climatic change is not new (Russian scientist Mikhail Budyko had proposed it in the 1970s), but this was the first time that a Nobel Prize-winning scientist with environmental credentials had made such a serious proposal. Crutzen also coined the word "Anthropocene," as the geologic epoch in which human beings are the primary force shaping the future of the Earth, to supersede "Holocene," the epoch in which most geology textbooks had said we live. He was not the first with this idea (the late nineteenth century geologist Antonio Stoppani spoke of the "Anthropozoic era"), but his nomenclature has coincided with humankind's obvious uncontrolled reeling into the Earth's climatic future via global warming, as well as the planetwide scope of proposed solutions, such as impregnating the atmosphere with sulfur. In 2016, the Commission on Stratigraphy (ICS) voted to make the new term official, after which every geology textbook on the planet became slightly obsolete.

There exists among advocates of the sulfur solution a paradoxical mixture of fear that humanity isn't up to the task of dealing with global warming, together with a large dose of techno-fixit hubris which assumes that squirting prodigious amounts of sulfur into the atmosphere will fix the problem. For the techno-fixers, I am left with a haunting thought from McKenzie Funk, author of *Windfall: The Booming Business of Global Warming*: "We should remember that we rarely recognize hubris before it is too late."

6.4.3 "Solar Radiation Management"

A fascinating array of people have stepped into this debate—Bill Gates (co-founder of Microsoft, which has funded geo-engineering research,) "with his near-mystical quest for energy, miracles' [tapping] into what may be our culture's most intoxicating narrative: the belief that technology is going to save us from the effects of our actions…our…most powerful form of magical thinking" (Klein, 2014, 255). This belief that we can save ourselves from ourselves has earned a pedestal in the hall of Hubris Hall of Fame. In *This Changes Everything: Capitalism and the Climate* (2014, 268), Lowell Wood, a co-inventor of the sky-hose idea, compared it to the idea of terraforming Mars, with its thin atmosphere, that is, 95% carbon dioxide. There exists, according to Wood, "a 50/50 chance that young children now alive will walk on Martian meadows…[and] will swim in Martian lakes" (Klein, 2014, 288). Cue to the Hubris-prone: don't buy your ticket to the Hubris Hall of Fame induction ceremony until you consider the hubris-worthy nature of assuming that you will step out of your space transporter expecting to pick peonies on Mars. Anyone who thinks that Mars is ripe for terraforming has been smoking something that doesn't grow on Mars, even in a hippie paradise that someone believes has been terraformed.

Will the oldest of our audience please remind the younger ones of who Newt Gingrich was? He also pitched sulfur's atmospheric virtues. A number of notable climate scientists have ventured the opinion that the situation is desperate enough to require serious consideration of the idea, one of whom has been Kenneth Caldeira, an environmental science professor at Stanford University's Carnegie Institution's Department of Global Ecology, who invented the phrase "ocean acid-ification." He also invented another phrase, describing the sulfur-as-savior solu-tion: "solar radiation management". Serious consideration is not the same as adopting sulfur as a solution, however. Caldeira also has said: "If we keep emitting greenhouse gases with the intent of offsetting global warming with ever-increasing loadings of particles in the stratosphere, we will be heading to a planet with extremely high greenhouse gases and a thick stratospheric haze that we would need to maintain more or less indefinitely. This leads to a dystopian world out of a sci-ence-fiction story". The fact that solar radiation management would add acid to the oceans also hasn't escaped Caldeira. Even so, he helped pitch Nathan Myhrvold's vision of a giant hose spewing sulfur into the stratosphere as a "senior inventor" (Klein, 2014, 264).

A tone of desperation is palpable in climate change science when well-known and usually seriously sane people propose that filling the stratosphere with sulfur dioxide may be the only way to stop runaway greenhouse warming. Do we really want to dump jumbo jet loads of a corrosive chemical, filling the stratosphere with sulfur to shroud the surface from warmth, then live in a perpetual acid mist?

When Crutzen advanced the sulfur shield idea in 2006, he cited a "grossly dis-appointing international political response" to increasing evidence of global warm-ing (Kerr, October 20, 2006, 401). Caldeira said that countries need to "undertake studies on what we might do" in a climate crisis, given the current trajectory of carbon concentration in the atmosphere. "Nobody likes the idea of engineering Earth's climate. Unfortunately, at some point, our other options may be even more unpleasant," said Caldeira (Eilperin, 2010).

"We should treat these ideas like any other research and get into the mindset of taking them seriously," said Ralph J. Cicerone, president of the National Academy of Sciences (Broad, 2006). Most of these proposals involve geo-engineering, large-scale rearranging of the Earth's environment to suit human needs. This idea "should not be taken as a license to go out and pollute," Cicerone said, emphasiz-ing that most scientists believe that reducing greenhouse gases in the atmosphere should be the top priority. He added, however, that "In my opinion, he [Crutzen] has written a brilliant paper" (Broad, 2006).

Wallace S. Broecker, a geo-engineering pioneer at Columbia University, also has proposed spreading tons of sulfur dioxide into the stratosphere, in imitation, he says, of erupting volcanoes. The injections, as calculated by him during the 1980s, would require a fleet of hundreds of jumbo jets in flight continuously (Broad, 2006). Again (repetition is called for here given the ridiculous nature of this proposal): has anyone calculated the amount of greenhouse gases that such flights would produce?

In the special issue of the scientific journal *Climatic Change* that was devoted to this subject, in August 2006, Crutzen discussed the theoretical basis of the idea, possible methodologies, as well as advantages and disadvantages. Several other authors also discussed the history of such proposals, the practical as well as ethical considerations of various approaches, and how best to evaluate them. The authors make it clear that geo-engineering climate is a less desirable potential solution to warming than controlling greenhouse emissions at their sources and that only if warming causes sufficiently harmful impacts would geo-engineering be a better choice (Editors' Choice, 2006, 387).

In a draft of his paper, Crutzen estimated the annual cost of his sulfur proposal at up to $50 billion, or about 5% of the world's annual military spending. "Climatic engineering, such as presented here, is the only option available to rapidly reduce temperature rises if international efforts fail to curb greenhouse gases," Crutzen wrote. "So far," he added, "there is little reason to be optimistic" (Broad, 2006). Supporters of this idea contend that any increase in sulfur at the Earth's surface would be small compared with the tons already being emitted from the smoke-stacks of coal-fueled power plants (Broad, 2006).

Bill Gates and the World's Leading Salesmen for the Sulfur Solution

David Keith, a Canadian who became a professor at Harvard and recipient (with others) of $6 million in grant money from the Bill Gates Foundation, since the 1990s became perhaps the world's most prominent single advocate of geo-engineering. Keith, who has an affinity for huge-scale, expensive techno-fixes, also promoted carbon capture and sequestration (CCS) in a report with four Canadian energy executives. By 2014, Keith was sponsoring a "field experiment aimed at understanding another geo-engineering technology that would use a balloon to release sun-blocking particles of sulfuric acid in the stratosphere" (Kintisch, 2013, 307). Keith's support for the "sulfur solution" is unequivocal: "It is possible to cool the planet by injecting reflective particles of sulfuric acid into the upper atmo-sphere…. To say that it's 'possible' understates the case: it is cheap and tech-nically easy…. for the price of a

Hollywood blockbuster" (Keith, 2013, ix). Keith writes with occasional humor and nuance and factors critics of his ideas into the discourse. For example, he proposes using "solar radiation manage-ment" to counter half of global warm-ing's effects to mitigate effects on precipitation distribution, most notably the Asian monsoon (Keith, 2013, 13–14).

Keith argues that the idea must be tested. He presents himself, as a friend of environmentalists who also subjects their assertions to intellectual rigor. He finds James Hansen to be something of an extremist, a "climate scientist turned activist" (2013, 23). Keith realizes global warming is serious business, however, even as he ridicules forecasts of "immi-nent doom" (2013, 23). Instead, Keith believes that alarmism incites "disaster fatigue," a response that numbs people and prevents action (2013, 170).

Despite the unequivocal nature of his rhetoric, Keith does agree with some of those whom he calls alarmists that "Whether or not we deploy geo-engineering, we must still *eventually* [one word with which James Hansen probably would disagree] decarbonize our energy system to reduce long-term climate risks" (2013, 100). He notes that geo-engineering can do little to stem the acidification of the oceans—a threat, eventually, to much of the marine food chain—which is related solely to the carbon dioxide level in the atmosphere, independent of its warming effects (2013, 103). For a person whom many regard as the world's leading salesman for geo-engineering, Keith displays a refreshing sense of self-reflection. He expects that "a world where geo-engineering is tested and available will be one that spends less on emissions" (2013, 131). And he can't quite agree with science journalist Eli Kintisch's assertion that "geo-engineering is a bad idea whose time has come" (2013, 173).

Keith's ideas have been called "barking mad" by geophysicist Ray Pierce-Humbert and an "utter political nightmare" by former British Royal Society head Martin Rees (Jones, 2013, 302). While advocating use of a fleet of jets to spray the stratosphere with sulfur to create an artificial volcano's cooling effect, even Keith admits that his plan could play a role in air pollution that "probably would cause thousands of deaths each year from asthma, heart disease, and lung cancer," perhaps skew monsoon rainfall patterns, damage the ozone layer, and aggravate ocean acidification (Jones, 2013, 302). However, Keith argues, this would be a small price versus the problems raised by accelerating temperature rise worldwide. Keith argues that "the potential upsides of geo-engi-

neering" require greater study of such ideas, which "may show that these technologies will not work." If so, he believes, "the sooner we find this out, the better" (Kintisch, 2013, 307). Our deepest condolences to everyone who suffers fatal diseases along the way. Is the road to hell really lined with good intentions? Mike Hulme commented in *Can Science Fix Climate Change* (2014, 51): "And these are not isolated results. Nearly all the modeling studies that have simulated the regional effects of stratospheric aerosol injection show that the existing mosaic of regional and local climates ends up being reconfigured. Stabilizing global temperature to avoid the danger zone of more than two degrees [Celsius] of warming only ends up *destabilizing* regional climates around the world."

Even so, Hulme describes a "slippery slope" down which humankind may slide that combines "technological and sociological 'lock-in'" with a bias toward innovation and "the vested interests of fortune, fear, fame, and fanaticism pushing the technology onwards," compelled by intensifying climate catastrophe (which he anticipates in acute detail) until "full-scale deployment [of stratospheric aerosol injection] eventually becomes unstoppable" (69, 76–81).

Looking back a decade or so from the slippery slope of geo-engineering, it's both a relief and a sense of consternation to review how many otherwise very intelligent individuals seriously sought salvation in sulfur. This idea seems to have vanished down Lewis Carroll's famous rabbit hole. Why not simply load up a half-dozen overdue active volcanoes around the world and let 'er rip? This, in preparation for another really hairbrained idea, which involves…

6.4.4   A Really *Big* Umbrella

If an acid bath for the upper atmosphere doesn't sound so good, how about a really *big* umbrella of fine particles to screen out sunlight—yet another perpetual haze? The Stratospheric Particle Injection for Climate Engineering (SPICE) has a cute acronym and a trial run $2.8 million price tag (cheap!), but as no one has yet stepped up to test run this sun-reflection scheme, "In a world struggling to control its greenhouse gas emissions, [this project] could also prove highly lucrative for inventors," wrote Daniel Cressey in *Nature* (2012, 429). The project initially was funded by the United Kingdom's government to investigate whether spurting reflective aerosols into the stratosphere could help to bounce some of the sun's warming rays back into space. As part of this project, SPICE had planned to test a possible delivery system: "pumping water up a 1-km-long hose to a balloon, where it would be sprayed into the sky," Cressey wrote. It was rejected in part over intellectual property rights to a patent application rather than concern regarding the feasibility of the technology or the fact that it does nothing to solve the fundamental, long-range problem, which is the rising level of greenhouse gases in Earth's atmosphere.

Project leader Matthew Watson, an Earth scientist at Great Britain's University of Bristol, said that modeling work on the project would continue. Meanwhile, the geo-engineering field has been troubled by its inability to adopt guiding principles, at a meeting in 2010. At the Asilomar Conference Center near Monterey, California. In 2011, a relatively small group adopted the Oxford Principles, stating that geo-engineering should be "regulated as a public good" (Cressey, 2012, 429). Some suggestions advocated an international control system for geo-engineering projects similar to that of atomic energy. David Keith agreed that legal restrictions should be placed on such a project and that "any technologies that could be controlled by a small number of people, yet have the capacity to rapidly alter our planet's climate, 'are deeply troubling'" (Cressey, 2012, 429).

Aside from squabbles over patents, scientific and environmental concerns over geo-engineering have intensified: "The risks of injecting aerosols into the stratosphere to combat global warming have been underestimated," said Klaus Keller and his team at Pennsylvania State University. Some have proposed that adding fine particles and liquid droplets, which reflect heat, to the atmosphere could postpone the need for deep cuts to carbon emissions. But by modeling the economic effects of substituting aerosol injections for carbon cuts, the researchers show that potentially damaging side effects of the aerosols—such as depletion of polar ozone and changes in precipitation patterns—could easily wipe out any benefit. And because aerosols disappear quickly whereas greenhouse gases linger, Earth could face even more abrupt and costly climatic change if any aerosol injections were interrupted by, for example, war or the breakdown of international pacts" (Geoengineering, 2011, 393).

Remember the junkie. Cut off the junk, and he's (or she's) *really* in trouble. No one wins by creating an addiction and then enabling it. Geo-engineering was banned by the Convention on Biological Diversity representing 193 nations on grounds of a threat to biodiversity late in 2010. The convention, adopted at a conference in Nagoya, Japan, states that "No climate-related geo-engineering activities that may affect biodiversity take place, until there is an adequate scientific basis on which to justify such activities and appropriate consideration of the associated risks for the environment and biodiversity and associated social, economic, and cultural impacts, with the exception of small-scale scientific research studies" under controlled circumstances (Eilperin, 2010). The United States did not ratify the convention. Also with an eye on side effects, the United Nations Convention on Biological Diversity in 2008 enacted a ban on iron fertilization. The ban was extended in 2010 over all schemes to alter temperatures via geo-engineering.

An increase in atmospheric particulate matter does not always exert a cooling influence, according to some reports. Researchers working with the US National Oceanic and Atmospheric Administration (NOAA) have assembled data indicating that periodic increases in atmospheric dust concentrations during glacial periods during the last 100,000 years may have resulted in significant regional warming and that this warming may have triggered some of the abrupt climatic changes observed in paleoclimate records.

Jonathan T. Overpeck, working with the Paleoclimatology Program at NOAA's National Geophysical Data Center in Boulder, Colorado, led a team of scientists who, during 1996, conducted global climate model simulations to examine the potential role of tropospheric dust in glacial climates. Comparing "modern dust" with "glacial dust" conditions, they found patterns of regional warming which increased at progressively higher latitudes. The warming was greatest (up to 4.4 °C) in regions with dust over areas that were covered with snow and ice (Abrupt Climate Change, 1996). Under some circumstances "aerosols can reduce cloud cover and thus significantly offset aerosol-induced radiative cooling at the top of the atmosphere on a regional scale" (Ackerman et al., 2000, 1042). Simply put, soot and other such pollution may not always mitigate other warming influences on climate as much as many of their proponent's assert.

6.4.5 Monumental Problems with Sulfur Shade

Filling the upper atmosphere with sulfur also may deplete stratospheric ozone and reduce overall precipitation, most notably during the African and Indian monsoons, which are crucial for hundreds of millions of subsistence farmers (Robock, 2008, 1166). Earth-wide, one-third of Earth's people depend on Asian and African monsoon rains for their staple foods. Reducing the intensity of sunlight also could reduce evaporation and rainfall.

Some scientists object to such a scheme on grounds that it would remove pressure to deal with the problem at its source, that is, to use energy sources other than fossil fuels. "I refuse to go down that road," said biochemist Meinrat Andreae of the Max Planck Institute for Chemistry in Mainz, Germany. "You're papering over the problem so people can keep inflicting damage on the climate system," he wrote (Kerr, October 20, 2006, 403). "The biggest risk of geo-engineering is that it eliminates pressure to decrease greenhouse gases," said Caldeira (Kerr, 2006).

Others argue that such a system could make the Earth dependent on a continuing human-provided sulfur "fix." Should the supply of sulfur falter, the Earth could heat up quickly within a few years. Some effects of increasing carbon dioxide levels, such as acidification of the oceans, would not be modified by the sulfur haze. The acidity of the oceans is influenced by the carbon dioxide level in the atmosphere, which the sulfur haze would not change, even if it cooled surface temperatures.

Additional strategies include the use of aircraft to maintain a dust cloud between Earth and Sun by making their engines less efficient, in other words, creating intentional air pollution. Roger P. Angel, an astronomer at the University of Arizona, has proposed a plan to put into orbit small lenses that would bend sunlight away from Earth—trillions of lenses, he calculated, each about 2 ft wide, extraordinarily thin, each weighing little more than a butterfly (Broad, 2006).

Some scientists believe that injection of sulfur into the atmosphere to counteract global warming may threaten the stratospheric ozone shield that protects the Earth and its flora and fauna from ultraviolet radiation. Sulfuric acid tends to attract chlorine atoms, creating a chemical combination that could assist chlorofluorocarbons (CFCs) in devouring stratospheric ozone.

Using a model of the sulfur load ejected in 1991 by the eruption of Mount Pinatubo, Simone Tilmes and colleagues wrote in *Science*: "We use an empirical relationship between ozone depletion and chlorine activation to estimate how this approach might influence polar ozone. An injection of sulfur large enough to compensate for surface warming caused by the doubling of atmospheric CO_2 would strongly increase the extent of Arctic ozone depletion during the present century for cold winters and would cause a considerable delay, between 30 and 70 years, in the expected recovery of the Antarctic ozone hole" (Tilmes et al., 2008, 1201). Adding sulfur to the stratosphere could lower temperatures there, promoting the formation of polar stratospheric clouds that provides surface area for creation of chlorine that destroys ozone.

Why not just quit burning carbon dioxide and methane? Is such a thing too prosaic for important people who would like to add "savior of the world" to their CVs? Thank you for not asking, because we have another dancing dog in the wings (no disrespect intended toward dancing dogs, of course). It has to do with seeding iron in the oceans to produce a population boom of phytoplankton, the basis of the oceanic food chain…

6.4.6   Iron Fertilization of the Oceans

Nearly half of the Earth's photosynthesis is performed by phytoplankton in the world's seas and oceans. This fact has led to proposals to geo-engineer a carbon dioxide sink using the same "biological pump" that is believed to have driven at least some of the Earth's past climate cycles. The chemistry of the oceans varies widely with regard to the amount of iron necessary to prime this pump and substantially increase carbon dioxide sequestration. In the equatorial Pacific and Southern ocean, wrote Sallie W. Chisholm, a marine biologist at the

Massachusetts Institute of Technology, it "is possible to stimulate the productivity of hundreds of square kilometers of ocean with a few barrels of fertilizer" (Chisholm, 2000, 686).

Should the oceans be seeded with large amounts of iron ore that will stimulate the growth of carbon dioxide-consuming phytoplankton? The idea has attracted some support among corporations and foundations looking for ways to minimize the effects of carbon dioxide without changing the world's basic energy generation mix. The idea is simple on its face: iron stimulates the growth of phytoplanktonic algae that are believed to be responsible for about half of the world's biologic absorption of carbon dioxide. By 2009, ocean fertilization had been largely rejected as a solution, in large part because, to succeed, it would disrupt large-scale ocean ecosystems in unpredictable ways, while holding little promise of significant reductions in atmospheric carbon dioxide (Strong et al., 2009, 347).

Ulf Riebesell, a marine biologist at the Alfred Wegener Institute for Polar and Marine Research in Bremerhaven, Germany, believes that ambitious iron seeding of the oceans could remove 3–5 billion tons of carbon dioxide per year, or about 10–20% of human-generated emissions, and on a factor of 10 times less expensive than planting forests to do the same thing (Schlermeier, 2003, 110). Patents have been issued for ocean fertilization and demonstration projects. One such project has been described and evaluated in *Nature* (Watson et al., 2000, 730–733).

In an experiment conducted between Tasmania and Antarctica, researchers confirmed that vast stretches of the world's southern oceans are primed to explode with photosynthesis, but lack only iron. As described by Brian Williams (2021), the researchers, who described their work in *Nature*, said it is too soon to start large-scale iron seeding because the new experiment raised as many questions as it answered. At best, they said, "iron seeding would absorb only a small amount of the carbon dioxide in the atmosphere." They also said that their "experimental bloom of plankton was not tracked long enough to determine whether the carbon harvested from the air sank into the deep sea or was again released into the environment as carbon dioxide gas" (Williams, 2021). Atsushi Tsuda and colleagues studied iron fertilization and found that, under some circumstances, it may dramatically increase phytoplankton mass (Tsuda et al., 2003, 958–961).

"There are still fundamental scientific questions that need to be addressed before anyone can responsibly promote iron fertilization as a climate control tactic," said Kenneth H. Coale, an oceanographer who has helped design studies of iron's effects in the tropical Pacific (Revkin, October 12, 2000, A-18). This seemingly simple proposition has some potential problems, however. First, there exists no way to measure the amount of carbon taken up by phytoplankton. Additionally, the algae produce dimethyl sulfide, which plays a role in cloud formation.

Additionally, phytoplankton increase the amount of sunlight as well as heat energy absorbed by ocean water. They also produce compounds such as methyl halides, which play a role in stratospheric ozone depletion. The iron also could promote the growth of toxic algae which may kill other marine life and change the chemistry of ocean water by removing oxygen. "The oceans are a tightly linked system, one part of which cannot be changed without resonating through the whole system," said Chisholm. "There is no free lunch" (Schlermeier, 2003, 110).

So much iron may be required to produce the desired effect that fertilization of this type will never [become] viable in a practical sense. "The experiments enabled us to make an initial determination about the amount of iron that would be required and the size of the area to be fertilized," said Ken O. Buesseler of the Woods Hole Oceanographic Institution, who co-authored a study of the idea. "Based on the studies to date, the amount of iron needed and [the] area of ocean that would be impacted is too large to support the commercial application of iron to the ocean as a solution to our greenhouse gas problem," he explained (Iron Link, 2003). "It may not be an inexpensive or practical option if what we have seen to date is true in further experiments on larger scales over longer time spans," Buesseler said. "The oceans are already naturally taking up human-produced carbon dioxide, so the changes to the system are already underway," he said. "We need to first ask will it work and then what are the environmental consequences?" (Iron Link, 2003).

One study asserted that, "To assess whether iron fertilization has potential as an effective sequestration strategy, we need to measure the ratio of iron added to the amount of carbon sequestered in the form of particulate organic carbon to the deep ocean in field studies" (Buesseler & Boyd, 2003, 67). To date, wrote Buesseler and Boyd, experiments of this type have "produced notable increases in biomass and associated decreases in dissolved inorganic carbon and macronutrients. However, evidence of sinking carbon particle carrying POC (particulate organic carbon) to the deep ocean was limited" (Buesseler & Boyd, 2003, 67). Small-scale experiments with iron fertilization indicate that it can do more than produce oxygen. It also may increase production of nitrous oxide, a greenhouse gas that is more potent than carbon dioxide (Williams, 2021).

Given the limits of present technology, this study estimates that an area a magnitude larger than the Southern Ocean (waters south of 50° S latitude) would have to be fertilized to remove 30% of the carbon that human activity presently injects into the atmosphere. Thus, according to this study, "ocean iron fertilization may not be a cheap and attractive option if impacts on carbon export and sequestration are as low as observed to date" (Buesseler & Boyd, 2003, 68).

Despite prospective problems, iron fertilization is considered possibly viable in some quarters. Scientists who fed tons of iron into the Southern Ocean reported evidence during 2004 that stimulating the growth of phytoplankton in this way may strengthen the oceans' viability as a carbon sink. Writing in *Science*, ocean biologists and chemists from more than 20 research centers said they triggered two huge blooms of phytoplankton that turned the ocean green for weeks and consumed hundreds, perhaps thousands of tons of carbon dioxide (Williams, 2021).

"These findings would be encouraging to those considering iron fertilization as a global geo-engineering strategy," said Ken Coale, a chief scientist at the Moss Landing Marine Laboratories. "But the scientists involved in this experiment realize that this looked only skin deep at the functioning of ocean ecosystems and much more needs to be understood before we recommend such a strategy on a global scale" (Hoffman, 2004). "From my work, I don't think this could solve a significant fraction of our greenhouse gas problem, while causing unknown ecological consequences," said Buesseler (Hoffman, 2004).

A scientific team led by Stéphane Blain reported observations of a phytoplankton bloom induced by natural iron fertilization, "an approach that offers the opportunity to overcome some of the limitations of short-term experiments. We found that a large phytoplankton bloom over the Kerguelen Plateau in the Southern Ocean was sustained by [a] supply of iron and major nutrients to surface waters from iron-rich deep water below. The efficiency of fertilization, defined as the ratio of the carbon export to the amount of iron supplied, was at least ten times higher than previous estimates from short-term blooms induced by iron addition experiments. This result sheds new light on the effect of long-term fertilization by iron and macronutrients on carbon sequestration, suggesting that changes in iron supply from below—as invoked in some palaeoclimatic and future climate change scenarios—may have a more significant effect on atmospheric carbon dioxide concentrations than previously thought" (Blain, 2007, 1070).

Iron fertilization faces another important problem—the ocean food chain. In 2008 and early 2009, tiny shrimplike crustaceans ate 159 miles^2 (about 300 km^2) of algae that were part of an iron fertilization experiment in the South Atlantic, according to the Alfred Wegener Institute of Germany. *National Geographic News* reported that working aboard the German research vessel *Polarstern*, German and Indian scientists mixed ten tons of ferrous sulfate with seawater. The team then pumped the artificially enhanced water back into the Atlantic outside ► Argentina's coastal waters. As expected, the experiment created a massive, CO_2-eating algae bloom. However, it was the wrong algae. The blooms were mostly tiny haptophytes, not the larger diatom algae the team had expected. The smaller algae variety is typically found only in coastal waters, and it's a favorite food of tiny shrimplike crustaceans called copepods. The copepods wolfed down the algae shortly after the new South Atlantic bloom appeared (Hearn, 2009).

The United Nations Convention on Biological Diversity meeting in Berlin during late May 2008, issued a moratorium, signed by representatives of 191 nations, on large-scale commercial projects to fertilize the oceans in pursuit of climate change mitigation. The ban will hold until scientists better understand the effects of such mitigation on the ocean food chain. Iron fertilization also could increase ocean acidity and reduce oxygen levels.

Delegates from 85 nations which comprise the London Convention Treaty (created in 1972 by the United Nations International Maritime Organization) on October 31, 2008, placed limits on iron fertilization, limiting it to "legitimate scientific research," a move to inhibit premature commercial application that exploits awards of carbon credits under diplomatic agreements such as the Kyoto Protocol. The body is charged with regulating pollution in international waters through such compacts as the London Protocol. The group acted after announcements of several plans to commercialize iron fertilization. Any research must consider carbon flux (such as emissions of nitrous oxides), impacts on the food web and oxygen levels, and the possibility that toxic species may grow (Kintisch, 2008, 835). Thus did the curtain fall on iron fertilization and, for the most part, on geo-engineering as a whole.

6.4.7   The Brave New World of "Smart Coral": Biology with a Genetic-Modification Chaser

We know we are in the Anthropocene when biologists develop "designer corals" that may be able to survive warmer and more acidic, and otherwise polluted ocean waters. They're propagating reefs that thrive in hot water so that they may become the new global normal as the rest fall victim to humankind's appetite for fossil fuels. Scientists now range the world seeking candidate corals, such as the *Acropora hyacinthus* near Ofu Island in American Samoa, which can withstand water temperatures of 35 °C in shallow pools under direct sunlight. These corals can be interbred with more fragile species that are being killed by warming waters. A fifth of corals have been lost since 1950 worldwide, and another 35% are in critical condition (Mascarelli, 2014). In some areas, such as the Caribbean Sea, coral mortality has reached 80%, provoking a sense of nearly messianic urgency among developers of designer reefs.

According to an account in *Nature* (Mascarelli, 2014, 444), the coral designers intend "to launch a program of 'human-assisted evolution,' creating resistant corals in controlled nurseries" hard-hit by changing conditions—all of them human made (warming, acidification, disease, overfishing, and pollution). "It's a brave new world of working with corals in this way," said Ruth Gates, a marine biologist at the University of Hawaii at Manoa who, along with coral geneticist Madeleine van Oppen at the Australian Institute of Marine Science in Townsville, is helping to pioneer the field.

"Smart corals" that can resist humankind's perils are being "outplanted" among vulnerable species and coaxed to take over. Corals that can survive pollution are being sought as well: "Sometimes we find reefs that are doing very, very well in places that you would least expect to find them," said Gates—such as a reef off Taiwan that lies below the wastewater outfall pipe of a nuclear power plant and experiences temperature fluctuations of between 6 and 8 °C/day. "By all of our understanding, we would expect those corals to all be dead. But they're not, they're flourishing".

In this brave new world, scientists give evolution a boost to assist survival in a world of warming water, rising ocean acidity, and increasing pollution, "handpick[ing] the hardiest, fastest-growing and most heat-resistant corals for their smart reef." The wisdom of reordering nature, or the necessity to amend conditions that kill weaker natural corals, seems not to be at issue. Paul Allen, co-founder of Microsoft, has given prizes for such work. "Ultimately," wrote Amanda Mascarelli in *Nature* (2014), "Gates and van Oppen hope to create a seed bank of gametes and fertilized embryos from extreme settings in which corals persist despite the odds—including the shallow reefs skirting Coconut Island, Hawaii, where both temperature and pH fluctuate drastically, reaching upper limits similar to those expected in the open ocean by 2050. In the old days, before the Anthropocene had a name, this might have been called "playing God." Now, it's simply called biology with a genetic-modification chaser.

Lithium Batteries: The Toxic Side of Electric Cars

Every source of transport energy aside from bicycles and walking shoes has an environmental cost. We know all about the deplorable cost of fossil fuels, but what about those wonderful electric cars that are supposed to solve this problem? The world's industrial base has been swiftly moving toward them. Reduce fossil fuel consumption, they do, but not without a cost. Lithium batteries, once used up, go to landfills leaching various noxious fluids into the soil and eventually into the water table. Who is working on ideas to repurpose these batteries?

Electric cars and their lithium batteries have received a lot of support in US President Joe Biden's plans to get the United States off fossil fuels. While only about 6% of new cars sold in the United States during 2021 were electric, Biden plans for 50% by 2030. By 2022, sales of electric and hybrid cars were taking off like proverbial rockets. In countries that are rich in lithium (Bolivia is the largest), a race is on to make an Earth-friendly dollar. "The amount of lithium we need in any of our climate goals is incredible," said Anna Shpitsberg, US deputy assistant secretary of state for energy transformation (Krauss, 2021, A-8). Lithium is a basic component of lithium-ion batteries that conduct electricity, "Because of the metal's light weight, long life, large storage capacity, and easy recharging, demand is expected to grow exponentially over the next decade to power an expanding fleet of vehicles produced by Tesla, Ford Motor, General Motors, General Motors, and other carmakers," wrote Clifford Krauss of the *New York Times* (2021, A-8).

The use of lithium-ion batteries is not the only potential problem with electric cars. Their energy source is only as clean as their sources of electricity. If your local power plant still burns coal, your power source is dirtier than gasoline. If the power plant has switched to 25% wind, 25% solar, and 50% coal, the mix is halfway there. Very few electric utilities have gotten this far (as of this writing). So, if you have an electric car and a power plant burning coal, you are destroying the planet faster than by driving a gas hog. Wipe that angelic smile off your face and get to work lobbying your local energy company to improve its sources.

While looking for an example of a state that is at the top of the United States in wind power, try our citadel of corn and milk cows, Iowa. This is no joke. Iowa is closing in on 100% wind energy, largely because Warren Buffett bought the state's public utility, MidAmerican Energy.

6

Treason Against Our Habitat: The *Real* High Crimes and Misdemeanors

Back when Donald J. Trump was impeached (twice—a first for the US Constitution), we heard a lot about high crimes and misdemeanors of the impeachable kind. The sordid, the corrupt, the inept, and the borderline insane became so ordinary that one wonders sometimes what our public life would be like without Russian election meddling, assaults on women by holders of high office, and so forth. What would we have done if we had tuned into the evening news and had suddenly found it overtaken by ordinary civility, sensitivity, self-control, decency, and competence?

We gorge on a diet of dysfunction with only occasional reference to the *real* issues, however. In a few decades, perhaps, history may recognize that the *real* high crimes and misdemeanors were hardly mentioned. It may be said that we amused ourselves to death with junk movies and television. In fact, our constitution has no words that recognize treason against our habitat as a transgression. Our legal system is barely beginning to catch up with the geophysical facts that entail high crimes against the Earth and future generations of human beings, other animals, and plants.

Our generation lives in what is, environmentally speaking, the most crucial time in the planet's history, when our actions (or lack thereof) will determine whether we will be able to establish, and then maintain, a habitable Earth with a sustainable energy supply. Until now, humanity has never faced such a task. To do so requires a departure from fossil fuels—*quickly*. Thermal inertia makes this race more urgent—and at a time when much of our political leadership does not even realize that such a threat exists. This threat has become background noise, easily ignored until a wildfire burns down *your* house.

In the long run, *these* are the most egregious crimes—most notably our continuing, foolish and ultimately fatal embrace of coal. We are deciding now whether we will condemn future generations to a hell on Earth. What a crime against nature this is! What a convincing argument for a constitutional amendment making the rape of mother Earth a criminal act, a form of treason. And lock up Trump, too, of course, if he ever becomes President of the United States again. Or, better yet, strip him of everything he owns, spend it on aid to refugees and climate research, and then issue him a potato sack and enough spare change to call Vladimir Putin. Then dump him at La Guardia International Airport in New York City with one-way exit visa. That, or dump him in New Delhi, India, on a May afternoon without air-conditioning or a gas mask, downwind of a coal-fired power plant. Last I checked, it was 112 °F there and viciously smoggy, in mid-May. The atmosphere in India today sounds like an environmental paradise designed by Trump and Co.

6.4.8 An "Artificial Leaf" May Someday Make Fossil Fuels Obsolete

Now, to our bedraggled readers comes some good news—possibly *really* good news. Once such a thing reaches the street, fossil fuels may eventually become past tense.

About 2010, for the first time, a team of researchers at the Massachusetts Institute of Technology (MIT) led by Daniel Nocera (in 2016 he was a professor of energy science at Harvard) created the prototype of an "artificial leaf" that can replicate the way that plants use sunlight to knit chemical bonds, possibly (with further development) providing a practical, inexpensive source of solar energy. The device was described by Robert F. Service in *Science* as "a silicon wafer about the size and shape of a playing card. Different catalysts coat each side of the wafer. The silicon absorbs sunlight and passes that energy to the catalysts to split water (H_2O) into molecules of hydrogen (H_2) and oxygen (O_2). The hydrogen then may be used in a fuel cell, reemerging as water.

By 2016, scientists were studying a form of artificial photosynthesis that, as was reported in *Science*: "When combined with solar photovoltaic cells, solar-to-chemical conversion rates should become nearly an order of magnitude more efficient than natural photosynthesis" (Liu et al., 2016, 1210). Liu and colleagues wrote that "on a previous artificial photosynthesis design, [we] combined the hydrogen-oxidizing bacterium *Ralstonia eutropha* with a cobalt-phosphorus water-splitting catalyst. This biocompatible self-healing electrode circumvented the toxicity challenges of previous designs and allowed it to operate aerobically."

Richard Martin reported in the *MIT Technology Review* in 2016 that Nocera and his colleague Pamela Silver "have devised a system that completes the process of making liquid fuel from sunlight, carbon dioxide, and water. And they've done it at an efficiency of 10%, using pure carbon dioxide—in other words, one-tenth of the energy in sunlight is captured and turned into fuel. That is much higher than natural photosynthesis, which converts about 1% of solar energy into the carbohydrates used by plants, and it could be a milestone in the shift away from fossil fuels…. Bill Gates has said that to solve our energy problems, someday we [will] need to do what photosynthesis does, and that someday we might be able to do it even more efficiently than plants," said Nocera. "That someday has arrived."

In other words, according to Martin, "Solar energy, water, and carbon dioxide [have now been used] to produce energy-dense liquid fuels. Nocera and Silver's system uses a pair of catalysts to split water into oxygen and hydrogen and feeds the hydrogen to bacteria along with carbon dioxide. The bacteria, a microorganism that has been bioengineered to specific characteristics, convert the carbon dioxide and hydrogen into liquid fuels." This new work "is really quite amazing," said Peidong Yang of the University of California, Berkeley, who had developed a similar system with much lower efficiency. "The high performance of this system is unparalleled" in any other artificial photosynthesis system reported to date" (Martin, 2016).

6.4.9 Some Tough Questions

Once every few years, usually on the eve of a major international climate change meeting, we enter the twilight world of climate change pledge season. In 2021, Joe Biden, as president of the United States, pledged to cut the United States' emissions of greenhouse gases in half compared to 2005 levels by the end of 2030. The good news is that we now had a president who knew something about climate science and was conscious enough of climate change's effects to advance a plan that may reduce them significantly. We would not have had such an experience 5 years earlier, when the official line emanating from the White House was that coal was good for us, and any whisper otherwise was a tree-hugging, hippie hoax. With Joe Biden, climate-wise, the United States is back in the game.

The other news is that a pledge has not yet been realized as reality—not yet, at any rate. The United States and other countries have signed plenty of these, including Kyoto in 1992 and Paris (2015). Trump pulled the United States out of the Paris Agreement, of course. Overall, talk about breaking the upward curve of greenhouse gases has had an effect, but a small one, in the United States and the European Union.

The extremely bad news is that China eclipsed United States' emissions in 2014, and kept on running up its total, doubling its emissions by 2020. You read that correctly: *in 6 years*. China added enough greenhouse gases to create a whole new, very large I country with greenhouse-gas emissions that spew twice as much effluent than as other, making two Chinas to one of the United States' as its effluent and that of the European Union were slowly falling. This was occurring as China also was becoming the world's largest producer and exporter of wind turbines and solar panels. China also remains the world's largest user of coal. It isn't alone. According to the World Energy Agency, the world's annual use of coal increased 4.5% in 2021, mainly because of increasing demand for coal-fired (and every other type of) electricity. Even with China's prodigious appetite for fossil fuels, the United States still uses much more *per capita*. The Chinese used 10.1 tons of greenhouse gases per person in 2019, as US citizens used 17.6 tons per capita. The European Union used 7.4 tons per capita and India 2.5. What ought to be obvious from such numbers is that no one is going to win any kind of climate change battle against greenhouse gases without China's full-throated cooperation.

The same goes for many other countries that we have not even mentioned yet in this context: Bangladesh, Brazil, Australia, Japan, Canada, Russia, India, and Indonesia, plus all of Africa. Some of these countries have sizable populations (India: 1.3 billion; Indonesia, 320 million; Bangladesh, 160 million; Nigeria, 140 million). Most of these are flooded with people who want only what most of us take for granted: enough food to survive, a small house (with electricity), a car, small truck or motorbike, perhaps, and fuel with which to propel it.

Many of these people live in the tropics, with hot, humid weather, where air-conditioning would be very nice but at present is an unaffordable luxury for most people. Anyone who has lived through a New Delhi pre-monsoon summer knows about this. To reduce worldwide emissions, are we going to give up our summer

291 **6**

6.5 · Is Global Warming Inevitable? A Contrary View

air-conditioning, or are all of us going to find new ways to supply electricity from renewable resources, such as sun and wind, rather than dirty coal?

6.5 Is Global Warming Inevitable? A Contrary View

6.5.1 The Coming Baby Bust, or The Imprecision of Predictability

Memo from the year 2121: Remember a century ago, just about the time when global warming (and its cousin, "climate change") became enough of a household phrase to scare many people that they would sizzle, drown, or starve in a fit of geophysical fury? According to many experts, by 2121 Earth was expected to be a scorched, useless ball of ashes in a century, if not sooner.

About the year 2021, a spate of studies and media reports had begun to tell our great-great grandparents that all this gloom and doom was wrong—that Earth's population would soon be falling, along with a concurrent decline in consumption of pressure on use of resources, including fossil fuels.

A newspaper as august as the *New York Times* had told your forebears in 2021 that many national populations already had begun to fall quickly, clinching a sustained long-term decline in Earth's birth rate for the first time in human history.

Now *there's* a news story: Stop the presses; hold the mail, Bruce Johansen (with some help from his media friends) is having a one-cent sale! Countries such as South Korea, Italy, parts of China, and others were already below replacement levels. The United States was hugging replacement level except for immigration. Australia and Canada were in a similar state.

6.5.2 What May Have Happened to 700 Million People?

Some villages in the northeastern part of China had emptied, totally. Remember 1.4 billion people? Those who say they were "in the know" soon realized that the number of people in China was about to be cut in half in less than a century.

As this came to pass, China's population was surpassed by Nigeria's, up from about 200 million in 2020. A baby boom continued in large parts of Africa and Mexico, long cradles of high birth rates, while much of the world fell to a replacement level of 2.1.

In Germany, hundreds of thousands of abandoned homes and other properties have been bulldozed and turned into parks. The drop in population represented by those empty houses came despite Germany's admission of several hundred refugees a few years after the beginning of the twenty-first century. Germans, as a whole, just have all but quit having babies.

So, what became of climate change in such a scenario? At first, not much. Anyone who has read this book by now knows about thermal inertia, which tells us that warming will continue until after we quit dumping heat into the environment for about 50 years on land and 150–200 years in the oceans, after the last belch of carbon dioxide enters the atmosphere. Thus, in 2121, the daily weather was warmer than in 2020, as it would have been even if no one had had a single new baby. Even a century after that, rising standards of living may keep temperatures rising even after we belch our last puff of fossil fuel emissions. Even without babies, this utopia may be a fantasy.

6.5.3 The Imprecision of Predictability

6

The proponents of the baby bust also have not figured in (and this is important), a propensity of human beings that we might call the imprecision of predictability. Or, as this fancy line is said in history, a field in which I have had some experience: "Any historian who tries to predict the future risks becoming a fool." Or in law: "Anyone who acts as his (or her) own lawyer has a fool for a client."

One other mitigating factor that rarely gets a murmur in population projections might be called the comfort paradigm. Birth rates may be falling in places with large populations and hot summers (India comes to mind), but energy consumption continues to rise. How many demographers check air-conditioner sales? What about the number of cars per thousand people? Material prosperity is a strong driver of energy consumption per person.

Who, a century ago, would have predicted that the Earth's population would rise from about 1.5 billion to about 8 billion in about 70 years? Who would have foreseen nuclear weapons or handheld computers that open a universe of information in several languages and also act as telephones? Or a time when these same wondrous machines could be used to build political and business empires based upon hate and lies by people who confuse information with verbal sewage? Social media's mean streak has made it a bastard child of our age, a breeding ground of both useful information and absolute lies that has ruined public discourse on a number of important issues, one of which is the veracity of climate change and the reality of such threats as viral disease. Note to Lewis Carroll: Just how do almost 9 billion human beings get out of this rabbit hole?

Conclusion: What We Should Do; Where Should We Go?

By way of conclusion, let's engage in some rudimentary numbers about a worldwide problem. How much fossil fuels must we quit using, and how much time do we have before weird, wild weather and sea-level rise make living on Earth very uncomfortable, if not nearly impossible?.

First, the numbers:

For that, we get the weather that we experience every day, described in the preceding pages—fires, floods, heat waves, and more. At 1.1°, climate change is more entertainment than crisis. If you read about it, or talk about it, or ignore it, your house has not yet burned, your beachfront villa probably hasn't washed out to sea (although you may have noticed the waves eating your beach). Temperatures are warm, even hot, but fall still comes, even if it is a couple of weeks later than it used to be.

At this stage, atmospheric science is not required reading for most of us. One may have heard that a large conference in Paris in 2015 pressed 195 nations to limit the rise in carbon dioxide to 1.5 °C, about four-tenths of a degree before real alarm bells begin to ring. This should be easy to do, one might think. However, some complications arise. One important problem is thermal inertia (about which you know quite a bit by now, we hope), the amount of time required to turn a given level of excess CO_2 into heat. You well know by now that this figure is about 50 years on land, and 100 to 150 years in the oceans. Having read this book, you have become acquainted with thermal inertia's effects.

However, here is a refresher and another number: Because of thermal inertia, the climate that we experience today is actually what a CO_2 level in 1971 would have done to us without a geophysical delay. We would have today the climate of Richard Nixon's time, before Watergate, and the end of the Vietnam War. More numbers: the carbon dioxide in Earth's atmosphere at the start of the industrial revolution was 285 parts per million. In 1971, it was 326; in 2020, it was 413. That's roughly 87 parts per million of weather mischief that is "in the pipeline," with potential to make the climate even more catastrophic at a level of 1.1 °C higher than the advent of the industrial age. At 1.5 °C, things get worse; at 2.0 °C, plus the 50–100 parts per million CO_2 that's still to come even if fossil fuel use stops cold at today's level, climate change is everyone's business, like it or not. At 3.0 °C, or 4.0, as was once said about the victims of nuclear war, the living will envy the dead.

There is a narrow path of escape from this trip into climatic hell. Get off fossil fuels, as quickly as possible, Many coal-fired power plants are being canceled, although worldwide use of coal is still rising. Wind and solar power are exploding in popularity, as electric cars and hybrids are selling at record rates.

6.6 Summary: How Do We Get Off the Road to Climatic Hell?

So, what do we do to get out of this trap? First, we need to take the science seriously. This is no "hoax." Donald Trump and other deniers are carbon dioxide's best friend. One of the best thumbnail descriptions of the situation was written by Brad Plumer in the *New York Times:* "The world's countries have begun to make meaningful progress in the fight against climate change in the past decade, data shows. Thanks to a rapid expansion of clean energy. Yet the planet is still on track for dangerous levels of warming in the years ahead unless those efforts accelerate rapidly" (Plumer, 2021b, A-10).

So, yes, we can preserve the habitability of the Earth—but it is a "war," so we have heard. All boots on the ground. It requires innovation and some degree of sacrifice. Ultimately, it is a war to save ourselves from the worst outcomes of the very bad habit of burning fossil fuels.

6.6.1 COPed Out: So How Do We Fix This Problem?

As of this writing, we had finished the Conference of Parties (COP26), a special meeting of the United Nations' membership (roughly 200 nations as of 2021), a group that started meeting in 1992, with a charge of solving the climate change problem, something which is easy to do in theory, but much tougher in practice. That is, each of the COP meetings has ended with professions of solutions, as the level of carbon dioxide keeps going up. If carbon dioxide had a sense of humor, it would be getting a huge laugh out of this situation. If the CO_2 had been watching the NBC Evening News on January 24, 2022, it would have seen reports of a large wildfire near Big Sur, in California. But wait: isn't this the *rainy* season in Northern California? Well, it used to be. Today, maybe, if you're lucky. Don't hope too hard, however. A deluge may follow a drought, and your rain may come as a mud-strewn wall of rocks over the highway that you use to get home, with your house sliding down the same muddy slope.

You may hear a guffaw from that devilish CO_2 already. The CO_2, it turns out, has a lot to laugh about. In the aisles of COP26 in Glasgow, word went around that Saudi Arabia's delegation was working to sabotage the talks by proposing that work cease by 6 PM each evening—not a world-endangering act unless the conference had only one day to come to an agreement on how to address the existential question on which all 197 national delegations had to agree. In the meantime, Mary Robinson, former president of Ireland and head of a group of senior political leaders on climate, told the Associated Press that Russia and Saudi Arabia were "pushing back hard" to block any mention in the final deal out of Glasgow of working to phase out coal or to reduce government subsidies for fossil fuels (Knickmeyer et al., 2021). On Thursday, a day before the conference was scheduled to end, United Nations Secretary-General António Guterres said that a proposal to limit future temperature increases to 1.5 °C was slipping away. The Earth, he said, with an undertone of frustration, was on "on life support," but "he hopes that world governments will step up to their pledges to slash their emissions of greenhouse gases" (UN Chief, 2021). Hope alone is not going to stop the ice caps from melting or the temperatures from rising. If you have gotten this far, you know the drill.

By November 2021, almost 30 years after the first COP meeting, the word on the street, among tens of thousands of protesters, most of them young, was that the human race does not have 30 more years to diddle around. At a level of 420–425 parts per million, the atmosphere's load of carbon dioxide has reached the level of the Pliocene, 2–3 million years ago, when temperatures were much higher (on the order of about about 4 °C than today), and the heat had melted any ice that the Earth may have had, enough to raise sea levels 25–30 ft in some places. That is where we are headed, as the atmosphere catches up with our present level of carbon dioxide—50 years on land, 150 years or so in the oceans. Do we need any starker warning? Or are some of Earth's nations making "pledges" that they are less than serious about fulfilling?

This is how the COP approach is failing us. Each meeting turns into squabbles between various nationalistic interest groups. Goals are set, but nothing is enforced.

This is like speeding at 200 miles/h the wrong way on an interstate highway without any police around. Our climate COPs need an ability to penalize nations, something that a worldwide body has never successfully done.

For example, Brazil committed at COP26 to stop illegal logging in 8 years, but 2 weeks after COP closed logging was at its highest toll in 15 years. (The country uses an August to July year to measure deforestation.) The steady rise in the Brazilian forest harvest raises huge questions about its intent and ability to meet a climate summit target. Brazil's 2020 deforestation total was on top of a 22% increase the previous year. These increases are supported by an infrastructure of tree cutters, millers, sellers, farmers, and more, all of whom Brazil's government has pledged to support, opposite the climate summit agreement that it had signed.

So how will we dispose of destructive nationalism, something else that the human race has never done, in the name of planetary survival? How much more evidence do we need to realize that the signposts on humanity's highway lead to the end of the world as anyone heretofore has ever known it? Please think it over and act. It will be the world that will be home to you, your children, and theirs.

6.7 Questions and Exercises

6.7.1 "Clean Coal" and Other Oxymoronic Orwellianisms

1. No matter what you spend in vulnerable coastal areas, you can't protect everything from every storm. Quite simply, it is impossible to stop land loss and guarantee the safety of people so long as the climate continues to change, the sea level continues to rise, and warming seas continue to create supercharged storms. It is impossible even if you spend lots of money, and do it the right way (Young, 2021, A-23).

   It may look dramatic to build gigantic levees around cities that are sinking amidst rising seas, as New Orleans did after Hurricane Katrina and before 2021s Hurricane Ida. Levees built for billions of dollars may even hold back the ocean for a while. However, water has a pernicious way of finding a way over, under, or around walls, especially if someone has built them of porous limestone. How wise has it been to build levees around cities (such as Miami and, most famously, New Orleans) which are sinking anyway? Run your eyes along a map of the Earth and identify large cities near coastlines that are on sinking ground (a dash of geology will be helpful here).

2. The IPCC's first reports today sound rather timid. Global warming had not even broached the debates of US presidential politics in 2000, when Al Gore tried to make an issue of it on this very public of stages. The rest of the field was stumbling along behind him, asking whether humankind had any role at all in the steady warming of the atmosphere that was becoming more evident. Republicans and others already were rallying a corps of denial. Every half decade or so, the case made by the IPC's member scientists became more urgent—not only that humankind has a role in the warming of the planet but also whether it was the major influence, with foreseeable consequences portending

6

very identifiable and unpleasant results. The debate over climate change thus became part of mainstream political debate and other discourse.

The daily weather has become more violent. Four-inch-diameter hail, 50 in. of rain from one hurricane, 4 ft of snow from one blizzard, 250-mile-an-hour wind gusts from one tornado, a foot of snow in Texas, a thaw on the same day in Alaska, and even a new type of disaster—a "firenado"—in Australia. Wildfires cover large parts of continents—Western North America, Brazil, Siberia, and many other places. "Thousand-year" rainstorm floods two or three times a decade. July 2021 became the hottest month in the history of human weather record keeping.

Calling upon personal experience, weather records, published accounts, and other sources, has weather actually become more violent and otherwise extreme?

3. This was not a tale for the squeamish by 2021, because those who dig into the science can find a warning even more ominous. Why are so many scientists sounding dire warnings that a decade or two of business-as-usual fossil fuel production and consumption will carry the Earth over various "tipping points" beyond which human ability to influence climate change may become irrelevant? Don't we have plenty of time for such a slow-motion crisis to unfold? Consider the two words "thermal inertia." You will be hearing them again.

Indeed, you are hearing about thermal inertia again, and it is worth a review at this point because it is a pernicious effect that postpones the effects of today's fossil fuels emissions between 50 years (on land) and 150–200 years (in the oceans). Reflect, please, on how such effects may mislead people who are oriented only toward present effects.

4. What we have read and discussed thus have been a climatic horror story, now made more monstrous by the drumbeat of the daily news, the hot, humid breath in our faces, when and where the story is no longer a question not of *if*, but *when*. What about solutions? Do we have any left? Yes, we do, in that "narrow window," described by the IPCC's 2021 report. In a world where solutions are mainly rooted in economics, one of the most important moments in recent environmental history was that glorious day when the cost curve of solar and wind energy for general use in everyday human life dipped below that of oil and coal.

5. Good news! Are we not starved for some of that these days? The fact that the costs of producing solar and wind power are falling below that of oil and coal ought to be the cause for reveling. There is work to do. What specific actions must be taken to remove oil, gas, and coal from our energy mix, where and how?

Just because wind and solar energy are cheaper to produce than oil and coal does not immediately solve the problem. It is a good opening gambit, however. Please estimate how long it will take to change world infrastructure to receive this gift and make use of it.

6. "Coal Capture and Sequestration (CCS): A Bad Idea Whose Time has Come—and Gone." With his low ranking of CCS as a good way to eliminate carbon dioxide from the atmosphere, is the author dismissing offhand a very good idea? Or not? What are the major reasons in support and opposition to CCS?

We now have a fair number of leaders in some of the Earth's most economically powerful countries advertising what they believe is magic that will make car-

bon dioxide vanish. Wow! Double wow! They did make an agreement at a meeting in mid-June 2014, to collectively cut carbon emissions in half by 2030. That sounds really good. But will it actually happen? Discuss this issue, pro and con.

7. "Coal is the single largest source of global carbon emissions, and this year's [2021] historically high level of coal power generation is a worrying sign of how far offtrack the world is in its efforts to put emissions into decline toward net zero," said IEA (International Energy Agency) Executive Director Fatih Birol (Gillespie, 2021, A-8). Rising prices of natural gas have been a factor, while talk about solar, wind, and other renewables has produced some progress, thus far, but very little in coal's big picture.

   What? Double what? After all this talk about oil and coal being condemned to a rancid hell as condemned former sources of fossil fuels, we see that many places are actually increasing their usage. Explain this, if you dare.

8. David Keith, a Canadian who became a professor at Harvard and recipient (with others) of $6 million in grant money from the Bill Gates Foundation, since the 1990s became perhaps the world's most prominent single advocate of geo-engineering. Keith, who has an affinity for huge-scale, expensive techno-fixes, also promoted carbon capture and sequestration (CCS) in a report with four Canadian energy executives. By 2014, Keith was sponsoring a "field experiment aimed at understanding another geo-engineering technology that would use a balloon to release sun-blocking particles of sulfuric acid in the stratosphere" (Kintisch, 2013, 307).... Geo-engineering was banned by the Convention on Biological Diversity representing 193 nations on grounds of a threat to biodiversity late in 2010.

   Does anyone out there still support the idea of "geo-engineering," to spray the atmosphere with sulfuric acid (as a large volcano would do) to bring down temperatures? Perhaps this once had been disregarded as a stupid or silly idea that could do more harm than good. Perhaps we can get a debate going on human intervention into other human affairs. Whom (other than volcanoes) puts sulfur into the sky to begin with?

9. *Well, then, if sulfuric acid won't do the job, how about iron ore? Seeding the oceans with iron ore to counteract growth of CO_2 production by human industry also has received a thumbs down by scientists.*

   Should the oceans be seeded with large amounts of iron ore that will stimulate the growth of carbon dioxide-consuming phytoplankton? The idea has attracted some support among corporations and foundations looking for ways to minimize the effects of carbon dioxide without changing the world's basic energy generation mix. The idea is simple on its face: iron stimulates the growth of phytoplanktonic algae that are believed to be responsible for about half of the world's biologic absorption of carbon dioxide. By 2009, ocean fertilization had been largely rejected as a solution, in large part because to succeed it would disrupt large-scale ocean ecosystems in unpredictable ways while holding little promise of significant success.

10. *Let us end this chapter's journey through possible solutions, including some real comedic losers with one that displays some potential, as well, as the usual temptation to*

place human folly over nature. In this case, we are testing corals' climatic endurance with "New World of "Smart Coral."' For more details, check ▶ Chap. 6.

You know that we are in the Anthropocene when biologists develop "designer corals" that may be able to survive warmer and more acidic and otherwise polluted ocean waters. They're propagating reefs that thrive in hot water so that they may become the new global normal as the rest fall victim to humankind's appetite for fossil fuels. Scientists now range the world seeking candidate corals, such as the *Acropora hyacinthus* near Ofu Island in American Samoa, which can withstand water temperatures of 35 °C in shallow pools under direct sunlight. These corals can be interbred with more fragile species that are being killed by warming waters. A fifth of corals have been lost since 1950 worldwide, and another 35% are in critical condition (Mascarelli, 2014). In some areas, such as the Caribbean Sea, coral mortality has reached 80%, provoking a sense of nearly messianic urgency among developers of designer reefs.

Is there any room for debate or discussion here, or should we stick with ideas that actually reduce greenhouse gases in the atmosphere? Shall we debate the use of lithium and nickel in electric car batteries, for example?

And finally: a contrarian point of view raised in the book: birth rates are falling in some parts of the world. Should such a trend continue, the human race may mitigate carbon emissions enough that, combined with other efforts, could help to solve the greenhouse gas problem with a minimum of pain and suffering. Now, here's a debate subject with which to conclude. Take a crack at forecasting the future.

References

Abrupt Climate Change. (1996, December 4). Abrupt climate change during last glacial period could be tied to dust-induced global warming. Press Release NOAA 96-78. Retrieved December 7, 2001, from http://www.noaa.gov/public-affairs/pr96/dec96/noaa96-78.html

Barringer, F. (2007, July 27). Navajos and environmentalists split on power plant. *New York Times.* Retrieved July 30, 2007, from http://www.nytimes.com/2007/07/27/us/27navajo.html

Beaty, A. (2011, April). *BP's 'Voices from the Gulf:' a rhetorical analysis of corporate propaganda.* MA thesis, Communication, University of Nebraska at Omaha.

Birnbaum, M. (2013, February 7). Europe consuming more coal. *Washington Post.* Retrieved March 1, 2013, from http://www.washingtonpost.com/world/europe-consuming-more-coal/2013/02/07/ec21026a-6bfe-11e2-bd36-c0fe61a205f6_print.html

Broad, W. J. (2006, June 27). How to cool a planet (maybe). *New York Times.* Retrieved July 25, 2006, from http://www.nytimes.com/2006/06/27/science/earth/27cool.html

Chisholm, S. W. (2000). Stirring times in the Southern Ocean. *Nature, 407,* 685–686.

Cressey, D. (2012). Cancelled project spurs debate over geo-engineering patents. *Nature, 485,* 429. Retrieved June 10, 2012, from http://www.nature.com/news/cancelled-project-spurs-debate-over-geoengineering-patents-1.10690

Crutzen, P. (2006). Albedo enhancement by stratospheric sulfur injections: A contribution to resolving policy dilemma? *Climatic Change, 77,* 211–220.

Dear Governor. (2008, May 29). Dear Governor Greenwash. Hansen: Governors aren't getting it. *Grist.* Retrieved June 12, 2008, from http://www.grist.org/article/dear-governor-greenwash/

Decade-Long. (2012, March 2). Decade-long grassroots campaign shuts two Chicago coal plants. *Environment NEWS Service*. Retrieved March 11, 2012, from http://www.ens-newswire.com/ens/mar2012/2012-03-02-01.html

Editors' Choice. (2006). *Science, 314*, 387.

Eilperin, J. (2010, October 29). Geo-engineering sparks international ban, first-ever congressional report. *Washington Post*. Retrieved November 11, 2010, from http://www.washingtonpost.com/wp-dyn/content/article/2010/10/29/AR2010102906365_pf.html

Fahrenthold, D. A. (2010, January 8). Scientists say mountaintop mining should be stopped. *Washington Post*, A-3.

Frosch, D. (2013, November 28). A part of Utah built on coal wonders what comes next. *New York Times*, A-16.

Gates, D. (2021, October 24). As climate concerns threaten air travel, aviation industry banks on technology solutions. *Seattle Times*, A-1.

Georgia Judge. (2008, June 30). Georgia Judge Yanks coal power permit on climate concerns. *Environmental News Service*. Retrieved July 9, 2008, from http://www.ens-newswire.com/ens/jun2008/2008-06-30-091.asp

Gillespie, T. (2021, December 18). World burning most coal ever to keep the lights on. *Bloomberg News in Omaha World-Herald*, A-8.

Hansen, J., Sato, M., & Ruedy, R. (1997). The missing climate forcing. *Philosophical Transactions of the Royal Society of London, B352*, 231–240.

Hansen, J. E., & Sato, M. (2021, October 14). September temperature update & COP 26. [Conference of Parties] personal website, no URL available.

Hearn, K. (2009, March 27). Huge man-made algae swarm devoured—Bad for climate? *National Geographic News*. Retrieved April 11, 2009, from http://news.nationalgeographic.com/news/2009/03/090327-iron-seeding.html

Hitt, M. A. (2012, April 26). Southwestern tribes lead 3-day March to move beyond coal. *Huffington Post*. Retrieved May 9, 2012, from http://www.huffingtonpost.com/mary-anne-hitt/coal_b_1456759.html

Hoffman, I. (2004, April 17). Iron curtain over global warming; ocean experiment suggests phytoplankton may cool climate. *Daily Review* (Hayward, CA), n.p. (in LEXIS).

Horowitz, J. (2021, August 22). Athens is only getting hotter. Its new 'Chief Heat Officer' hopes to cool it down. *New York Times*. Retrieved September 2, 2021, from https://www.nytimes.com/2021/08/21/world/europe/athens-is-only-getting-hotter-its-new-chief-heat-officer-hopes-to-cool-it-down.html/

Hulme, M. (2014) *Can science fix climate change? A case against climate engineering*. Polity Press.

Iron Link. (2003, April 10). Iron link to CO2 reductions weakened. *Environment News Service*. Retrieved April 22, 2003, from http://ens-news.com/ens/apr2003/2003-04-10-09.asp#anchor8

Johansen, B. E. (2010, July–August). The climatic consequences of coal: Nebraska's role. *Nebraska Report*, 1, 3, 10.

Jones, N. (2013). Geo-engineering: One cool solution. *Nature, 502*, 302.

Keith, D. (2013). *A case for climate engineering*. MIT/Boston Review.

Kerr, R. A. (2006). Pollute the planet for climate's sake? *Science, 314*, 401–403.

Kintisch, E. (2008). Rules for ocean fertilization could repel companies. *Science, 322*, 835.

Knickmeyer, H. A., Jordans, F., Ghosal, A., & Borenstein, S. (2021, November 12). Saudi Arabia denies playing saboteur at Glasgow. Associated Press. Retrieved November 24, 2021, from https://apnews.com/article/climate-science-business-glasgow-united-nations-41a155b329be0567e944fe912ad4ccb5?user_email=&utm/

Krauss, C. (2021, December 16). Energy riches in Bolivia lure global suitors. *New York Times*, A-1, A-8, A-9.

Kruger, T. (2014). Stratospheric folly [Review: Hulme, Mike. *Can science fix climate change? a case against climate engineering*. Cambridge, UK: Polity Press, 2014]. *Nature, 508*, 457.

Liu, C., Colón, B. C., Ziesack, M., Silver, P. A., & Nocera, D. G. (2016). Water splitting–biosynthetic system with CO2 reduction efficiencies exceeding photosynthesis. *Science, 352*, 1210–1213.

Mann, M., & Hassol, S. J. (2021, June 30). Climate ignited the heat dome frying the northwest. *New York Times*, A-22.

Martin, R. (2016, June 7). A big leap for an artificial leaf: A new system for making liquid fuel from sunlight, water, and air is a promising step for solar fuels. *MIT Technology Review*. Retrieved June 15, 2016, from https://www.technologyreview.com/s/601641/a-big-leap-for-an-artificial-leaf

Mascarelli, A. (2014). Climate-change adaptation: Designer reefs. *Nature, 508*, 444–446. Retrieved April 30, 2014, from http://www.nature.com/news/climate-change-adaptation-designer-reefs-1.15073

Mazzei, P. (2021, June 3). Sea wall looms as Miami plans for rising seas. *New York Times*, A-1, A-18.

Moapa Paiute March. (2012, April 27). Moapa Paiute March 50 miles in anti-coal protest. *Indian Country Today Media Network*. Retrieved May 10, 2012, from http://indiancountrytodaymedianetwork.com/article/moapa-paiute-march-50-miles-in-anti-coal-protest-110450

Moapa Paiute Sue. (2013, August 9). Moapa Paiute Sue over coal plant contaminants. *Indian Country Today Media Network*. Retrieved August 20, 2013, from http://indiancountrytodaymedianetwork.com/2013/08/09/moapa-paiute-sue-over-coal-plant-contaminants-150806

Morton, O. (2007). Is this what it takes to save the world? *Nature, 447*, 132–136.

Mufson, S. (2007, September 4). Coal rush reverses, power firms follow plans for new plants stalled by growing opposition. *Washington Post*, D-1. Retrieved September 10, 2007, from http://www.washingtonpost.com/wp-dyn/content/article/2007/09/03/AR2007090301119_pf.html

Newman, A. (2021, September 3). Dozens are dead as record deluge stuns northeast. *New York Times*, A-1.

Palmer, M. A., Bernhardt, E. S., Schlesinger, W. H., Eshleman, K. N., Foufoula-Georgiou, E., Hendryx, M. S., Lemly, A. D., Likens, G. E., Loucks, O. L., Power, M. E., White, P. S., & Wilcock, P. R. (2010). Mountaintop mining consequences. *Science, 327*, 148–149.

Phillips, M. (2021, June 11). How a tiny green fund turned the Exxon tanker. *New York Times*, B-3.

Plumer, B. (2021a, September 3). Devastating night, and a warning: Big storms are packing a more powerful punch. *New York Times*, A-15.

Plumer, B. (2021b, October 27). In war on climate change, cause for hope and alarm. *New York Times*, A-10.

Plumer, B., & Fountain, H. (2021, August 9). A hotter future is now inevitable. *New York Times*, A-1, A-8.

Powell, J. L. (2011). *The inquisition of climate science*. Columbia University Press.

Revkin, A. C. (2000, October 12). Antarctic test raises hope on a global-warming gas. *New York Times*, A-18.

Robock, A. (2008). Whither geo-engineering? *Science, 320*, 1166–1167.

Rosenthal, E. (2010, November 21). Nations that debate coal use export it to feed China's need. *New York Times*. Retrieved December 2, 2010, from http://www.nytimes.com/2010/11/22/world/asia/22fossil.htm

Ruff, J. (2010, June 17). UP target of global warming protest. *Omaha World-Herald*. Retrieved June 18, 2010, from http://www.omaha.com/article/20100617/MONEY/100619732

Schlermeier, Q. (2003). The oarsmen. *Nature, 421*, 109–110.

Service, R. F. (2011a). Artificial leaf turns sunlight into a cheap energy source. *Science, 332*, 25.

Strong, A., Chisholm, S., Miller, C., & Cullen, J. (2009). Ocean fertilization: Time to move on. *Nature, 461*, 347–348.

Tilmes, S., Müller, R., & Salawitch, R. (2008). The sensitivity of polar ozone depletion to proposed geo-engineering schemes. *Science, 320*, 1201–1204.

Torres, C., & Eilperin, J. (2011, July 21). Mayor Bloomberg gives $50 million to fight coal-fired power plants. *Washington Post*. Retrieved July 30, 2011, from http://www.washingtonpost.com/national/health-science/mayor-bloomberg-gives-50-million-to-fight-coal-fired-power-plants/2011/07/20/gIQAEKKURI_print.html

Tsuda, A., Takeda, S., Saito, H., Nishioka, J., Nojiri, Y., Kudo, I., Kiyosawa, H., Shiomoto, A., Imai, K., Ono, T., Shimamoto, A., Tsumune, D., Yoshimura, T., Aono, T., Hinuma, A., Kinugasa, M., Suzuki, K., Sohrin, Y., Noiri, Y., Tani, H., Deguchi, Y., Tsurushima, N., Ogawa, H., Fukami, K., Kuma, K., & Saino, T. (2003). A mesoscale iron enrichment in the western subarctic Pacific induces a large centric diatom bloom. *Science, 300*, 958–961.

UN Chief. (2021, November 12). UN Chief: Climate goal on 'life support'. Associated Press in *Omaha World-Herald*, A-3.

6

UN Climate. (2013, November 19). UN climate chief tells coal industry 'Leave It in the Ground'. *Environment News Service.* Retrieved November 27, 2013, from http://ens-newswire.com/2013/11/19/un-climate-chief-tells-coal-industry-leave-it-in-the-ground

Warner, M. (2009, February 15). Is America ready to quit coal? *New York Times.* Retrieved February 17, 2009, from http://www.nytimes.com/2009/02/15/business/15coal.html

Watson, A. J., Bakker, D. C. E., Ridgwell, A. J., Boyd, P. W., & Law, C. S. (2000). Effect of iron supply on Southern Ocean CO2 uptake and implications for glacial atmospheric CO2. *Nature, 407*(6805), 730–733.

Watson, T. (2009, March 9). Companies rethink coal plants. *USA Today,* 9-A.

Williams, B. (2021, November 25). Problems with ocean fertilization. Retrieved from https://www.briangwilliams.us/global-warming-6/problems-with-ocean-iron-fertilization.html

Woodard, S. (2012, August 17). Moapa Paiutes find solar solution amid coal ash plague. *Indian Country Today Media Network.* Retrieved September 2, 2012, from http://indiancountrytodaymedianetwork.com/article/moapa-paiutes-find-solar-solution-amid-coal-ash-plague-129554/

Young, R. S. (2021, September 3). We can't defend against every hurricane. *New York Times,* A-23.

Yukhananov, A., & Volcovici, V. (2013, July 17). World Bank to limit financing of coal-fired plants. *Reuters.* Retrieved July 25, 2013, from http://uk.reuters.com/article/2013/07/16/us-worldbank-climate-coal-idUKBRE96F19U20130716

Zhong, R. (2022, February 23). Climate scientists warn of a 'Global Wildfire Crisis'. *New York Times.* Retrieved February 23, 2022, from https://www.nytimes.com/2022/02/23/climate/climate-change-un-wildfire-report.html/

Further Reading

Ackerman, A. S., Toon, O. B., Stevens, D. E., Heymsfield, A. J., Ramanathan, V., & Welton, E. J. (2000). Reduction of tropical cloudiness by soot. *Science, 288,* 1042–1047.

Archer, D. (2009). *The long thaw: How humans are changing the next 100,000 years of earth's climate.* Princeton University Press.

Blain, S., Quéguiner, B., Armand, L., Belviso, S., Bombled, B., Bopp, L., Bowie, A., Brunet, C., Brussaard, C., Carlotti, F., Christaki, U., Corbière, A., Durand, I., Ebersbach, F., Fuda, J.-L., Garcia, N., Gerringa, L., Griffiths, B., Guigue, C., Guillerm, C., Jacquet, S., Jeandel, C., Laan, P., Lefèvre, D., Monaco, C. L., Malits, A., Mosseri, J., Obernosterer, I., Park, Y.-H., Picheral, M., Pondaven, P., Remenyi, T., Sandroni, V., Sarthou, G., Savoye, N., Scouarnec, L., Souhaut, M., Thuiller, D., Timmermans, K., Trull, T., Uitz, J., van Beek, P., Veldhuis, M., Vincent, D., Viollier, E., Vong, L., & Wagener, T. (2007). Effect of natural iron fertilization on carbon sequestration in the Southern Ocean. *Nature, 446,* 1070–1074.

Brown, K. (2021, August 4). The west is on fire. It's past time to act on climate change. *New York Times,* A-24.

Buesseler, K. O., & Boyd, P. W. (2003). Will ocean fertilization work? *Science, 300,* 67–68.

Ewing, J., & Boudette, N. E. (2022, February 8). Electric cars set for breakthrough as sales soar. *New York Times,* A-1, A-11.

Geoengineering. (2011). Aerosols pose climate dangers. *Nature, 472,* 393.

Goodell, J. (2011). *How to cool the planet: Geo-engineering and the audacious quest to fix earth's climate.* Houghton-Mifflin Harcourt.

Kintisch, E. (2010). *Hack the planet: Science's best hope—or worst nightmare—for averting climate catastrophe.* Wiley.

Kintisch, E. (2013). Dr. Cool. *Science, 342,* 307–309.

Klein, N. (2014). *This changes everything: Capitalism and the climate.* Simon & Schuster.

Kolbert, E. (2009, November 16). Hosed: Is there a quick fix for the climate? *The New Yorker,* 75–77.

Kolbert, E. (2020, July 27). The catastrophist: NASA's climate expert delivers the news that no one wants to hear. *The New Yorker,* 24–29.

GPSR Compliance

The European Union's (EU) General Product Safety Regulation (GPSR) is a set of rules that requires consumer products to be safe and our obligations to ensure this.

If you have any concerns about our products, you can contact us on ProductSafety@springernature.com

In case Publisher is established outside the EU, the EU authorized representative is:

Springer Nature Customer Service Center GmbH
Europaplatz 3
69115 Heidelberg, Germany

The manufacturer's authorised representative in the EU is Springer
Nature Customer Service Centre GmbH, Europaplatz 3, 69115 Heidelberg,
Germany. If you have any concerns regarding our products, please
contact ProductSafety@springernature.com

Printed and bound by CPI Group (UK) Ltd, Croydon, CR0 4YY

24/04/2026

02096317-0009